10 0621551 2
UNIVERSITY OF NOTTINGHAM
WITHDRAWN
FROM THE LIBRARY

D1338075

Dairy Fats and Related Products

The Society of Dairy Technology (SDT) has joined with Wiley-Blackwell to produce a series of technical dairy-related handbooks providing an invaluable resource for all those involved in the dairy industry, from practitioners to technologists, working in both traditional and modern large-scale dairy operations.

For information regarding the SDT, please contact Maurice Walton, Executive Director, Society of Dairy Technology, P.O. Box 12, Appleby in Westmorland, CA16 6YJ, UK. email: execdirector@sdt.org

Other volumes in the Society of Dairy Technology book series:

Probiotic Dairy Products (ISBN 978 1 4051 2124 8)
Fermented Milks (ISBN 978 0 6320 6458 8)
Brined Cheeses (ISBN 978 1 4051 2460 7)
Structure of Dairy Products (ISBN 978 1 4051 2975 6)
Cleaning-in-Place (ISBN 978 1 4051 5503 8)
Milk Processing and Quality Management (ISBN 978 1 4051 4530 5)
Dairy Powders and Concentrated Products (ISBN 978 1 4051 5764 3)

Dairy Fats and Related Products

Edited by

Dr A.Y. Tamime
Dairy Science and Technology Consultant
Ayr, UK

A John Wiley & Sons, Ltd., Publication

This edition first published 2009

Blackwell Publishing was acquired by John Wiley & Sons in February 2007. Blackwell's publishing programme has been merged with Wiley's global Scientific, Technical, and Medical business to form Wiley-Blackwell.

Registered office
John Wiley & Sons Ltd, The Atrium, Southern Gate, Chichester, West Sussex, PO19 8SQ, United Kingdom

Editorial offices
9600 Garsington Road, Oxford, OX4 2DQ, United Kingdom
2121 State Avenue, Ames, Iowa 50014-8300, USA

For details of our global editorial offices, for customer services and for information about how to apply for permission to reuse the copyright material in this book please see our website at www.wiley.com/wiley-blackwell.

Library of Congress Cataloging-in-Publication Data:

Dairy fats and related products / edited by Adnan Tamime.
p. cm.
Includes bibliographical references and index.
ISBN 978-1-4051-5090-3 (hardback : alk. paper) 1. Oils and fats, Edible. 2. Dairy products.
I. Tamime, A. Y.
TP670.D25 2009
637–dc22

2008047429

A catalogue record for this title is available from the British Library

Typeset in 10/12.5 Times-Roman by Laserwords Private Limited, Chennai, India
Printed and bound in Singapore by Ho Printing Singapore Pte Ltd

Contents

Preface to Technical Series

For more than 60 years, the Society of Dairy Technology (SDT) has sought to provide education and training in the dairy field, disseminating knowledge and fostering personal development through symposia, conferences, residential courses, publications, and its journal, the *International Journal of Dairy Technology* (previously known as the *Journal of the Society of Dairy Technology*).

In recent years, there have been significant advances in our understanding of milk systems, probably the most complex natural food available to man. Improvements in process technology have been accompanied by massive changes in the scale of many milk processing operations, and the manufacture a wide range of dairy and other related products.

The Society has now embarked on a project with Blackwell Publishing to produce a Technical Series of dairy-related books to provide an invaluable source of information for practicing dairy scientists and technologists, covering the range from small enterprises to modern large-scale operation. This eighth volume in the series, on *Dairy fats* under the editorship of Dr A.Y. Tamime, provides a timely and comprehensive update on the principles and practices involved in producing these products. Milk fats have provided a valuable and pleasurable source of energy over the centuries though in recent years fashion has moved away from their consumption. However, recent research has found more evidence of their benefits within a balanced diet. This book provides a timely and valuable review of the progress being made in the provision of these products.

R.A. Wilbey
Chairman of the Publications Committee, SDT

Preface

Reviews of butter and, to a lesser extent, other fat-rich products have been included in general dairy technology books though few have been published in the last 10 years. Specialist monographs such as the *Cream Processing Manual* (2nd Edn) and *Milk Fat – Production Technology and Utilization*, which had been published by the SDT in 1989 and 1991 respectively, are now largely out-of-date.

Improvements in hygienic design, efficiency and in automation and control systems have resulted from changes in the industry. The annual world production of butter alone is just below 7 million tonnes.

While milk fat has always been appreciated for its flavour, the market had suffered from concerns over cardiovascular diseases associated with the consumption of animal fats. However, recent clinical studies have indicated benefits, particularly in relation to conjugated linoleic acids (CLA), in the prevention of certain diseases. The range of spreads has also increased, including the addition of probiotic organisms and/or plant extracts to reduce serum cholesterol levels.

The primary aim of this publication will be to detail the state-of-the-art manufacturing methods for:

- Cream
- Butter
- Yellow fat spreads, both pure milk fat based and mixtures with other fats
- Anhydrous milk fat and its derivatives
- Human health benefits associated with dairy fat components.

The manufacturing technology will be complemented by coverage of nutrition issues and analytical methods.

The authors, who are all specialists in their fields in respect to these products, have been chosen from around the world. There is no doubt that the book will receive international recognition by dairy scientists and technologists within both the industry and those with similar processing requirements, researchers and students, thus becoming an important component of the Technical Series promoted by the SDT.

A. Y. Tamime

This book is dedicated to the memory of Dr Richard Robinson, who generously devoted much time and effort to checking the text of the volumes in the SDT technical series prior to publication.

Contributors

Editor

Dr A.Y. Tamime
Dairy Science & Technology Consultant
24 Queens Terrace
Ayr KA7 1DX
Scotland – United Kingdom
Tel. +44 (0)1292 265498
Fax +44 (0)1292 265498
Mobile +44 (0)7980 278950
E-mail: adnan@tamime.morfsnet.co.uk

Contributors

Dr D. Allersma
NIZO Food Research
PO Box 20
6710 BA Ede
The Netherlands
Tel. +31-318 659 578
Fax +31-318 650 400
E-mail: durita.allersma@nizo.nl

Dr H.C. Deeth
School of Land and Food Sciences
The University of Queensland
Brisbane Qld 4072
Australia
Tel. + 61 (0)7 3346 9191
Fax + 61 (0)7 3365 1177
E-mail: h.deeth@uq.edu.au

Dr G. Fitzgerald
University College Cork
Department of Microbiology
Cork
Ireland
E-mail: g.fitzgerald@ucc.ie

Prof P.F. Fox
Department of Food and Nutritional Sciences
University College Cork
Cork
Ireland
Tel. +353-21-4902362
Fax +353-21-4270001
E-mail: pff@ucc.ie

Dr T.P. Guinee
Moorepark Food Research Centre
Teagasc Moorepark
Fermoy
Co. Cork
Ireland
Tel. + 353 25 42204
Fax: + 35325 42340
E-mail: tim.guinee@teagasc.ie

Dr M. Gunsing
NIZO Food Research
PO Box 20
6710 BA Ede
The Netherlands
Tel. +31-318 659 658
Fax +31-318 650 400
E-mail: michiel.gunsing@nizo.nl

Mr M. Hickey
Michael Hickey Associates Food
Consultancy Derryreigh
Creggane
Charleville
Co. Cork
Ireland
Tel. +353 (0)63 89392
Mobile: +353 (0)87 2385653
E-mail: mfhickey@oceanfree.net

Dr H.C. van der Horst
NIZO Food Research
PO Box 20
6710 BA Ede
The Netherlands
Tel. +31-318 659 588
Fax +31-318 650 400
E-mail: caroline.van.der.horst@nizo.nl

Dr T. Huppertz
NIZO Food Research
P.O. Box 20
6710BA Ede
The Netherlands
Tel. +31 318 659600
Fax +31 318 650400
E-mail: Thom.Huppertz@nizo.nl

Mr D. Illingworth
5 Lancewood Lane
Palmerston North
New Zealand
Tel. +64 6 354 8623
Mobile +64 21 1330713
E-mail: david.illingworth@clear.net.nz

Dr P. de Jong
NIZO Food Research B.V.
Principle Scientist
Division Manager Processing
PO Box 20
6710 BA Ede
The Netherlands
Kernhemseweg 2
6718 ZB Ede
The Netherlands
Tel. +31(0)318 659 575
Fax +31 318 650 400
E-mail: peter.de.jong@nizo.nl

Dr A.L. Kelly
Dean of Graduate Studies
Department of Food and Nutritional Sciences
University College Cork
Cork
Ireland
Tel. +353-21-4903405 and 4902810
Fax +353-21-4270001
E-mail: a.kelly@ucc.ie

Dr S. Mills
Moorepark Food Research Centre
Teagasc Biotechnology Department
Fermoy
Co. Cork
Ireland
E-mail: Susan.mills@teagasc.ie

Dr B.K. Mortensen
Bytoften 16
DK 8660 Skanderborg
Denmark
Tel. +45 86 52 5528
E-mail: borge.k.mortensen@privat.dk

Dr P.W. Parodi
Human Nutrition and Health Research
Dairy Australia
9 Hanbury Street
Chermside QLD 4032
Australia
Tel. +61733596110
Fax +61733591848
E-mail: peterparodi@uq.net.au

Dr G.R. Patil
Joint Director (Academic)
National Dairy Research Institute
Karnal - 132 001 (Haryana)
India
Tel. +91 (0)184 2254751 (Office)
+91 (0)184 2283133 (Residence)
Fax +91 (0)184 2250042
Mobile +91 (0)94661 49003
E-mail: grpndri@yahoo.co.in

Dr R.P. Ross
Moorepark Food Research Centre
Teagasc Biotechnology Department
Fermoy
Co. Cork
Ireland
Tel. +353 (0)25 42229
Fax +353 (0)25 42340
E-mail: Paul.Ross@tegasc.ie

Ms M.A. Smiddy
NIZO Food Research
P.O. Box 20
6710BA Ede
The Netherlands
Tel. +31-318-659511
Fax +31-318-650400
E-mail: mary.smiddy@nizo.nl

Dr C. Stanton
Moorepark Food Research Centre
Teagasc Biotechnology Department
Fermoy
Co. Cork
Ireland
Tel. +353 (0)25 42229
Fax +353 (0)25 42340
E-mail: cstanton@moorepark.teagasc.ie

Mr B.B.C. Wedding
Senior Food Industry Consultant
Innovative Food Technologies
Department of Primary Industries and Fisheries
P.O. Box 652
Cairns Qld 4870
Australia
Tel. +61 7 40441604
Fax +61 7 40355474
E-mail: brett.wedding@dpi.qld.gov.au

Dr R.A. Wilbey
University of Reading
School of Biosciences
P. O. Box 226
Reading RG6 6AP
United Kingdom
Tel.
Fax
E-mail: r.a.wilbey@reading.ac.uk

1 Milk Lipids – Composition, Origin and Properties

T. Huppertz, A.L. Kelly and P.F. Fox

1.1 Introduction

Milk is the fluid secreted by the female of all mammalian species, primarily to meet the complete nutritional requirements for the neonate, such as energy, essential amino acids and fatty acids, vitamins, minerals and water. Milk is an effective and balanced source of lipids, proteins (caseins and whey proteins), carbohydrates (mainly lactose), minerals (e.g. calcium and phosphate), enzymes, vitamins and trace elements.

The primary role of the lipids in milk is to provide a source of energy to the neonate. From a practical viewpoint, milk lipids derive a high level of importance from the distinctive nutritional, textural and sensory properties they confer on a wide variety of dairy products such as liquid milk, cheese, ice cream, butter and yoghurt. Milk lipids were long regarded as the most economically valuable constituent of milk and, as a result, the milk price paid to farmers was for many years determined primarily by the concentration of lipids in the milk; only more recently has the concentration of milk proteins attained an equal, or even higher, weighting factor in the determination of milk price. In this chapter, the aim is to provide an overview of the composition, origin and properties of milk lipids. The focus is primarily on the lipids of cow's milk, but comparisons with milk from other major dairy species are made where appropriate.

1.2 Composition of milk lipids

Lipids are esters of fatty acids and related compounds that are soluble in apolar organic solvents and (nearly) insoluble in water and may be divided into three groups:

- Neutral lipids (tri-, di- and monoacylglycerols)
- Polar lipids (phospholipids and glycolipids)
- Miscellaneous lipids (sterols, carotenoids and vitamins)

The term 'lipids' is often readily interchanged with 'fat', but this is incorrect since the latter represents only one subgroup of neutral lipids, the triacylglycerols; hence, the term lipids will be used throughout this chapter. Cow's milk contains ~45 g lipids L^{-1} on average, but this can range from 30 to 60 g L^{-1}, depending on the breed, diet, stage of lactation and health of the cow. There are very large inter-species differences in lipid content between mammals, and the concentration can reach >500 g L^{-1} milk for some species

Table 1.1 Concentration of lipids in the milk of different species.

Species	Lipid content (g L^{-1})	Species	Lipid content (g L^{-1})
Cow	33–47	Marmoset	77
Buffalo	47	Rabbit	183
Sheep	40–99	Guinea pig	39
Goat	41–45	Mink	71
Musk ox	109	Snowshoe hare	134
Dall sheep	32–206	Chinchilla	117
Moose	39–105	Rat	103
Antelope	93	Red kangaroo	9–119
Elephant	85–190	Dolphin	62–330
Human	38	Manatee	55–215
Horse	19	Pygmy sperm whale	153
Monkey	22–85	Harp seal	502–532
Lemur	8–33	Bear	108–331
Pig	68		

Data compiled from Christie (1995).

(Table 1.1). In cow's milk, >98% of lipids are triacylglycerols, but diacylglycerols and monoacylglycerols, free fatty acids, phospholipids, sterols, carotenoids, fat-soluble vitamins and flavour compounds are also found (Table 1.2). In this section, an overview of the various classes of lipids found in milk is provided. More extensive reviews on this subject were compiled by Christie (1995), Jensen (2002), Fox & Kelly (2006) and MacGibbon & Taylor (2006).

Table 1.2 Primary classes of lipids in cow's milk.

Lipid class	% of total
Triacylglycerols	98.3
Diacylglycerols	0.3
Monoacylglycerols	0.03
Free fatty acids	0.1
Phospholipids	0.8
Sterols	0.3
Carotenoids	Trace
Fat-soluble vitamins	Trace
Flavour compounds	Trace

Data compiled from Walstra *et al.* (1999).

1.2.1 *Fatty acids*

A fatty acid is a carboxylic (organic) acid, often with a long aliphatic tail. The fatty acid composition of lipids is particularly important in determining their physical, chemical and nutritional properties. The cow's milk lipids are among the most complex naturally occurring groups of lipids, because of the large number of fatty acids they contain. The fatty acids arise from two sources, synthesis *de novo* in the mammary gland and plasma lipids originating from the feed, as discussed further in Section 1.3. Approximately 400 fatty acids have been identified in cow's milk fat to date, an extensive review of which is provided by Jensen (2002). Primary distinguishing variables among fatty acids are as follows:

- Chain length – An overview of the major fatty acids in cow's milk lipids is given in Table 1.3, from which it is clear that palmitic, oleic, stearic and myristic acids are the principal fatty acids in cow's milk lipids. Short- and medium-chain fatty acids occur in lower amounts, at least when expressed on a weight basis; these have the interesting characteristic that, unlike long-chain acids, they are absorbed into the blood stream in non-esterified form and are metabolised rapidly (Noble, 1978). Furthermore, the chain length influences the melting characteristics of lipids (Section 1.6).
- Degree of saturation – Saturated fatty acids contain no double bonds along the chain; the term *saturated* refers to hydrogen, in that all carbon atoms, apart from the carboxylic acid group, contain as many hydrogen atoms as possible. Saturated fatty acids contain an alkane chain of only single-bonded carbon atoms (–C–C–), whereas

Table 1.3 Major fatty acids in bovine milk lipids.

Number of carbon atoms	Number of double bonds	Shorthand designation	Systematic name	Trivial name	Average range (g 100 g^{-1})
4	0	$C_{4:0}$	Butanoic acid	Butyric acid	2–5
6	0	$C_{6:0}$	Hexanoic acid	Caproic acid	1–5
8	0	$C_{8:0}$	Octanoic acid	Caprylic acid	1–3
10	0	$C_{10:0}$	Decanoic acid	Capric acid	2–4
12	0	$C_{12:0}$	Dodecanoic acid	Lauric acid	2–5
14	0	$C_{14:0}$	Tetradecanoic acid	Myristic acid	8–14
15	0	$C_{15:0}$	Pentadecanoic acid	–	1–2
16	0	$C_{16:0}$	Hexadecanoic acid	Palmitic acid	22–35
16	1	$C_{16:1}$	9-Hexadecanoic acid	Palmitoleic acid	1–3
17	0	$C_{17:0}$	Heptadecanoic acid	Margaric acid	0.5–1.5
18	0	$C_{18:0}$	Octadecanoic acid	Stearic acid	9–14
18	1	$C_{18:1}$	9-Octadecanoic acid	Oleic acid	20–30
18	2	$C_{18:2}$	9,12-Octadecadienoic acid	Linoleic acid	1–3
18	3	$C_{18:3}$	9,12,15-Octadecatrienoic acid	Linolenic acid	0.5–2

Data compiled from Jensen (2002).

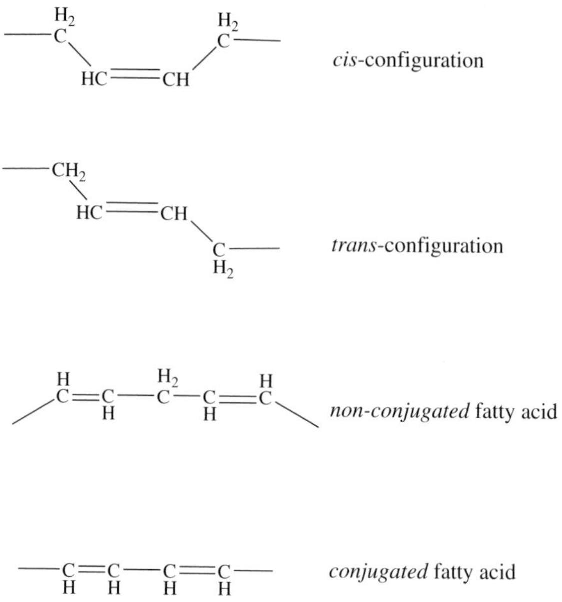

cis-configuration

trans-configuration

non-conjugated fatty acid

conjugated fatty acid

Fig. 1.1 Examples of *cis* and *trans* configuration in unsaturated fatty acids and *non-conjugated* and *conjugated* polyunsaturated fatty acids.

unsaturated fatty acids contain at least one alkene group of double-bonded carbon atoms (–C=C–).

- Configuration of double bonds – The carbon atoms in the chain on either side of the double bond can occur in a *cis* or *trans* configuration (Figure 1.1). In the *cis* configuration, which is most common in nature (>95% of unsaturated fatty acids), the two carbons are on the same side of the double bond. Owing to the rigidity of the double bond, the *cis* isomer causes the chain to bend. In the *trans* configuration, the carbon atoms on either side of the double bond are orientated to the opposite sides of the bond (Figure 1.1) and do not cause the chain to bend much; as a result, their shape is similar to the more linear saturated fatty acids.
- Conjugation of double bonds – For polyunsaturated fatty acids, the term *non-conjugated* indicates that two double bonds in the fatty acid carbon chain are separated by a methylene group ($-CH_2-$), whereas, in a *conjugated* fatty acid, the double bonds are separated by only one single bond (Figure 1.1); most of the naturally occurring fatty acids are non-conjugated. Particularly in the last decade, conjugated linoleic acid (CLA) has gained major interest in human nutrition, as discussed in Chapter 2 and by Bauman & Lock (2006).

1.2.2 *Triacylglycerols*

A triacylglycerol (also known as a *triglyceride* or *triacylglyceride*) is a glyceride in which glycerol is esterified to three fatty acids; likewise, in a mono- or diacylglycerol, the glycerol is esterified to one or two fatty acids, respectively. The composition of triacylglycerols is

Table 1.4 Total carbon number of triacylglycerols in cow's milk lipids.

Total fatty acid carbon number	% of total
C26	0.1–1.0
C28	0.3–1.3
C30	0.7–1.5
C32	1.8–4.0
C34	4–8
C36	9–14
C38	10–15
C40	9–13
C42	6–7
C44	5–7.5
C46	5–7
C48	7–11
C50	8–12
C52	7–11
C54	1–5

Data compiled from Jensen (2002).

defined in terms of the kinds and amounts of fatty acids present and can be expressed as the total carbon number, that is, the sum of the number of carbon atoms in the three fatty acids. The proportions of triacylglycerols according to total carbon number in cow's milk are given in Table 1.4. Stereospecific analysis has enabled the distribution of the fatty acids on the *sn*-1, *sn*-2 and *sn*-3 positions of glycerol to be determined (Figure 1.2) and shown that the distribution of fatty acids in the triacylglycerols of milk is highly specific (Table 1.5). The short-chain fatty acids, butyric and caproic acids, are esterified almost exclusively at the *sn*-3 position, while the medium-chain acids, lauric and myristic acids are esterified preferentially at the *sn*-2 position. Palmitic acid is esterified preferentially at the *sn*-1 and *sn*-2 positions,

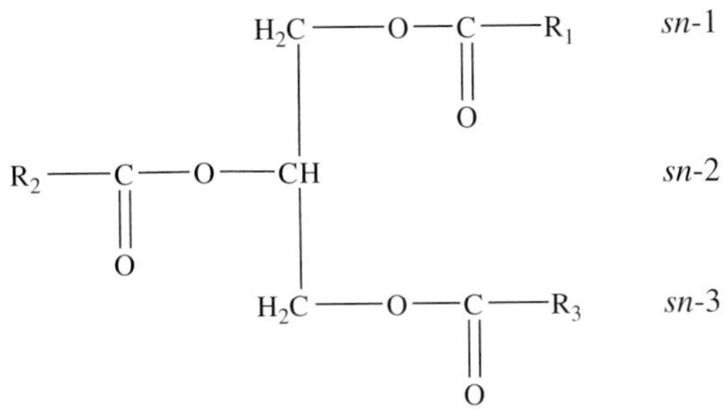

Fig. 1.2 Schematic diagram of a triacylglycerol showing the stereospecific *sn*-1, *sn*-2 and *sn*-3 positions.

Table 1.5 Positional distribution of the major fatty acids in the bovine triacylglycerols.

Fatty acid	Fatty acid composition (mol%)		
	sn-1	*sn*-2	*sn*-3
4:0	–	0.4	30.6
6:0	–	0.7	13.8
8:0	0.3	3.5	4.2
10:0	1.4	8.1	7.5
12:0	3.5	9.5	4.5
14:0	13.1	25.6	6.9
16:0	43.8	38.9	9.3
18:0	17.6	4.6	6.0
18:1	19.7	8.4	17.1

Data compiled from MacGibbon & Taylor (2006).

stearic acid at the *sn*-1 position and oleic acid at the *sn*-1 and *sn*-3 positions (MacGibbon & Taylor, 2006).

1.2.3 *Mono- and diacylglycerols and free fatty acids*

Immediately after milking, milk contains only small amounts of di- and monoacylglycols and free fatty acids, but these levels can increase considerably during storage, due to enzymatic hydrolysis of the ester bonds in triacylglycerols. Enzymatic hydrolysis of triacylglycerols is referred to as lipolysis, and may arise from the action of the indigenous milk enzyme lipoprotein lipase or of bacterial lipases. The diacylglycerols in freshly drawn milk are unlikely to be the products of lipolysis since they are mostly non-esterified at the *sn*-3 position, whereas lipases preferentially attack the *sn*-1 and *sn*-3 position and should thus result in a mixture of diacylglycerols with non-esterified *sn*-1 and *sn*-3 positions. More likely, the diacylglyerols in freshly drawn milk are intermediates in the biosynthesis of triacylglycerols, since the *sn*-3 position is the last to be esterified. The profile of free fatty acids in freshly drawn milk also differs from that of the fatty acids esterified in the triacylglycerols, making it unlikely that free fatty acids are the products of lipolysis. Diacylglycerols have physical properties similar to those of triacylglycerols, but monoacylglycerols, particularly those containing a long-chain fatty acid, are amphiphilic and are thus surface active (Taylor & MacGibbon, 2002; Walstra *et al.*, 2006).

1.2.4 *Phospholipids*

Phospholipids are a class of lipids formed from four components: a backbone (glycerol or sphingosine), fatty acids, a negatively charged phosphate group and a nitrogen-containing compound or sugar. Phospholipids with a glycerol backbone are known as *glycerophospholipids* and have a fatty acid at the *sn*-1 and *sn*-2 positions and a phosphate and a polar head group

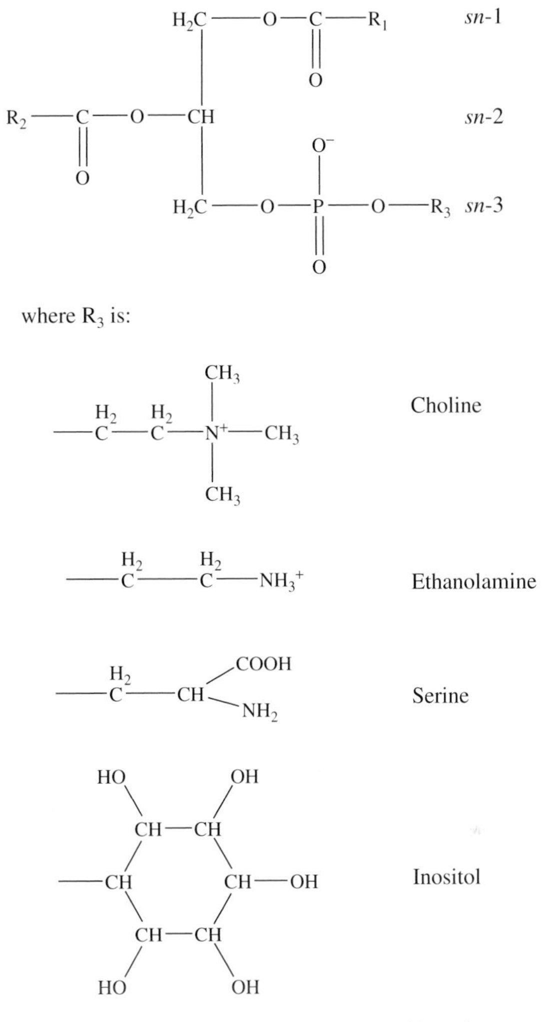

H_2C—O—C(=O)—R_1 *sn*-1

R_2—C(=O)—O—CH *sn*-2

H_2C—O—P(=O)(O^-)—O—R_3 *sn*-3

where R_3 is:

—CH_2—CH_2—$N^+(CH_3)_3$ Choline

—CH_2—CH_2—NH_3^+ Ethanolamine

—CH_2—CH(COOH)(NH_2) Serine

—CH(CHOH)₅ ring (HO, OH, OH, HO, OH) Inositol

Fig. 1.3 Schematic projection of a glycerophospholipid and its possible polar groups.

(choline, ethanolamide, serine or inositol) at the *sn*-3 position (MacGibbon & Taylor, 2006); a general projection of a glycerophospholipid is given in Figure 1.3. Phosphatidylcholine is commonly referred to as *lecithin*. The major fatty acids in glycerophospholipids are palmitic, stearic, oleic and linoleic acids, with very few shorter chain fatty acids (Bitman & Wood, 1990). Sphingolipids consist of a long-chain amino alcohol, sphingosine, to which a fatty acid is attached through an amide linkage to yield a ceramide. Linkage of phosphorylcholine group to the terminal alcohol group of a ceramide yields sphingomyelin, a sphingophospholipid (Figure 1.4); glycosphingolipids have one or more hexose units (e.g. glucose) attached to the terminal alcohol group of a ceramide (MacGibbon & Taylor, 2002).

The proportion of phospholipids in milk lipids is typically 0.9 g 100 g^{-1} milk lipids, but this proportion is considerably higher in skimmed milk (25 g 100 g^{-1}) and buttermilk (22 g 100 g^{-1}); in contrast, the proportion of phospholipids in 40 g 100 g^{-1} fat cream is lower

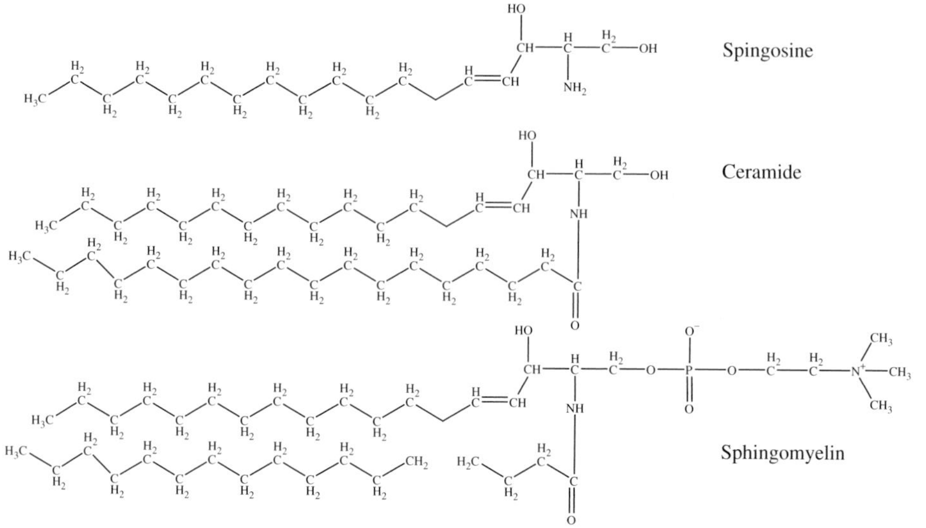

Fig. 1.4 Schematic projection of sphingosine, a ceramide and sphingomyelin.

(~0.5 g 100 g^{-1} milk lipids) than in whole milk (Mulder & Walstra, 1974). Phosphatidylcholine, phosphatidylethanolamine and sphingomyelin are the principal phospholipids in cow's milk, each representing ~25–35% of the total phospholipids. Phospholipids are important in dairy products because they are good emulsifiers. Hence, it is not surprising that the majority of phospholipids (60–65%) are present in the membrane surrounding the milk lipid globules (MLGs), whereas the remainder are found in the milk plasma, primarily in soluble fragments of the lipid globule membrane (Huang & Kuksis, 1967). The role of phospholipids in the milk lipid globule membrane (MLGM) is discussed in Section 1.4. Some phospholipids have a number of health-beneficial functions; sphingomyelin shows a strong anti-tumour activity, influences the metabolism of cholesterol and exhibits an anti-infective activity; glycerol phospholipids protect against mucosal damage (Parodi, 2004, 2006).

1.2.5 Minor constituents

Sterols or steroic alcohols

These constituents are minor components, accounting for only ~0.3% of total milk lipids; cholesterol accounts for >95% of total sterols in milk (Walstra *et al.*, 1999). Since cholesterol is found primarily in the MLGM, the proportion of cholesterol in milk lipids is, as for phospholipids, considerably higher in skimmed milk and buttermilk than in whole milk and cream (Russel & Gray, 1979). The cholesterol content of various dairy products is summarised in Table 1.6. The association of dietary cholesterol with coronary heart disease has led to consumer preference for food products, including dairy products, containing a low amount of cholesterol. This has led to considerable research into the development of methods for the removal of cholesterol from dairy products; for an overview of biological, chemical and physical processes used for this purpose see Sieber & Eyer (2002).

Table 1.6 Cholesterol content of dairy products.

Dairy product	Lipid content (g 100 g^{-1} product)	Cholesterol content (mg 100 g^{-1} product)
Skimmed milk	0.3	2
Whole milk	3.3	14
Medium cream	25.0	88
Skimmed milk powder	0.8	20
Cream cheese	34.9	110
Ice cream	11.0	44
Cheese		
Cheddar	33.1	105
Brie	27.7	100
Swiss	27.5	92
Butter	81.1	219

Data compiled from Jensen (2002).

Carotenoids occur only in trace amounts (μg g^{-1} lipids) in cow's milk lipids; β-carotene accounts for >95% of total carotenoids in cow's milk. The level of β-carotene is highly variable, depending both on the concentration of carotene in the feed, as well as on the breed of the cow. β-Carotene is responsible for the yellow colour of milk fat (Walstra *et al.*, 2006).

The lipid-soluble vitamins (i.e. vitamins A, D, E and K) are also found in the lipid fraction of milk; milk is a significant source of vitamin A, but the concentrations of vitamins D, E and K are too low to make significant contributions to the consumers' vitamin requirements.

Finally, milk lipids contain a large number of flavour compounds, especially lactones, fatty acids, aldehydes and methyl ketones, which contribute to the overall organoleptic properties of milk. The concentrations of flavour compounds are influenced primarily by the feed of the cow. For a description of the flavour compounds in milk lipids, the reader is referred to Schieberle *et al.* (1993).

1.3 Origin of milk lipids

1.3.1 *Biosynthesis and origin of the fatty acids in milk lipids*

The fatty acids of milk arise from two sources: (a) synthesis *de novo* in the mammary gland and (b) uptake from the circulating blood. The composition of the fatty acids derived from the two sources differs markedly. The fatty acids produced *de novo* are all those with a carbon chain of 4–14 atoms and a proportion of the C_{16} fatty acids, whereas the remainder of the C_{16} fatty acids and almost all of the C_{18} acids arise from the blood. In this section, a brief overview of the two sources of fatty acids for bovine lipids are given; for more extensive reviews, the reader is referred to Hawke & Taylor (1995), Barber *et al.* (1997) and Palmquist (2006).

1.3.2 De novo synthesis of fatty acids

Ruminants use acetate (C_2) and β-hydroxybutyrate (C_4) that is synthesised by bacteria in the rumen as the carbon sources for fatty acid synthesis; acetate contributes ~92% of the total carbon in milk fatty acids. Monogastric animals use glucose as the principal source of carbon source for fatty acid synthesis (Palmquist, 2006). In ruminants, the first step of fatty acid synthesis is the conversion of acetate, derived from the blood, to acetyl coenzyme-A by the cytosolic enzyme acetyl coenzyme-A synthase, according to Figure 1.5.

$$acetate + coenzyme\ A + ATP \rightarrow acetyl\text{-}coenzyme\ A + ADP.$$

Acetyl-coenzyme A is converted to malonyl coenzyme A (Figure 1.5):

$$acetyl\text{-}coenzyme\ A + HCO_3^- + ATP \rightarrow malonyl\text{-}coenzyme\ A + ADP + P_i$$

The synthesis of fatty acids subsequently occurs under the influence of the multi-enzyme system 'fatty acid synthase', according to Figure 1.5.

$$acetyl\text{-}coenzyme\ A + n\ malonyl\text{-}coenzyme\ A + 2n\ NADPH + 2nH^+ \rightarrow$$
$$fatty\ acid + nCO_2 + (n+1)\ coenzyme\ A + 2n\ NADP^+ + (n-1)\ H_2O$$

where the number of carbon atoms in the fatty acid is $(2n+2)$. Activation of β-hydroxybutyrate leads to the formation of β-hydroxybutyl-coenzyme A, which contributes equally to acetyl-coenzyme A as the primer for fatty acid formation by fatty acid synthase, but does not contribute to chain elongation. The synthesis of fatty acids by fatty acid synthase is

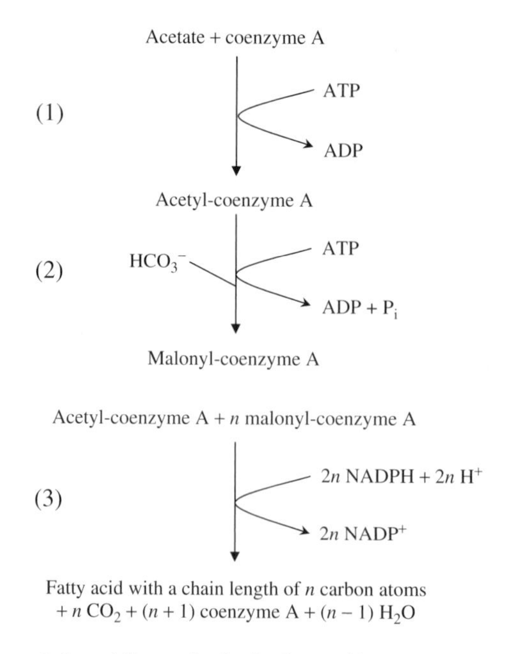

Fig. 1.5 Schematic representation of the synthesis of a fatty acid.

described in detail by Smith *et al.* (2003). Cycling through the fatty acid synthase system, with the addition of C_2 units from malonyl-coenzyme A, continues until the nascent fatty acid reaches a chain length of 6 to 16 carbon atoms, when a thioesterase specific for that chain length releases the fatty acid and terminates the cycle (Palmquist, 2006).

1.3.3 *Uptake of fatty acids from the blood*

Lipids taken up by the mammary gland from the blood can originate from the digestive tract or from mobilised body-fat reserves. To overcome their insolubility in aqueous media, dietary triacylglycerols are transported in the form of lipoproteins or, more specifically, a subclass of lipoproteins, the very-low density lipoproteins. In the mammary gland, fatty acids are de-esterified from triacylglycerols by lipoprotein lipase, a process described by Barber *et al.* (1997). Fatty acids released from body-fat reserves by the action of a hormone-sensitive lipase are also taken up by the mammary gland, although their contribution is thought to be limited. As mentioned previously, almost all C_{18} fatty acids are taken up from the blood, but $C_{16:0}$ is derived from the blood and by *de novo* synthesis. When the level of dietary lipid intake is low, almost all $C_{16:0}$ is synthesised *de novo*, but the proportion of $C_{16:0}$ synthesised *de novo* can decrease by $>70\%$ when the uptake from the blood increases (Palmquist, 2006).

1.3.4 *Desaturation of fatty acids*

From Section 1.3.1, it is clear that only saturated fatty acids are synthesised *de novo* in the mammary gland; furthermore, due to the very low redox potential in the rumen, extensive hydrogenation of unsaturated fatty acids in the diet occurs. Hence, the initial concentration of unsaturated fatty acids in the mammary gland is low and an extensive desaturation must occur to achieve the levels of unsaturated fatty acids found commonly in bovine milk. The rate-limiting enzyme in this desaturation process is stearoyl-coenzyme A desaturase, the activity of which is particularly high in the lactating mammary gland, but low in non-lactating mammary gland tissue. Stearyl-coenzyme A is located primarily in the endoplasmic reticulum (ER) and its principal substrates are stearoyl-coenzyme A and palmitoyl-coenzyme A (Palmquist, 2006).

1.3.5 *Synthesis of triacylglycerols*

The glycerol-3-phosphate pathway is the primary route for triacylglycerol synthesis in the mammary gland and is presented in Figure 1.6. In this multistep enzyme-catalysed pathway, the *sn*-1, *sn*-2 and *sn*-3 positions of glycerol are esterified consecutively. The resulting triacylglycerols associate into spherical droplets, which become covered with a membrane layer, as described in Section 1.5.

1.4 Factors affecting the composition of milk lipids

When considering the composition and origin of milk lipids, it is important to consider that the composition of milk lipids is not a static, but a dynamic, phenomenon, which is influenced

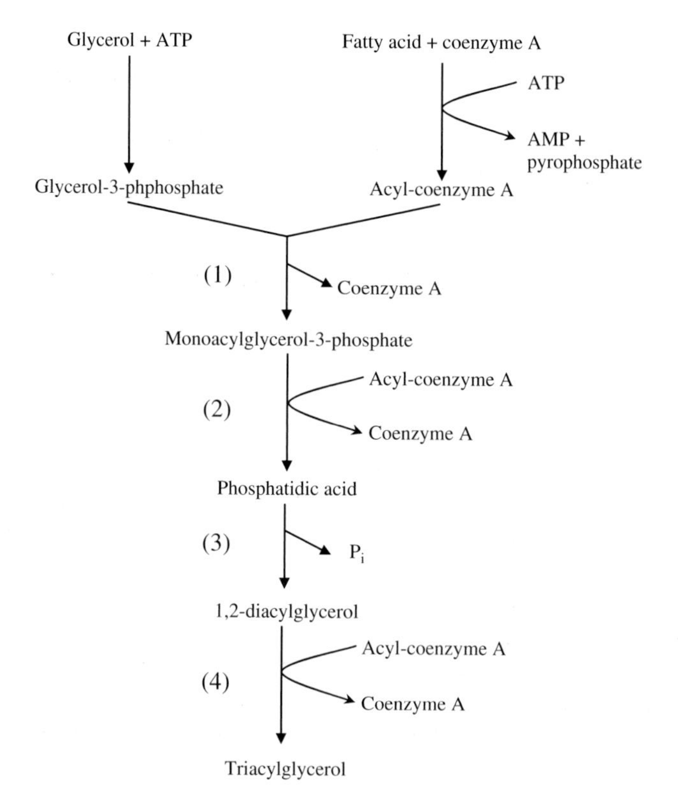

Fig. 1.6 Schematic representation of the synthesis of triacylglycerols in the mammary gland.

by both physiological and nutritional factors. Physiological factors that influence milk lipid composition include other species of mammals, as well different breeds of a species. Milk lipid composition is also affected by the stage of lactation. Colostrum is rich in C_{12}, C_{14} and C_{16} fatty acids, but the relative proportions of C_4–C_{10} and C_{18} acids increase rapidly afterwards and the relative proportions of fatty acids generally stabilise 1 week postpartum. During the subsequent lactation, the relative proportions of fatty acids synthesised *de novo* increases, whereas the proportion of dietary fatty acids decreases concomitantly (Palmquist *et al.,* 1993; Palmquist, 2006).

The fatty acid profile of milk lipids is also influenced strongly by dietary factors, which have been reviewed extensively (Grummer, 1991; Palmquist *et al.,* 1993; Palmquist, 2006), and only a few points are mentioned here:

- Low-fat diets greatly reduce the proportion and yield of C_{18} acids in milk lipids, whereas increasing the level of C_{18} acids in a low-fat diet linearly increases the level of C_{18} acids in milk.

- The incorporation of high levels of unsaturated fatty acids in the diet has little effect on the degree of unsaturation of milk lipids, due to extensive biohydrogenation in the rumen.
- The concentrations of C_{16} and C_{18} acids in milk lipids can be increased by increasing the dietary uptake of these fatty acids.

1.5 Intracellular origin of milk lipid globules and the milk lipid globule membrane

1.5.1 *Secretion of milk lipid globules*

The secretion of MLG is represented schematically in Figure 1.7. The precursors for the MLG originate in the ER, as the so-called micro lipid droplets (MLD), which have a diameter <0.5 μm and consist of a triacylglycerol-rich core surrounded by a coat consisting of proteins and polar lipids (Keenan & Mather, 2006). Since the size distribution of MLG is considerably broader than that of the MLD, considerable growth must occur after micro lipid droplet formation. Whether droplet growth is a controlled or random process is unknown. Although there are several potential mechanisms through which the droplet growth can occur, fusion of droplets appears to be the only mechanism for which there is both biochemical and morphological evidence (Keenan & Mather, 2006). Such fusion is restricted to the small droplets, which can fuse with each other or with larger droplets; apparently, large

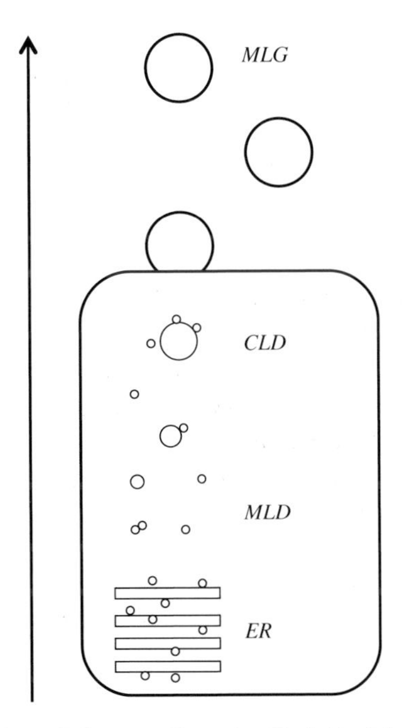

Fig. 1.7 Schematic representation of the secretion of milk lipid globules. ER, endoplasmic reticulum; MLD, micro lipid droplet; CLD, cytoplasmic lipid droplet; MLG, milk lipid globule. The upward arrow indicates the direction in which the secretion process progresses.

droplets (>2 μm) cannot fuse. The larger droplets resulting from the fusion of MLD are termed *cytoplasmic lipid droplets*. The micro and cytoplasmic lipid droplets move through the cell to its apical region. Upon arrival at the apical membrane, the droplets come into contact with a dense proteinaceous coat on the inner face of the plasma membrane and are gradually surrounded by plasma membrane and subsequently released into the lumen as MLG (Figure 1.7). The composition of the membrane surrounding the MLG is described in the following section.

1.5.2 *The milk lipid globule membrane*

The MLGM, which is also referred to as the milk fat globule membrane (MFGM), has been the subject of considerable research and was reviewed by King (1955), Patton & Keenan (1975), Keenan & Dylewski (1994) and Keenan & Mather (2002, 2006).

As described above, the MLGM originates from the membranes of the milk-secreting cells. The droplets released from the endoplasmatic reticulum are coated primarily by proteins and phospholipids; this surface layer has no discernable bilayer characteristics (Keenan & Dylewski, 1995; Keenan & Mather, 2006). On arrival at the apical region, the droplets interact with the plasma membrane prior to secretion. During this process, the droplets are enveloped with plasma membrane. As a result, the composition of the MLGM, which is given in Table 1.7, closely resembles that of the cell membrane from which it is derived. The current view of the molecular organisation of the MLGM is that it comprises a true bilayer membrane with a thick (10–50 nm) dense proteinaceous coat on the inner surface of the bilayer membrane, plus an innermost layer that existed before secretion of the globules (Keenan & Mather, 2006). From Table 1.7, it is apparent that lipids and proteins, including enzymes, are the principal constituents of the MLGM and these are described in Sections 1.5.3, 1.5.4 and 1.5.5. The MLGM material has attracted considerable attention from both nutritional and technological viewpoints in recent years and was reviewed by Ward *et al.* (2006).

Table 1.7 Gross composition of the milk lipid globule membrane.

Constituent class	Amount
Proteins	25–60 g 100 g^{-1}
Total lipids	0.5–1.1 mg mg^{-1} protein
Neutral lipids	0.25–0.88 mg mg^{-1} protein
Phospholipids	0.13–0.34 mg mg^{-1} protein
Glycosphingolipids	13 μg mg^{-1} protein
Hexoses	108 μg mg^{-1} protein
Hexosamines	66 μg mg^{-1} protein
Sialic acids	20 μg mg^{-1} protein
RNA	20 μg mg^{-1} protein
Glycosaminoglycans	0.1 μg mg^{-1} protein

Data compiled from Keenan & Mather (2002).

1.5.3 *Lipids of the milk lipid globule membrane*

The principal classes of lipids in the MLGM are summarised in Table 1.8, from which it is apparent that triacylglycerols and phospholipids, in a ratio of ~2:1, are the most abundant lipid classes. The triacylglycerols of the MLGM contain a higher proportion of long-chain fatty acids than the triacylglycerols of the milk lipid globule core. Di- and monoacylglycerols are also present, but it is unclear whether they are true membrane constituents or the products of lipolysis (Keenan & Mather, 2006). Of the total phospholipids in milk, ~60% are present in the MLGM. The distribution and fatty acid composition of phospholipids in the MLGM are similar to those in skimmed milk, suggesting that they are derived from a common source (Keenan & Dylewski, 1995).

1.5.4 *Proteins of the milk lipid globule membrane*

Proteins account for 25–60% of the mass of the MLGM (Table 1.8); these proteins are extremely diverse and were reviewed by Mather (2000). Electrophoresis on the basis of molecular mass reveals eight major protein bands, which are discussed briefly. Most of these proteins are known under several different names; the nomenclature proposed by Mather (2000) is followed here.

Butyrophilin (*BTN*) is the most abundant protein of the MLGM, comprising ~40% of total membrane protein in the milk of Holstein cows. BTN is a transmembrane protein with an externally oriented N-terminus; its association with the membrane is probably stabilised by disulphide bonds. The protein is expressed specifically in the lactating mammary gland and is concentrated at the apical cell membrane and on the MLG. BTN is a glycosylated protein with a molecular mass of ~66 kDa. A comprehensive review of the molecular and cellular biology of BTN is provided by Mather & Jack (1993).

Table 1.8 Lipids of the milk lipid globule membrane.

Lipid class	% of total lipids
Triacylglycerols	62
Diacylglycerols	9
Monoacylglycerols	0–0.5
Unesterified fatty acids	0.6–6.0
Phospholipids	26–31
Sphingomyelin	22[a]
Phosphatidyl choline	36[a]
Phosphatidyl ethanolamine	27[a]
Phosphatidyl inositol	11[a]
Phosphatidyl serine	2[a]

[a]Concentration expressed as a percentage of total phospholipids in milk.
Data compiled from Keenan & Mather (2002).

Xanthine dehydrogenase/xanthine oxidoreductase (XDH/XO), which is an internally disposed constituent of the proteinaceous coat of the MLGM, is the most abundant protein with a known enzyme activity in the MLGM and represents ~20% of membrane-associated protein. XDH/XO is an iron- and molybdenum-containing oxidoreductase, which can exist as a dehydrogenase or an oxidase; the structure and properties of the oxidase form that predominates in milk were reviewed by Harrison (2006). XDH/XO, a homodimer with a monomeric mass of ~147 kDa, plays a key role in the terminal steps of the purine metabolism, catalysing the oxidation of hypoxanthine to xanthine and uric acid. In milk, XDH/XO may play a role by providing hydrogen peroxide for the antimicrobial hydrogen peroxide–lactoperoxidase–thiocyanate system in milk.

Mucin 1 (*MUC-1*) is a heavily glycosylated (~50% carbohydrate) protein that displays allelic polymorphism and has a mass of ~56 kDa. MUC-1 occurs in milk at a concentration of ~40 mg L^{-1} and is incorporated into the lipid globules during the contact of the lipid droplet with the plasma membrane. Its biological function is unclear but, since MUC-1 is a transmembrane protein with an externally oriented N-terminal and appears to be present in filamentous structures extending up to 1 μm from the membrane, it may provide the membrane with steric stabilisation and protect it against physical damage and invasive microorganisms (Mather, 2000).

Periodic acid Schiff 6/7 (PAS 6/7) is an externally disposed, extrinsic MLGM protein; it appears as a doublet with a molecular mass in the range of 48 to 54 kDa on electrophoretic separation on the basis of molecular mass; this rather broad size range is due primarily to post-translational modification, particularly glycosylation. The specific functions of PAS 6/7 in milk lipid globule secretion or lactation are unknown (Mather, 2000).

Periodic acid Schiff III (PAS III) is a rather poorly characterised glycoprotein with a molecular mass of ~100 kDa; it is located on the apical surfaces of the epithelial cells in the mammary gland.

Cluster of differentiation 36 (CD 36) has a molecular mass of ~77 kDa and is heavily glycosylated; carbohydrates, excluding sialic acids, account for ~24% of its mass. It is an integral protein of the MLGM and remains associated with the membrane upon destabilisation of the globules. CD36 is a transmembrane protein with short N- and C-terminal segments internally disposed and a large hairpin loop disposed externally (Mather, 2000).

Adipophilin (*ADPH*) was recognised as a constituent of the MLGM only recently, because its molecular mass (~52 kDa) overlaps with that of PAS 6/7, and it has a relatively low solubility using conventional methods for the preparation of samples for electrophoresis. ADPH is not glycosylated, but may be acetylated with long-chain fatty acids, and may play a role in the secretion of MLG (Mather, 2000).

Fatty acid binding protein (*FABP*) has a molecular mass of ~13 kDa, and is the smallest of the major proteins of the MLGM; FABP is not glycosylated, but may be modified posttranslationally. Within the mammary epithelial cells, FABP may serve as a transporter of fatty acids, but its role in milk lipid globule formation and secretion is unknown.

1.5.5 *Enzymes of the milk lipid globule membrane*

The MLGM is a significant source of indigenous milk enzymes, although, except for XDH/XO, whose concentrations are too low to be considered as major constituents.

Table 1.9 Enzymes found in the milk lipid globule membrane.

Xanthine oxidoreductase	Alkaline phosphatase	β-Galactosidase
Lipoamide dehydrogenase	Acid phosphatase	Plasmin
NADPH oxidase	Phosphatic acid phosphatase	Inorganic pyrophosphatase
NADP oxidase	5′-Nucleotidase	Adenosine triphosphatase
Sulphydryl oxidase	Glucose-6-phosphatase	Nucleotide pyrophosphatase
Catalase	Phosphodiesterase I	Aldolase
γ-Glutamyl transpeptidase	Ribonuclease I	Acetyl-coenzyme-A carboxylase
Galactosyl transferase	UDP-glycosyl hydrolases	
Choline esterase	β-Glucosidase	

A summary of enzymatic activities found in the MLGM is given in Table 1.9. More than half of the enzymatic activities in the MLGM are hydrolases, with oxidoreductases and transferases also representing major proportions. Biological roles of enzymes of the MLGM remain largely unstudied, and studies on their functional significance have focussed largely on effects on the processing characteristics and organoleptic properties of milk and dairy products (Keenan & Mather, 2006).

1.6 Physicochemical stability of milk lipid globules

As will be described in more detail in the later chapters of this book, the fact that milk lipids are present primarily in the form of globules has implications for many properties of dairy products. Hence, the physicochemical stability of MLG is of major importance to the dairy industry. An excellent and unrivalled overview of this area was provided by Mulder & Walstra (1974); the more recent reviews of Walstra (1995) and Huppertz & Kelly (2006) update knowledge in this area. This section will provide a summary of the work reviewed in the aforementioned publications.

1.6.1 *Size distribution of milk lipid globules*

As mentioned previously, the lipids in milk are present predominantly in the form of globules, which in cow's milk range from $\sim$0.2 to 15 μm in diameter. Cow's milk contains $>10^{10}$ lipid globules mL^{-1}, of which $\sim$80% have a diameter <1 μm; however, these small globules comprise only $\sim$3% of the total mass of the lipids in milk. The majority ($>$90%) of milk lipids are in globules with a diameter in the range 1 to 10 μm, with a small proportion of globules having a diameter >10 μm (Walstra, 1969, 1995).

The most common parameters used to express average globule size are derived from the so-called 'moments' of the particle size distribution, which are particularly useful auxiliary parameters. The *n*th moment of a particle size distribution is given by

$$S_n = \sum d_i^n N_i$$

Table 1.10 Common parameters used to express average milk lipid globule size.

Name	Abbreviation	Calculation	Average value for cow's milk (μm)
Number mean diameter	d_n or $d_{1,0}$	S_1/S_0	0.81[a]
Volume mean diameter	d_v or $d_{3,0}$	$(S_3/S_1)^{1/3}$	1.8[a]
Volume surface-weighted mean diameter	d_{vs} or $d_{3,2}$	S_3/S_2	3.3[a]
Volume moment-weighted mean diameter	d_{vm} or $d_{4,3}$	S_4/S_3	3.5[b]

[a]Data compiled from Walstra (1969).
[b]Data compiled from Huppertz *et al.* (2003).

where N_i and d_i are the number and diameter of the particles in size class i (Walstra, 2003). By calculating the various moments of the particle size distribution, usually S_1, S_2, S_3 and S_4, a number of parameters representing average particle size can be derived, as summarised in Table 1.10. The size of MLG can be determined by a variety of methods, including dynamic or static light scattering, Coulter counting, electroacoustics, ultrasonic spectroscopy and light or electron microscopy.

1.6.2 *Colloidal stability of milk lipid globules*

Colloidal interactions form the basis of the stability or instability of the MLG emulsion and will, for instance, govern whether the droplets remain as discrete entities or aggregate. For MLG, colloidal stability is governed by a balance between attractive forces (van der Waals attractions) and repulsive forces (electrostatic and steric repulsions) (Walstra, 1995; Huppertz & Kelly, 2006). Electrostatic repulsion occurs between molecules with a permanent electrical charge. For MLGs, repulsion arises from the net negative charge on the globule membrane. However, even when electrostatic repulsion is minimised, MLGs remain as discrete entities, indicating that considerable stability is derived from steric repulsion. In the case of MFGs, such steric stabilisation is provided by various glycoproteins of the MLGM (Section 1.5.4); hydrolysis of these proteins, for example, by papain, causes aggregation of MFGs (Shimizu *et al.*, 1980).

All food emulsions, including those containing MFGs, are physically unstable over time. Such instability may be evident as gravitational separation or droplet aggregation, as outlined in Figure 1.8. Gravity separation of MLGs is a result of a density difference between the globules and the milk plasma and may occur under the influence of a gravitational or centrifugal force. As MLGs have a lower density than milk plasma, the predominant type of gravitational separation of MLGs is creaming, which is described in Section 1.6.3. Sedimentation of MLGs occurs only under extreme circumstances, for example, when the protein load on the globule membrane is sufficiently high to increase the density of the globule above that of the milk plasma. Aggregation of droplets occurs when they stay together for a time longer than can be accounted for in the absence of colloidal interactions. Flocculation implies the aggregation of droplets to give three-dimensional structures, but is rarely observed in MLGs. Coalescence is the process by which two or more droplets merge to form a larger

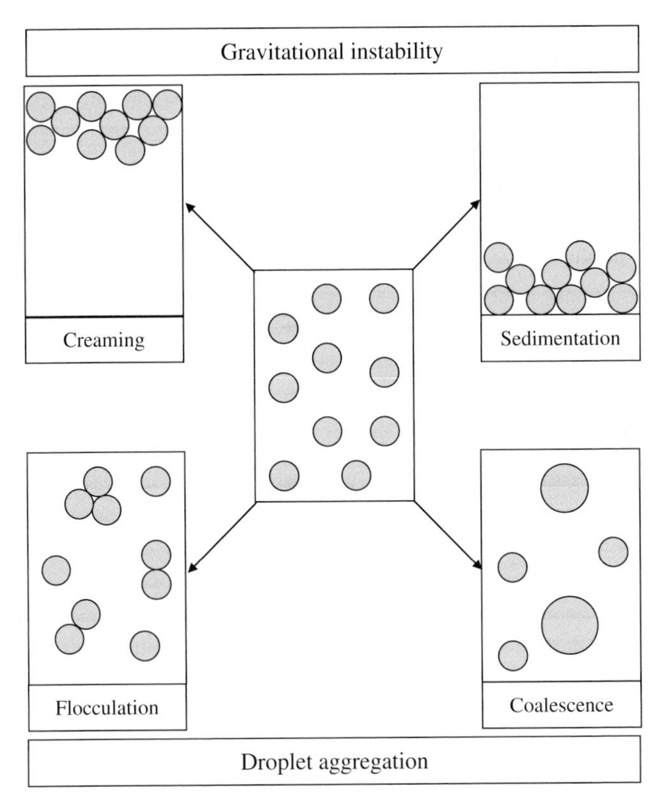

Fig. 1.8 Schematic representation of types of physical instability of emulsions.

droplet; it may be partial when anisometrically shaped conglomerates are formed because true coalescence is prevented. True and partial coalescence are described in Section 1.6.4.

1.6.3 *Creaming of milk lipid globules*

Creaming of MLGs implies the rise under gravitational or centrifugal force due to a difference in density between the globules and the milk plasma. For perfectly spherical globules, the rate of rise of globules, v, is given by Stokes' Law:

$$v = a\frac{(\rho_p - \rho_l) \cdot d^2}{18 \cdot \eta_p}$$

where d is the diameter of the globules, ρ and η are the density and viscosity, respectively, the subscripts p and l refer to the density of the plasma and lipid phase of milk, respectively, and a is the acceleration, which is 9.8 m s^{-2} for gravitational separation. For centrifugal separation, $a = R\omega^2$, where R is the centrifugal radius and ω is the angular velocity, which is equal to $2\pi n/60$, where n is the number of revolutions per minute (rpm). Centrifugal creaming of milk forms the basis of industrial separation of the cream and skimmed milk phases of milk and is described in Chapter 4.

Considerable creaming of cow's milk occurs under quiescent conditions; in fact, when raw, unhomogenised milk is stored at refrigeration temperature, the rate of rise of MFGs is much higher than can be accounted for by Stokes' Law for individual MLGs. This is due to the fact that the globules in such milk rise as large clusters, up to 1 mm in diameter, which are formed as a result of cold agglutination of MLG (Dunkley & Sommer, 1944). Cold agglutination of MLGs requires three components, that is, MLGs, immunoglobulin M (IgM) and lipoproteins present in the milk plasma; the latter fraction is often referred to as the *skimmed milk membrane* (*SMM*). On cold agglutination, IgM interacts with both the globules and SMM, whereas the latter cannot interact with each other (Euber & Brunner, 1984). Heating milk at a temperature ~62°C impairs cold agglutination due to the denaturation of IgM, a heat-labile component, whereas homogenisation disrupts the SMM, a homogenisation-labile component, and thereby impairs cold agglutination (Walstra, 1995; Huppertz & Kelly, 2006). Cold agglutination is almost unique to bovine milk; it does not occur, or if it does, to a limited extent, in sheep, goat, buffalo and camel milk (Huppertz & Kelly, 2006).

1.6.4 *Coalescence of milk lipid globules*

As mentioned in Section 1.6.2, coalescence is the process whereby lipid globules merge to form a single, larger globule. This process stabilises the emulsion thermodynamically, as it reduces the contact area between the globule membrane and the milk plasma. Although the general process of coalescence is not fully understood, it is clear that coalescence can be induced by collisions and prolonged contact of the lipid globules (Walstra, 2003). Coalescence of MLGs, which can be induced by the enzymatic removal of the polar head of phospholipids in the MLGM, is of limited importance in milk and dairy products (Shimizu *et al.*, 1980).

Partial coalescence is of far greater importance for milk and dairy products, particularly in the preparation of ice cream, whipped cream (Chapter 4) and butter (Chapter 5). Partial coalescence occurs when two lipid globules, in which the lipids are partially crystalline, come into contact. The aggregate will partially retain the shape of the individual globules, because the presence of a solid lipid phase prevents complete merger. Partial coalescence is enhanced by applying a velocity gradient, increasing the lipid content, carefully manipulating the concentration of solid lipids and minimising the repulsion provided by the globule membrane (Walstra *et al.*, 1999, 2006). Partial coalescence plays a major role in the textural defect 'bitty' or 'broken' cream, which is highlighted by the presence of large cream particles floating on top of milk or cream. Partial coalescence also plays a major role in re-bodying, a phenomenon whereby cooled cream, when warmed to ~30°C and subsequently recooled, becomes extremely viscous or even solid-like. These textural defects are described in more detail in Chapter 4.

1.6.5 *Homogenisation and properties of homogenised milk lipid globules*

As outlined in Section 1.6.3, MLGs cream readily, which is an undesirable feature of liquid milk products. From Stokes' Law, it is apparent that the rate of rise of MLGs can be reduced by a number of measures, of which a reduction in size is the most effective. For this purpose,

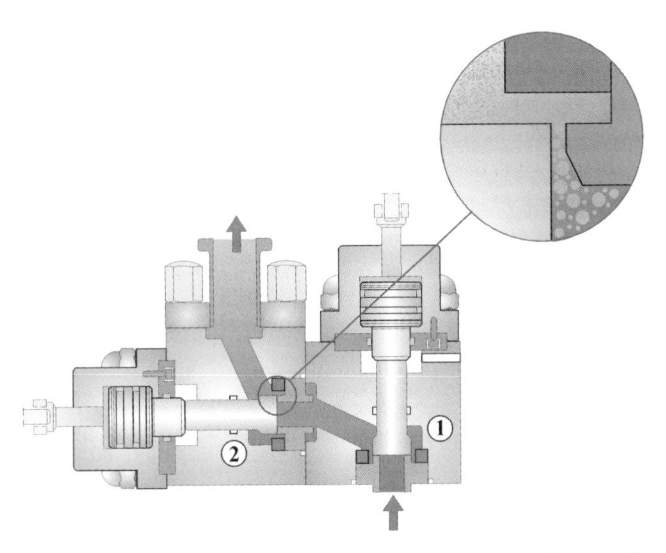

Fig. 1.9 Cross-sectional view of a homogeniser. Reproduced by permission of Tetra Pak, Lund, Sweden.

a process called homogenisation has been commonly applied in the dairy industry for a century. Homogenisation involves forcing milk through a small orifice at high pressure; the principle of operation is shown in Figure 1.9. The effectiveness of a homogeniser in reducing particle size depends on a number of factors, including homogenisation pressure, valve geometry and the number of passes through the homogeniser (Mulder & Walstra, 1974). The relationship between homogenisation pressure (P_h, in MPa) and average particle size is given by

$$\log d_{3,2} = k - 0.6 \cdot \log P_h$$

where k is a constant which generally varies between −2 and −2.5 (Walstra, 1975). Conventionally, a pressure in the range 10 to 20 MPa (100–200 bar or 14 500–29 000 psi) is used. In recent years, several so-called high-pressure homogenisers have been developed, which are capable of reaching pressures up to 400 MPa. The influence of both conventional and high-pressure homogenisation on the size distribution of MLGs is shown in Figure 1.10. Homogenisation is most effective when the milk lipids are in a liquid state, so prewarming the milk to $>40^\circ$C prior to homogenisation is required.

During homogenisation, MFGs are deformed and disruption occurs if the force of deformation is larger than the resistance to deformation, which depends on the Laplace pressure in the globule and the difference in viscosity between the plasma and lipid phases. Following disruption of the lipid globules in the homogenising valve, the surface of the lipid globules far exceeds that which the amount of original globule membrane material can cover, so adsorption of casein micelles and fragments thereof, and also, at high temperature, of some whey proteins, on the lipid-milk plasma interface occurs (Huppertz & Kelly, 2006). Freshly homogenised MLGs are susceptible to clustering if the globule surface is not covered sufficiently rapidly and, as a result, share casein micelles on their interface. The formation of homogenisation clusters is prevented by the application of a second homogenisation stage,

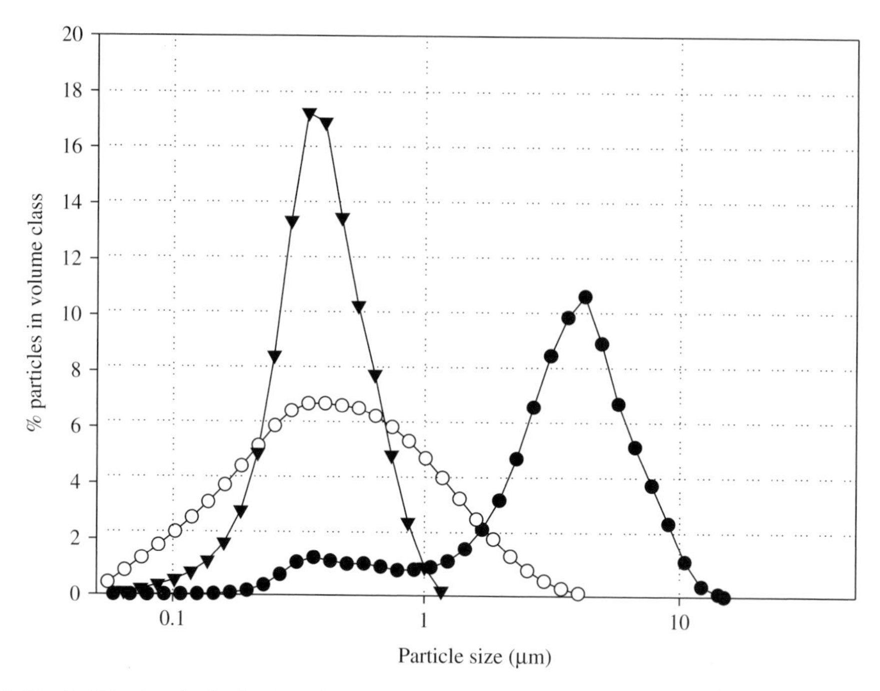

Fig. 1.10 Particle size distribution in unhomogenised milk (•) or in milk homogenised using a conventional homogeniser at 20 MPa (○) or using a high-pressure homogeniser at 200 MPa (▼). Data from T. Huppertz.

at a considerably lower pressure than the first stage (Kiesner *et al.*, 1997). Homogenisation clusters are particularly evident in products containing a high level of lipids, for example, homogenised cream, where the casein:lipid ratio is too low to provide sufficient surface coverage following homogenisation.

As a result of the altered composition of the lipid globule membrane, homogenised lipid globules behave differently than natural globules. Due to their coverage by caseins, the homogenised lipid globules behave like casein micelles, which can be beneficial in the case of yoghurt manufacture; however, the rennet-induced coagulation of milk and the stability of milk against heat-induced coagulation are impaired by homogenisation (Walstra, 1995; Huppertz & Kelly, 2006).

1.6.6 *Temperature-induced changes in milk lipid globules*

The MLGM is influenced considerably by temperature. Cooling of milk induces the transfer of phospholipids from the membrane to the milk plasma. Freezing and subsequent thawing of milk, and particularly cream, can result in the clumping of the lipid globules. Heat treatment of milk increases the amount of protein associated with the MLGM; proteins associated with the membrane as a result of heat treatment are mainly whey proteins, which presumably associate with membrane proteins via sulphydryl-disulphide interchange reactions. Heat treatment can also reduce the level of triacylglycerols in the MLGM, but the influence on the phospholipid content is unclear, with both heat-induced increases and decreases being reported (Mulder & Walstra, 1974; Walstra, 1995; Huppertz & Kelly, 2006).

1.7 Crystallisation and melting of milk triacylglycerols

The crystallisation behaviour of the triacylglycerols greatly affects the consistency of high-fat dairy products (e.g. butter), as well as the occurrence and rate of partial coalescence of MLGs; hence, knowledge of the crystallisation behaviour of the triacylglycerols of milk is very important for achieving and maintaining the desired product characteristics. Crystallisation of triacylglycerols is a complicated subject and the extremely wide and variable composition of milk triacylglycerols adds further complications. In the following section, a brief description of crystallisation of milk triacylglycerols and some factors affecting it are presented. For more detailed information, the reader is referred to the excellent reviews by Mulder & Walstra (1974), Walstra *et al.* (1995) and Wright & Marangoni (2006).

As milk fat is composed of hundreds of different triacylglycerols, it has a very broad melting range, rather than a discrete melting temperature. Milk fat is not completely solid until it reaches a temperature below – 40°C, and must be warmed to +40°C to ensure complete melting; the solid fat content in the range 0–40°C is of industrial significance and is shown as a function of temperature in Figure 1.11 for a typical milk fat.

For crystallisation of fat to occur, nucleation is required, which occurs when the molecules are 'supercooled', that is, cooled to a temperature below their melting temperature. Supercooling is the thermodynamic driving force that initiates crystallisation. On supercooling, triacylglycerols aggregate continuously into very small clusters, most of which redissolve rapidly. Only when a certain size is reached does the cluster become stable and can it start to act as a nucleus for crystallisation. The minimum size required for cluster stability decreases

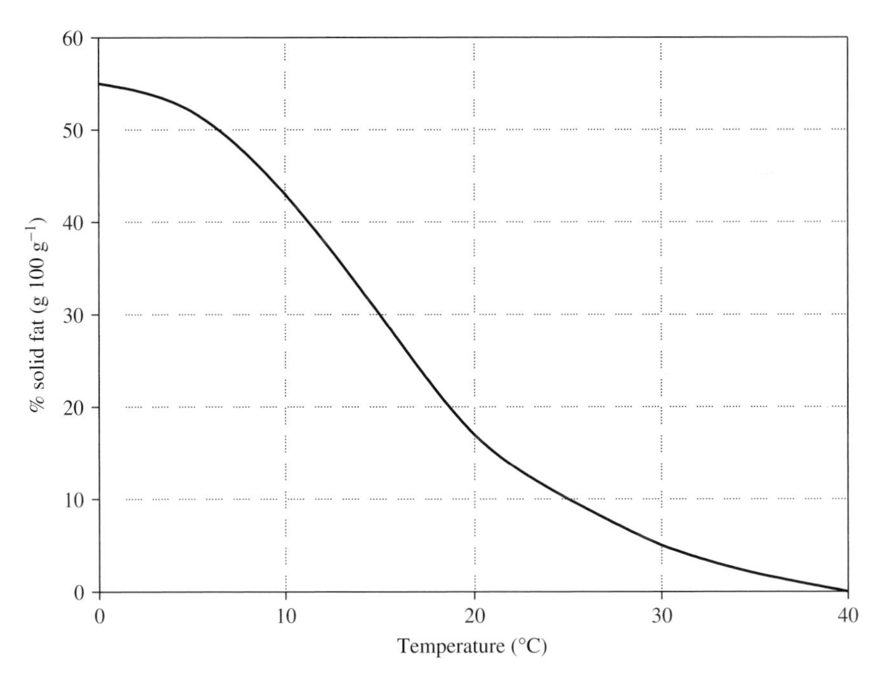

Fig. 1.11 Solid fat content of typical milk fat as a function of temperature. Redrawn using the data of Wright & Marangino (2006).

with temperature. Three types of nucleation can generally occur in fats (Walstra *et al.*, 1995; Wright & Marangoni, 2006):

- *Primary homogeneous nucleation* occurs in the absence of foreign materials and interfaces. This process generally requires very deep supercooling and is very rare in milk fat because sub-zero temperatures are required for this process to occur.
- *Primary heterogeneous nucleation* is much more common for milk fat and is initiated at the surface of catalytic impurities. The degree of primary heterogeneous nucleation decreases with decreasing temperature. For milk fat, micelles formed by monoacylglycerides may act as catalytic impurities.
- *Secondary nucleation* occurs at the interface of fat crystals that have formed during cooling and is also important in milk fat.

Following nucleation, growth of crystals can occur to a degree which depends on the degree of supersaturation, the rate of molecular diffusion to the crystal surface and the time required for triacylglycerols to fit into the growing crystal lattice. Crystals of milk fat can occur in three polymorphic modifications, designated the α, β′ and β forms. The α crystals have the simplest and least densely packed structure and are often formed first on rapid cooling; subsequently, the metastable α crystals may undergo molecular rearrangements to form the more thermodynamically stable β′ and ultimately β forms, as depicted in Figure 1.12; *vice versa* transitions, that is from β to β′ to α do not occur. In milk fat, the majority of fat crystals remain in the β′ form, even after prolonged storage (Wright & Marangoni, 2006).

In a complex mixture of triacylglycerols, like milk fat, impure crystals are formed. Milk fat may also form mixed or compound crystals, which contain two or more molecular species; the probability of these compound crystals being formed is greater when the compounds are more alike. Mixed crystals are formed most readily in the α form, where the packing density is not too dense and there is some conformational freedom to fit different molecules into the crystal lattice. Compositional variation is considerably more restricted in the β′ and, particularly, in the β crystal form (Walstra *et al.*, 1995). The presence of mixed crystals has important consequences for the influence of temperature on the solid fat phase (Mulder & Walstra, 1974; Walstra *et al.*, 1995):

- The melting range of the fat is narrower than it would be if the crystals contained only one molecular species.

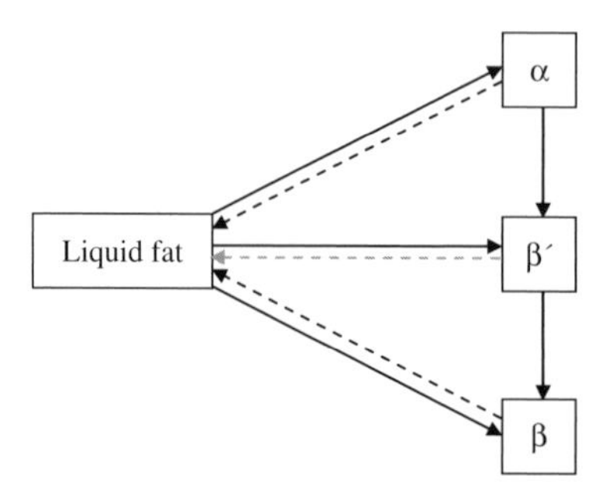

Fig. 1.12 Polymorphic transitions in fats.

- The temperature at which most of the fat melts depends on the temperature at which crystallisation occurred.
- Stepwise or slow cooling produces less crystalline fat than rapid cooling.
- Precooling to a lower temperature followed by bringing the fat to the final temperature results in greater crystallisation than direct cooling to the final temperature.

From the above, it is apparent that whether a quantity of milk fat is present as a continuous mass (e.g. anhydrous milk fat (AMF) or butter oil) or in numerous small globules (e.g. as in milk or cream) has a considerable influence on its crystallisation behaviour. Some reasons why crystallisation of fat in globules may differ from that in bulk milk fat are as follows (Mulder & Walstra, 1974; Huppertz & Kelly, 2006):

- Because of the lower thermal conductivity of bulk fat and the fact that bulk fat cannot be agitated efficiently, heat dissipation in bulk fat is considerably slower than in milk or cream.
- Not all lipid globules may contain the catalytic impurities required to initiate heterogeneous nucleation, so that nuclei would have to form spontaneously in those globules.
- The surface layer of the lipid globule may act as a catalytic impurity, for example, when it contains, mono- or diglycerides with long-chain fatty acid residues.
- The composition of bulk fat is uniform, but there are inter-globular differences in milk lipid composition, resulting in differences in crystallisation behaviour.

1.8 Conclusions

Many desirable organoleptic attributes of dairy products are due to the lipids they contain. As a result, milk lipids have always been valued highly and have been the subject of scientific interest for more than a century. As summarised in this chapter, extensive research has provided considerable insight into the lipids of milk. The >400 different fatty acids in milk, which are derived from the feed as well as *de novo* synthesis, are esterified into tri-, di- and monoacylglycerols and phospholipids and exist in emulsified state, that is, in MLG, which are surrounded by a membrane. The properties and stability of the MLGs are of crucial importance for many dairy products, as described in the later chapters of this book. To optimise the functionality and value of milk lipids, the fundamentals described in this chapter need to be considered carefully and explored further. The great English physicist Michael Faraday (1791–1867) is reported to have said 'Nothing is too wonderful to be true if it be consistent with the laws of nature, and in such things as these, experiment is the best test of such consistency' and while he did not refer to milk lipids, it is undoubtedly true that pursuit of fundamental knowledge in this area can yield surprising and fascinating new insights that will increase further the value of milk lipids.

References

Barber, M.C., Clegg, R.A., Travers, M.T. & Vernon, R.G. (1997) Lipid metabolism in the lactating mammary gland. *Biochimica et Biophyisica Acta*, **1347**, 101–126.

Bauman, D. E. & Lock, A. L. (2006) Conjugated linoleic acid: biosynthesis and nutritional significance. *Advanced Dairy Chemistry 2: Lipids*, (eds. P. F. Fox & P. L. H. McSweeney), 3rd edn., pp. 93–136, Springer, New York.

Bitman, J. & Wood, D. L. (1990) Changes in milk phospholipids during lactation. *Journal of Dairy Science*, **73**, 1208–1216.

Christie, W. W. (1995) Composition and structure of milk lipids. *Advanced Dairy Chemistry 2: Lipids*, (ed. P. F. Fox), 2nd edn., pp. 1–36, Chapman & Hall, London.

Dunkley, W. L. & Sommer, H. H. (1944). *The Creaming of Milk*, Research Bulletin No. 151, *Agricultural Experiment Station*, University of Wisconsin, Madison.

Euber, J. R. & Brunner, J. R. (1984) Reexamination of fat globule clustering and creaming in cow milk. *Journal of Dairy Science*, **67**, 2821–2832.

Fox, P. F. & Kelly, A. L. (2006) Chemistry and biochemistry of milk constituents. *Food Biochemistry and Food Processing*, (ed. Y. H. Hui), pp. 425–452, Blackwell Publishing, Oxford.

Grummer, R. R. (1991) Effect of feed on the composition of milk fat. *Journal of Dairy Science*, **74**, 3244–3257.

Harrison, R. (2006) Milk xanthine oxidase: properties and physiological roles. *International Dairy Journal*, **16**, 546–554.

Hawke, J. C. & Taylor, M. W. (1995) Influence of nutritional factors on the yield, composition and physical properties of milk fat. *Advanced Dairy Chemistry 2: Lipids*, (ed. P. F. Fox), 2nd edn., pp. 37–88, Chapman & Hall, London.

Huang, T. C. & Kuksis, A. (1967) A comparative study of the lipids of globular membrane and fat core and of the milk serum of cows. *Lipids*, **2**, 453–470.

Huppertz, T., Fox, P. F. & Kelly, A. L. (2003) High pressure-induced changes in the creaming properties of bovine milk. *Innovative Food Science and Emerging Technologies*, **4**, 349–359.

Huppertz, T. & Kelly, A. L. (2006) Physical chemistry of milk fat globules. *Advanced Dairy Chemistry 2: Lipids*, (eds. P. F. Fox & P. L. H. McSweeney), 3rd edn., pp. 173–212, Springer, New York.

Jensen, R. G. (2002) The composition of bovine milk lipids: January 1995 to December 2000. *Journal of Dairy Science*, **85**, 295–350.

Keenan, T. W. & Dylewski, D. P. (1995) Intracellular origin of milk lipid globules and the nature and structure of the milk lipid globule membrane. *Advanced Dairy Chemistry 2: Lipids*, (eds. P. F. Fox), 2nd edn., pp. 89–130, Chapman & Hall, London.

Keenan, T. W. & Mather, I. H. (2002) Milk fat globule membrane. *Encyclopedia of Dairy Sciences*, (eds. H. Roginski, J. W. Fuquay & P. F. Fox), pp. 1568–1576, Academic Press, Amsterdam.

Keenan, T. W. & Mather, I. H. (2006) Intracellular origin of milk fat globules and the nature of the milk fat globule membrane. *Advanced Dairy Chemistry 2: Lipids*, (eds. P. F. Fox & P. L. H. McSweeney), 3rd edn., pp. 137–171, Springer, New York.

Kiesner, C., Hinrichsen, M. & Jahnke, S. (1997) Mehrstufiger Druckabbau beim Homogenisieren von Rahm. *Kieler Milchwirtschaftliche Forschungsberichte*, **49**, 197–206.

King, N. (1955) *The Milk Fat Globule Membrane and Associated Phenomena*. Commonwealth Agricultural Bureau, Farnham Royal.

MacGibbon, A. K. H. & Taylor, M. W. (2002) Phospholipids. *Encyclopedia of Dairy Sciences*, (eds. H. Roginski, J. W. Fuquay & P. F. Fox), pp. 1559–1563, Academic Press, Amsterdam.

MacGibbon, A. K. H. & Taylor, M. W. (2006) Composition and structure of bovine milk lipids. *Advanced Dairy Chemistry 2: Lipids*, (eds. P. F. Fox & P. L. H. McSweeney), 3rd edn., pp. 1–42, Springer, New York.

Mather, I. H. (2000) A review and proposed nomenclature for major proteins on the milk-fat globule membrane. *Journal of Dairy Science*, **83**, 203–247.

Mather, I. H. & Jack, L. J. W. (1993) A review of the molecular and cellular biology of butyrophilin, the major protein of the bovine milk fat globule membrane. *Journal of Dairy Science*, **76**, 3832–3850.

Mulder, H. & Walstra, P. (1974) *The Fat Globule: Emulsion Science as Applied to Milk Products and Comparable Foods*, Centre for Agricultural Publishing and Documentation, Wageningen.

Noble, R. C. (1978) Digestion, absorption and transport of lipids in ruminant animals. *Progress in Lipids Research*, **17**, 55–91.

Palmquist, D. L. (2006) Milk fat: origin of fatty acids and influence of nutritional factors thereon. *Advanced Dairy Chemistry 2: Lipids*, (eds. P. F. Fox & P. L. H. McSweeney), 3rd edn., pp. 43–92, Springer, New York.

Palmquist, D. L., Beaulieu, A. D. & Barbano, D. M. (1993) Feed and animal factors influencing milk fat composition. *Journal of Dairy Science*, **76**, 1753–1771.

Parodi, P. W. (2004) Milk fat in human nutrition. *Australian Journal of Dairy Technology*, **59**, 3–59

Parodi, P. W. (2006) Nutritional significance of milk lipids. *Advanced Dairy Chemistry 2: Lipids*, (eds. P. F. Fox & P. L. H. McSweeney), 3rd edn., pp. 601–639, Springer, New York.

Patton, S. & Keenan, T. W. (1975) The milk fat globule membrane. *Biochimica et Biophysica Acta*, **415**, 273–309.

Russel, C. E. & Gray, I. K. (1979) The cholesterol content of dairy products. *New Zealand Journal of Dairy Science and Technology*, **14**, 281–289.

Schieberle, P., Gassenmeier, K., Guth, H., Sen, A. & Grosch, W. (1993) Character impact odour compounds of different kinds of butter. *Lebensmittel Wisssenschaft und Technologie*, **26**, 347–356.

Shimizu, M., Yamauchi, K. & Kanno, C. (1980) Effect of proteolytic digestion of milk fat globule membrane proteins on stability of the globules. *Milchwissenschaft*, **35**, 9–12.

Sieber, R. & Eyer, H. (2002) Cholesterol removal from dairy products. *Encyclopedia of Dairy Sciences*, (eds. H. Roginski, J. W. Fuquay & P. F. Fox), pp. 1611–1617, Academic Press, Amsterdam.

Smith, S., Witkowski, A. & Joshi, A. K. (2003) Structural and functional organization of the animal fatty acid synthase. *Progress in Lipid Research*, **42**, 289–317.

Taylor, M. W. & MacGibbon, A. K. H. (2002) Lipids–general characteristics. *Encyclopedia of Dairy Sciences*, (eds. H. Roginski, J. W. Fuquay & P. F. Fox), pp. 1544–1550, Academic Press, Amsterdam.

Walstra, P. (1969) Studies on milk fat dispersion. II. The globule-size distribution of cow's milk. *Netherlands Milk and Dairy Journal*, **23**, 99–110.

Walstra, P. (1975) Effect of homogenization on the fat globule size distribution in milk. *Netherlands Milk and Dairy Journal*, **29**, 297–294.

Walstra, P. (1995) Physical chemistry of milk fat globules. *Advanced Dairy Chemistry 2: Lipids*, (ed. P. F. Fox), 2nd edn., pp. 131–178, Chapman & Hall, London.

Walstra, P. (2003) *Physical Chemistry of Foods*, Marcel Dekker, Inc., New York.

Walstra, P., van Vliet, T. & Kloek, W. (1995) Crystallization and rheological properties of milk fat. *Advanced Dairy Chemistry 2: Lipids*, (ed. P. F. Fox), 2nd edn., pp. 179–211, Chapman & Hall, London.

Walstra, P., Geurts, T. J., Noomen, A., Jellema, A. & van Boekel, M. A. J. S. (1999) *Dairy Technology, Principles of Milk Properties*, Marcel Dekker Inc., New York.

Walstra, P., Wouters, J. T. M. & Geurts, T. J. (2006) *Dairy Technology*, 2nd edn., Marcel Dekker Inc., New York.

Ward, R. E., German, J. B. & Corredig, M. (2006) Composition, application, fractionation, technological and nutritional significance of the milk fat globule membrane material. *Advanced Dairy Chemistry 2: Lipids*, (eds. P. F. Fox & P. L. H. McSweeney), 3rd edn., pp. 213–244, Springer, New York.

Wright, A. J. & Marangoni, A. G. (2006) Crystallization and rheological properties of milk fat. *Advanced Dairy Chemistry 2: Lipids*, (eds. P. F. Fox & P. L. H. McSweeney), 3rd edn., pp. 245–291, Springer, New York.

2 Milk Fat Nutrition

P.W. Parodi

2.1 Introduction

Milk from domesticated animals has provided food for humans for more than 8500 years (Patton, 2004). Milk contains a wide range of readily bioavailable nutrients, which enables this nutrient-dense product to be the sole food for neonates and infants during the first stage of growth and development. Moreover, milk and dairy products make a significant contribution to the total supply of nutrients for adolescents and adults (Gurr, 1998).

From the early 1900s, public health authorities encouraged the consumption of milk and milk products to improve the nutritional status of the population, especially children. However, the image of milk and its products was blemished during the past few decades, mainly because they contain saturated fatty acids that can increase the level of serum cholesterol, which is considered a risk factor for coronary heart disease (CHD). More recently, fat in general has been linked to the emerging obesity epidemic and to cancer at some sites (Parodi, 2004, 2006).

Recommendations by public health authorities to limit milk fat intake means that the diet is deprived of several bioactive components that may help prevent disease throughout the life span. This review discusses the nutrition of these components and examines the role of milk fat in coronary artery disease (CAD), obesity and cancer. The composition of milk fat components is found in reviews by Jensen (1995, 2002) and Parodi (2003, 2004).

2.2 Conjugated linoleic acid

Perhaps the most important bioactive component in milk fat is conjugated linoleic acid (CLA), which is the descriptor for all possible positional and geometric isomers of octadecadienoic acid with conjugated double bonds. Milk fat is the richest natural source of CLA with *cis*-9, *trans*-11-18:2, now named rumenic acid (RA), representing more than 90% of the CLA isomers present. Some 20 minor isomers have been reported, the predominant one being *trans*-7, *cis*-9-18:2; however consumption of a low-fibre diet can increase the level of *trans*-10, *cis*-12-18:2 (Parodi, 2003; Lock & Bauman, 2004).

2.2.1 *Origin of rumenic acid*

In part, RA is produced in the rumen from pasture and feed polyunsaturated fatty acids as a consequence of certain biohydrogenation-associated reactions. A linoleate *cis*-12, *trans*-11 isomerase from the rumen bacteria *Butyrivibrio fibrisolvens* transposes the cis-double bond at carbon-12 of dietary linoleic acid (*cis*-9, *cis*-12-18:2) to carbon-11, and in doing so assumes the trans-configuration to produce RA, a transitory intermediate that is subsequently hydrogenated to *trans*-11-18:1 (vaccenic acid, VA). The major pasture fatty acid, α-linolenic (*cis*-9, *cis*-12, *cis*-15-18:3) cannot be converted directly to RA; it is rather hydrogenated to VA. A portion of the VA produced escapes further hydrogenation and passes from the rumen, is absorbed from the digestive tract and transported in the circulation to peripheral adipose tissue and the lactating mammary gland where Δ^9-desaturases convert it to RA (Lock & Bauman, 2004).

2.2.2 *CLA nutrition*

During the past two decades, multiple animal studies showed that CLA has anticarcinogenic, antiatherogenic, antidiabetogenic, antiadipogenic, immunomodulating and bone growth–enhancing properties (Pariza *et al.*, 2001; Roche, *et al.*, 2001; Belury, 2002a; Parodi, 2002; Wahle *et al.*, 2004). Most of these studies, especially the early ones, used a synthetic mixture of many CLA isomers that usually contained about 70 to 80% of near equal quantities of RA and *trans*-10, *cis*-12-18:2. Now it is realised that these two isomers may exert disparate, common or synergistic effects (Pariza *et al.*, 2001). Thus, it is not possible to identify the biological effects of RA from most of these studies. In this review, mixed isomers will be denoted by CLA whereas individual isomers will be separately indicated.

2.2.3 *CLA as an anticancer agent*

Pariza & Hargraves (1985) were the first to report an anticancer action for CLA when they showed that a mutagenesis inhibitor from beef, later identified as CLA, inhibited chemically induced murine epidermal tumour development. Since then, many studies have demonstrated that CLA at physiological concentrations suppressed the growth of an extensive range of human and animal cultured cancer cells that included colon, liver, bladder, lung, breast, ovarian, prostate, mesothelioma, glioblastoma and leukaemia. This inhibition by CLA was in contrast to linoleic acid that usually stimulated growth in these cell lines (Scimeca, 1999; Parodi, 2002, 2004).

Dietary CLA at levels of around 1% inhibited the growth of chemically induced skin, stomach, colon and mammary gland tumours. Further, the growth and metastasis from human prostate or breast cancer cells transplanted into mice was inhibited by dietary CLA (Scimeca, 1999; Belury, 2002a, 2002b; Parodi, 2002, 2004). In particular, CLA is a potent inhibitor of mammary tumourigenesis. In a seminal study, Ip *et al.* (1991) fed rats a basal diet or a diet supplemented with 0.5, 1.0 and 1.5 g 100 g^{-1} CLA for 2 weeks before and following administration of a chemical carcinogen. At the conclusion of the trial, the total number of mammary adenocarcinomas in the rats fed 0.5, 1.0 and 1.5 g 100 g^{-1} CLA was reduced by 32, 56 and 60%, respectively, compared to the rats that were fed the basal diet. Tumour incidence (percentage of rats with tumours), tumour multiplicity (number of tumours per rat) and total tumour weight were reduced similarly.

A number of subsequent studies by this group confirmed that dietary CLA inhibited mammary tumourigenesis and extended our understanding of its action (Scimeca, 1999; Ip *et al.*, 2003; Parodi, 2004). When the dose of carcinogen was halved, and the tumours took longer to develop, CLA at levels of 0.05 to 0.5 g 100 g^{-1} was effective for tumour inhibition. CLA was equally effective with a direct-acting carcinogen or with a carcinogen that required prior endogenous metabolic activation. The amount and type of fat in the diet of rodents can influence tumourigenesis. However, tumour inhibition was similar when CLA was part of a low-fat (5 g 100 g^{-1}) or high-fat (20 g 100 g^{-1}) diet. Inhibition of tumour development was also similar when CLA was included in a 20 g 100 g^{-1} unsaturated fat diet as corn oil or a 20 g 100 g^{-1} saturated fat diet as lard. Although a diet containing 12 g 100 g^{-1} linoleic acid produced more mammary tumours than a 2 g 100 g^{-1} linoleic acid diet, CLA inclusion reduced tumour development to the same extent. CLA was preferentially incorporated into the neutral lipids of mammary tissue. When CLA was removed from the diet its disappearance from the neutral lipids paralleled the rate of appearance of new tumours. The breast is abundant in adipose tissue, which can store dietary CLA and may influence future challenge from carcinogens.

The age of the rat at the time of CLA administration can influence its outcome. When rats were fed CLA from weaning at 21 days of age until day 51 only, and then administered a carcinogen at day 57, they were protected from subsequent tumour development. However, when CLA was included in the diet for the same period after carcinogen administration and when the animals were older, there was no protection against tumour development. A continuous intake of CLA was then necessary to obtain equivalent protection. The period from age 21 to 51 days corresponds to development of the mammary gland to adult stage morphology. Further studies showed that during the pubescent period, CLA reduced the development and branching of the expanding mammary ductal tree. There was also a reduction in the density and rate of proliferation of terminal-end bud (TEB) cells. TEBs are the least differentiated and most actively growing glandular ductal structures, which are most abundant from weaning to puberty and are the site of chemically induced tumours (Ip *et al.*, 2003). These observations could be important because there is evidence that in humans the risk of breast cancer associated with certain events increased with decreasing age at which exposure occurred or commenced (Parodi, 2005).

2.2.4 *Rumenic acid and mammary tumour prevention*

The initial difficulty in preparing pure isomers of RA and *trans*-10, *cis*-12-18:2, and then their high cost, hampered studies to elucidate their individual properties. However, Ip *et al.* (2002) later demonstrated that both isomers were equally effective in suppressing chemically induced tumour development in rats. With cell culture studies, it has been found that both isomers generally inhibit cell proliferation, but the relative efficiency can vary with cell line. Also, the individual isomers may modulate different anticancer pathways (Chujo *et al.*, 2003).

Because of the potential health benefits for RA, animal scientists developed techniques to increase its level in milk by dietary manipulation. Ip *et al.* (1999) fed an RA-enriched butter produced by these techniques to rats during the time of pubescent mammary gland development. Compared to control butter with normal RA levels, the RA-enriched product reduced mammary epithelial mass, decreased the size of the TEB population, suppressed the proliferation of TEB cells and decreased tumour incidence and total tumour numbers.

The extent of these benefits was similar to those for the animals that were fed an equivalent amount of synthetic RA or *trans*-10, *cis*-12-18:2. There was a consistently higher level of RA in the liver, mammary fat pad, peritoneal fat and blood serum of rats that were fed RA-enriched butter than in synthetic RA-fed rats. This elevated RA resulted, no doubt, from endogenous Δ^9-desaturation of VA, which was concurrently increased in the RA-enriched butter.

Studies in animals and humans showed that dietary VA, the predominant trans-monounsaturated fatty acid of milk fat, could be converted endogenously to RA by Δ^9-desaturases. About 50 g 100 g^{-1} of the VA in tissues of mice was converted to RA. A small study in humans indicated that about 20 g 100 g^{-1} of dietary VA was converted to RA, but individual variation was large (Parodi, 2004; Palmquist *et al.*, 2005). Follow-up anticancer studies showed that 2 g 100 g^{-1} dietary VA inhibited chemically induced premalignant lesions in the mammary gland to the same extent as 1 g 100 g^{-1} RA (Banni *et al.*, 2001). Corl *et al.* (2003) produced a high VA-enriched butter that was used to produce diets low in RA, but with increasing levels of VA. Increasing the VA content of the diet resulted in a dose-dependent accumulation of RA in the mammary fat pad, which was accompanied by a corresponding decrease in mammary tumour development. However, Lock *et al.* (2005a) demonstrated that this antitumour effect of VA was blocked by a Δ^9-desaturase inhibitor and concluded that the antitumorigenic effect is predominantly, perhaps exclusively, mediated through its conversion to RA.

Mechanisms for the anticancer action of CLA and RA have been studied in detail, but are still poorly understood. However, they can influence numerous events in various pathways associated with carcinogenesis, but a discussion is beyond the scope of this chapter and more detail can be obtained from reviews by Scimeca (1999), Belury (2000a, 2000b), Ip *et al.* (2003), Parodi (2002, 2004) and Wahle *et al.* (2004). Nevertheless, it is important to point out here that tumours develop when the processes of cell proliferation and apoptosis are out of balance. Apoptosis or programmed cell death is the process where the body eliminates cells where DNA damage is irreparable. RA can reduce cell proliferation and induce apoptosis. RA can also inhibit angiogenesis. Angiogenesis is the process for generation of new blood capillaries through sprouting of pre-existing vessels and is an absolute requirement for tumour growth. In addition, RA can inhibit the activation of cyclooxygenase (COX)-2 that catalyses the conversion of arachidonic acid to prostaglandins (PG) that are associated with carcinogenesis. COX-2 is overexpressed in many forms of cancer and animal studies have demonstrated that COX-2 inhibitors can suppress tumour development.

Epidemiology

Evidence from epidemiological studies for an association between RA intake and the risk of breast cancer are inconsistent. The initial case–control study by Aro *et al.* (2000) found an inverse association between dietary RA, serum RA and VA and the risk of breast cancer in Finnish postmenopausal women, but not in premenopausal women. A New York study found that although RA was not related to overall breast cancer risk, the risk of oestrogen receptor–negative breast cancer, which is the most aggressive form, in premenopausal women was reduced with increasing intake (McCann *et al.*, 2004). Three other studies did not suggest an association between RA consumption and breast cancer risk. Methodological inadequacies in these studies have been discussed (Parodi, 2004).

2.2.5 *CLA, RA and colon tumour prevention*

The role of CLA in the prevention of intestinal tumours has not been studied as extensively as for mammary tumour prevention. CLA effectively inhibited the growth of a range of human colon cancer cell lines, and a few studies showed that it could suppress the development of chemically induced colon tumour and premalignant lesions induced by mutagenic heterocyclic amines in cooked food. Some studies, however, did not detect any benefit for CLA (Scimeca, 1999; Parodi, 2004; Wahle *et al.*, 2004). Studies with RA are sparse. RA can inhibit growth in cultures of human colon cancer cell lines, and Chen *et al.* (2003) reported that RA significantly inhibited chemically induced mouse forestomach tumours. Recently, a large Swedish epidemiological study of prospective format suggested that a high intake of RA and also high-fat dairy food reduced the risk of colorectal cancer (Larsson *et al.*, 2005).

2.2.6 *Rumenic acid and the prevention of atherosclerosis*

Early studies with rabbits and hamsters, common models for the study of atherosclerosis, showed that CLA added at a level of 1 g 100 g^{-1} or less to an atherogenic diet could reduce atherosclerotic lesion formation. In these studies the effect of CLA on serum lipoproteins was variable (Roche *et al.*, 2001; Kritchevsky *et al.*, 2004; Parodi, 2004). There are now several reports on the antiatherogenic action of RA. Kritchevsky *et al.* (2004) demonstrated that RA not only inhibited the development of cholesterol-induced atherosclerosis in rabbits but also reduced the severity of pre-existing lesions.

The hamster is a popular model for the study of atherogenesis because, unlike most other models, its serum lipoprotein profile resembles that of humans. However, different strains of hamsters respond differently to dietary perturbations. Lock *et al.* (2005b) showed that a naturally RA-enriched butter (0.46 g RA kg^{-1} diet) significantly reduced serum total cholesterol, LDL-cholesterol and triglycerides and increased HDL-cholesterol by 22% in cholesterol-fed hamsters compared to an equivalent dietary level of a conventional RA butter (0.07 g RA kg^{-1} diet).

Mitchell *et al.* (2005) found that RA reduced the frequency of atherosclerotic lesions in the hamster aorta compared to a control diet supplemented with an equivalent level of linoleic acid. In this study, there was no significant effect on serum lipoprotein profile. In another study, hamsters were fed a butter fat diet with or without 1 g 100 g^{-1} RA or fish oil. The RA-supplemented diet produced the lowest aortic lipid deposition with little change in serum lipoproteins (Valeille *et al.*, 2005). The studies of Mitchell *et al.* (2005) and Valeille *et al.* (2005) were severe tests for the effect of RA on atherosclerosis since the control diets contained equivalent levels of linoleic acid and fish oil, respectively, which are considered antiatherogenic compounds.

Extrapolation of data from animal models to humans is often criticised. Recently, gene deletion technology has allowed the development of a range of transgenic or knockout mice suitable for atherosclerosis research. A particularly popular model is the apoprotein E–deficient mouse (apo$E^{-/-}$), which on a standard chow diet has the propensity to spontaneously develop atherosclerotic lesions that are similar to those that develop in humans. Toomey *et al.* (2006) fed apo$E^{-/-}$ mice a normal diet or a diet supplemented with 1% RA. Inclusion of RA in the diet not only retarded further development of atherosclerotic lesions but also induced regression in established lesions.

A review of human studies indicates that CLA has no major effect on serum total, LDL- and HDL-cholesterol levels (Terpstra, 2004). Recently, Tricon *et al.* (2004) reported the first human study to compare the effect of RA and *trans*-10, *cis*-12-18:2 on blood lipids. They found that the two isomers had divergent effects. *Trans*-10, *cis*-12-18:2 increased the ratios of serum low-density lipoprotein to high-density lipoprotein (LDL:HDL) cholesterol and total to HDL cholesterol, and triglycerides levels whereas RA decreased them. No human studies have investigated the effect of RA or CLA on arterial pathology. In the animal studies, RA-induced improvement in atherosclerosis could not be associated with changes in serum lipid profile. Thus, other factors may be partially or even totally responsible for the antiatherosclerotic properties of RA.

It is now generally accepted that atherosclerosis is a chronic inflammatory disease. Inflammatory responses are involved in every stage of the disease, that is, initiation, progression, rupture of the fibrous plaque and subsequent thromboses (Libby *et al.*, 2002). These mechanisms are exceedingly complex and involve a large number of chemokines, cytokines, growth factors, enzymes and vasoregulatory molecules, which are expressed by monocytes, macrophages, T-lymphocytes and smooth muscle cells as a result of immune responses.

There is emerging evidence from animal studies that RA-induced changes in atherosclerosis development are associated with negative regulatory expression of pro-inflammatory genes (Mitchell *et al.*, 2005; Toomey *et al.*, 2006; Valeille *et al.*, 2005).

2.2.7 *Trans fatty acids and coronary heart disease*

The role of trans fatty acids (TFAs) in CHD is of current concern for the dairy industry. In the early 1990s, a number of clinical studies showed that consumption of TFAs increased serum levels of LDL cholesterol and decreased levels of HDL cholesterol, which are markers associated with increased risk of CHD. Later, a number of large prospective epidemiological studies found a positive association between the intake of TFAs and the risk of CHD (see the reviews by Parodi, 2004, 2006).

Because of these observations, a number of countries have introduced, or are considering introducing, regulations designed to reduce trans fatty acid intake, including nutritional labelling of the trans fatty acid content of food items and limiting the use of industrially prepared trans fats (Lock *et al.*, 2005c). However, five of the six epidemiological studies that assessed separately the risk of CHD for TFAs from animal fat and hydrogenated vegetable oil did not find that animal TFAs were associated with risk. TFAs from both sources contain a range of positional isomers. In the case of animal fat, a large proportion of its TFAs are represented by VA, which has a double bond at position 11 (Δ^{11}) whereas in the case of hydrogenated oil, the double bond of the TFAs exhibit a gaussian type distribution centred on Δ^9 to Δ^{12}. As discussed previously, VA can be bioconverted to RA and RA itself is a major contributor to the total TFA content of milk and other animal fat (Parodi, 2004; Lock *et al.*, 2005c). The inhibition of atherosclerosis in animal models by RA and the fact that addition of RA and other dairy TFAs to human diets did not adversely affect serum lipoproteins (Tricon *et al.*, 2004; Desroches *et al.*, 2005) provides a plausible biological explanation for the divergent results in epidemiological studies for animal- and vegetable oil-derived TFAs and the risk of CHD.

2.2.8 *Rumenic acid and immunomodulation*

Many *in vitro*, *ex vivo* and animal studies with a number of species show that CLA can modulate aspects of both the innate and adaptive immune systems, although results were not always consistent (O'Shea *et al*., 2004). Furthermore, recent studies with RA and the *trans*-10, *cis*-12 isomer suggest they can exert different effects on the immune system (O'Shea *et al*., 2004; de Roos *et al*., 2005). *In vitro* and *ex vivo* studies with RA (Yu *et al*., 2002; de Roos *et al*., 2005; Loscher *et al*., 2005) show that it can decrease the production of a number of pro-inflammatory factors including the following:

- COX-2 and PGE_2
- Inducible nitric oxide synthase (NOS) and nitric oxide (NO), which is a potent vaso-relaxant secreted by the endothelium that helps maintain blood vessel integrity
- Cytokines (tumour necrosis factor (TNF)-α, interleukin (IL)-1, IL-6 and IL-12)
- Macrophage migration inhibitory factor

In addition, RA can increase the production of the anti-inflammatory cytokine IL-10 and different forms of heat shock protein. These factors are important in disease states associated with inflammation such as allergy, arthritis, atherosclerosis and cancer.

The beneficial effect of RA on inflammatory activity results, at least in part, from the inactivation of the transcription factor nuclear factor-κB (NF-κB) that promotes genes encoding proinflammatory cytokines (Loscher *et al*., 2005). Suppression of NF-κB in turn results from upstream activation of peroxisome proliferator-activated receptor (PPAR)γ, a nuclear receptor for which RA is a potent agonist (Belury, 2002a). PPARγ is widely expressed in immune cells and also in other RA target cells. PPARs are promiscuous receptors that are activated by a number of ligands like triazolidinediones used to treat diabetes and the fibrate group of drugs for treating dyslipidemias. Like RA, these agonists also exert multiple health benefits (Parodi, 2002).

Results from the few human studies with RA have proved unremarkable for the dose and limited range of immune markers studied.

2.2.9 *Rumenic acid and type 2 diabetes mellitus*

The incidence of type 2 diabetes (non-insulin-dependent diabetes mellitus) is approaching epidemic proportions worldwide and is a major cause of death. The underlying abnormality is insulin resistance (i.e. a situation in which the normal amount of secreted insulin fails to elicit adequate glucose uptake and glycogen synthesis in target tissues such as skeletal muscle and adipose tissue) with progression to impaired glucose tolerance, which are important risk factors for CHD. In the Zucker Diabetic Fatty rat, a model for obesity and type 2 diabetes, supplementation of diets with 1.5 g 100 g^{-1} CLA normalised impaired glucose tolerance and reduced serum insulin, triglycerides and nonesterified fatty acid (NEFA) levels. The magnitude of these reductions was similar to littermates supplemented with troglitazone, a member of the thiazolidinedione family of antidiabetic drugs. A subsequent study that compared CLA with RA from an enriched butter suggested that it is *trans*-10, *cis*-12-18:2 that possesses antidiabetogenic properties (Belury, 2000a).

In contrast, Roche *et al*. (2002) found *trans*-10, *cis*-12-18:2, but not RA, promoted insulin resistance and significantly increased serum glucose and insulin levels in obese *ob/ob* mice. *Trans*-10, *cis*-12-18:2 also promoted insulin resistance in apo$E^{-/-}$ mice, whereas RA beneficially modulated indices and serum markers of insulin resistance compared to the mice that were fed a control diet (de Roos *et al.*, 2005). It is not known if these divergent outcomes reflect species differences or the diabetic state of the test animals.

A few human studies are available from which isomer-specific effects on insulin resistance can be deduced. There were no differences in serum triglycerides, NEFAs, glycerol and insulin levels in healthy men and women supplemented for 8 weeks with either CLA or an RA-rich mixture compared to a linoleic acid control (Noone *et al.*, 2002). On the other hand, Riserus *et al.* (2002a) found that supplementation with *trans*-10, *cis*-12-18:2 for 12 weeks decreased insulin sensitivity and increased serum glucose and insulin levels compared to CLA and a placebo in obese men with signs of the metabolic syndrome (abdominal obesity, dyslipidemia and hypertension). It is interesting to note that in this study *trans*-10, *cis*-12-18:2 increased serum levels of C-reactive protein (CRP) compared to placebo, whereas CLA significantly decreased CRP levels (Riserus *et al.*, 2002b). CRP is a marker of systemic inflammation and is a powerful predictor of CHD (Libby *et al.*, 2002). These results suggest that RA has a neutral or perhaps beneficial effect on insulin resistance. A small follow-up study where obese men were supplemented with RA did not show a significant increase in insulin resistance after adjustment for lipid peroxidation (Riserus *et al.*, 2004).

There is evidence from *in vitro* studies with human adipocytes that *trans*-10, *cis*-12-18:2 down-regulates the expression of PPARγ and promotes NF-κB activation that leads to impaired glucose uptake and insulin resistance (Chung *et al.*, 2005).

2.2.10 *Rumenic acid as a growth factor*

CLA has been shown to reduce body fat and increase lean body mass in growing animals, and dietary supplements are widely promoted for weight loss. However, it is the *trans*-10, *cis*-12-isomer and not RA that modulates body composition (Pariza *et al.*, 2001). Nevertheless, RA may have growth-promoting properties.

Bone formation

There is tentative evidence that RA can promote bone mineral density and bone formation. Watkins *et al.* (1997) showed that growing chicks fed butter fat had a higher rate of bone growth in the tibia relative to soya bean oil. Increased bone formation was associated with decreased *ex vivo* production of PGE_2, a known bone resorption factor, and an increased level of insulin-like growth factor-1, which is an established bone growth factor. Later, Li & Watkins (1998) fed rats CLA and found that for all tissues analysed, CLA, and particularly the RA isomer, was preferentially incorporated in bone lipids. A recent epidemiological study suggested that RA intake might benefit hip and forearm bone mineral density in postmenopausal women (Brownbill *et al.*, 2005). At physiological concentrations, PPAR activators can induce the maturation of bone forming osteoblasts *in vitro* and bone formation *in vivo* (Jackson & Demer, 2000). This action may explain the bone growth properties of RA, since RA is a potent activator of PPARs (Belury, 2002a).

Infant growth

Pariza *et al.* (2001) presented data from a small study with weanling mice fed a control diet or a diet supplemented with RA or *trans*-10, *cis*-12-18:2. The RA diet enhanced body weight gain and feed efficiency compared to the other diets, whereas *trans*-10, *cis*-12-18:2 reduced body fat levels, but did not enhance body growth or feed efficiency.

2.3 Sphingolipids

Sphingolipids are a diverse family of compounds based on a long-chain amino alcohol called sphingosine. If a fatty acid is linked to the -NH_2 group of sphingosine, a ceramide is produced. Further, addition of a phosphocholine group to the primary hydroxyl group of sphingosine results in the phospholipid sphingomyelin, whereas attachment of one or more sugar residues produces glycolipids of various complexity including cerebrosides and gangliosides. Sphingolipids are associated with the outer surface of membranes; thus, the milk fat globule membrane, which is enhanced in buttermilk and whey lipids, is a rich source. Sphingolipid content and composition of milk is reported in Parodi (2004) and Rombaut & Dewettinck (2006).

Sphingomyelin (SM) is the predominant sphingolipid. In cells, extracellular agonists such as IL-1, TNF-α and 1,25-dihydroxyvitamin D_3 [1,25$(OH)_2$ D_3], acting through their membrane receptors, activate sphingomyelinase that produces ceramide. Other agonists can activate ceramidase to generate sphingosine. Ceramide and sphingosine act as second messengers for the agonists in signalling cascades that regulate cell proliferation, differentiation and apoptosis, and have been referred to as tumour suppressor lipids (Parodi, 2004; Schmelz, 2004).

Sphingolipids are digested throughout the small intestine and colon to produce ceramide and sphingosine, which are absorbed by enterocytes where they may be utilised to re-synthesise sphingolipids. Thus, diet can subject the intestinal tract to a ready supply of the tumour suppressor lipids with the potential to prevent colon cancer (Schmelz, 2004; Duan, 2005).

2.3.1 *Sphingolipids in colon cancer prevention*

In a series of animal studies, Schmelz (2004) and Schmelz *et al.* (2006) showed that SM and other sphingolipids did indeed inhibit tumour development in the intestinal tract, and they concluded the following:

- Mice fed 0.025 to 0.1 g 100 g^{-1} of milk-derived SM had less than half the incidence of chemically induced colon tumours than non-supplemented mice.
- The ganglioside GM_1 fed at a level of 0.025 g 100 g^{-1} produced significantly fewer chemically induced colonic aberrant crypt foci (ACF; microscopically determined pre-neoplastic lesions that can develop into colon tumours) than control-fed mice.
- There was also a significant reduction in ACF in mice that were fed up to 0.1 g 100 g^{-1} SM.
- In another study, supplementation with SM did not reduce tumour incidence, but the tumours in the control-fed mice were all invasive adenocarcinomas whereas in the SM-fed mice there were many benign adenomas.
- A synthetic SM produced a comparable suppression of ACF to the milk-derived products, which indicates that the benefits ascribed to milk-derived SM in previous experiments were not due to a contaminant co-extracted from milk.

- Milk-derived complex sphingolipids glucosylceramide, lactosylceramide and the ganglioside GD_3 all reduced the level of ACF development to the same extent as SM.
- *Apc*$^{min/+}$ mice have a germline mutation in the adenomatous polyposis coli (*Apc*) gene and they spontaneously develop adenomas (polyps) throughout the intestinal tract, but preferentially in the small intestine. In humans, germline mutations in the tumour suppressor *APC* gene are responsible for familial adenomatous polyposis (FAP), a syndrome in which patients develop multiple adenomas in the colon. If not resected, a proportion of these adenomas may progress to become malignant. In human colon cancer, mutations in the *APC* gene initiate the development of most tumours. *Apc*$^{min/+}$ mice were fed a control diet or a diet supplemented with 0.1 g 100 g^{-1} ceramide, a mixture of sphingolipids similar to their composition in milk and this mixture plus ceramide (60:40). All test diets dramatically reduced spontaneous tumour development in all regions of the intestine.
- SM was equally effective in inhibiting tumour development when supplementation commenced before administration of the chemical carcinogen or after tumour initiation, which suggests both a chemoprotective and chemotherapeutic benefit.
- Prevention of tumour development by sphingolipids was associated with normalisation of cell proliferation and apoptosis.

The rate of sphingolipid digestion in the intestinal tract reflects the distribution pattern of alkaline sphingomyelinase and ceramidase in the intestine. In humans, the activity of alkaline sphingomyelinase was 75% less in tissue from human colon carcinomas than in normal colonic tissue. In FAP patients, there was a 90% decrease in alkaline sphingomyelinase activity in colonic adenomas and in surrounding mucosa compared to tissue from healthy controls. Human colon carcinomas had a 50% decrease in cellular ceramide content compared to normal colon mucosa. In addition, SM can protect colonic cells against the toxic action of bile acids, which can increase proliferation and increase the risk of carcinoma (Duan, 2005; Parodi, 2006).

Administration of SM was shown to enhance the effect of the anticancer drug 5-fluorouracil in mice bearing human colon cancer cells (Modrak *et al.*, 2000). A recent preliminary study suggests that sphingolipid metabolites may have a role in chemoprevention and chemotherapy of ovarian cancer (Schmelz *et al.*, 2006).

2.3.2 *Sphingomyelin and cholesterol absorption*

Animal studies showed that dietary SM could inhibit the absorption of cholesterol, which in turn may be responsible for lower serum cholesterol levels. The chow-fed mice absorbed 31.4% of its cholesterol. Supplementation with 0.1, 0.5 and 5 g 100 g^{-1} milk-derived SM resulted in a decrease in cholesterol absorption by 20, 54 and 85%, respectively. This reduced absorption was greater than the values obtained with dipalmitoyl phosphatidylcholine and much greater than with egg yolk phosphatidylcholine (Eckhardt *et al.*, 2002).

The interaction of SM and cholesterol is considered to be due to the ability of SM to form a hydrogen bond between the amide group of SM and the hydroxyl group of cholesterol (Noh & Koo, 2004). Dietary SM reduced intestinal absorption of fat as well as that of cholesterol. Again, milk-derived SM was more effective than egg SM, which suggested that chain length as well as saturation of the acyl chain of SM is the determining factor for inhibition of lipid absorption (Noh & Koo, 2004). In a long-term study, rats fed 1 g 100 g^{-1} SM throughout two generations had a decrease in serum cholesterol of 30%. This reduction was accomplished without affecting the SM concentration in serum phospholipids (Kobayashi *et al*., 1997).

2.3.3 *Sphingomyelin and the immune system*

Ceramide-dependent signalling in the SM pathway described previously also occurs in most cells comprising the immune system and influences their development, activation and regulation (Ballou *et al*., 1996; Cinque *et al*., 2003).

In addition to the customary role in cell proliferation, differentiation and apoptosis, ceramide signalling plays an important role in efficient phagocytosis and B- and T-lymphocyte function, particularly antigen processing. In T cells, the SM pathway can influence the production of cytokines such as IL-1 to IL-6, IL-10, TNF-α and interferon-γ that are critical in the inflammatory process. These cytokines can by autocrine and paracrine mechanisms influence important downstream targets like COX-2, NOS, NF-κB, PGE_2, matrix metalloproteinase and adhesion molecules in endothelial cells (Ballou *et al*., 1996; Cinque *et al*., 2003). This is an emerging and difficult area of study that requires much research to identify the various interweaving pathways.

2.3.4 *Sphingolipids and intestinal diseases*

Milk sphingolipids may play an important role in the maintenance of intestinal health, especially in infants. In the intestine, sphingolipids act as membrane receptors for various bacteria, viruses and protein toxins (Hanada, 2005). Soluble synthetic sphingolipids bound to the human immunodeficiency virus (HIV) *in vitro* and prevented entry to immune cells, the natural cellular target for HIV (Fantini *et al*., 1997). Thus, dietary sphingolipids, accentuated by their slow digestion in the intestine, may act as 'false intestinal receptors' for some enteric pathogens and toxins (Rueda *et al*., 1998). Rueda *et al*. (1998) showed that a ganglioside supplemented milk formula not only reduced the intestinal content of *Escherichia coli* in preterm newborn infants, but also acted as a prebiotic in increasing the beneficial bifidobacteria population of the intestine.

Digestion products of sphingolipids, lysosphingomyelins and sphingosine were powerful bactericidal agents *in vitro* that inhibited the growth of *E. coli*, *Salmonella enteritidis*, *Campylobacter jejuni*, *Listeria monocytogenes* and *Clostridium perfringens* (Sprong *et al*., 2001). Further, a milk-derived SM-supplemented artificial formula accelerated enzymatic and morphological maturation of the intestine in young rats (Motouri *et al*., 2003).

Total milk phospholipids protected against ethanol- and aspirin-induced gastric mucosal damage in rats. In humans, milk phospholipid supplementation protected the gastric mucosa against aspirin provocation. A common, yet potentially serious, intestinal pathogen is *Helicobacter pylori* that colonises the gastric mucosa. Infected individuals can exhibit chronic gastric inflammation, peptic ulcer disease, and in some subjects the pathogen leads to mucosa-associated lymphoid tissue lymphomas and gastric adenocarcinomas. An *in vitro* study suggests that bovine colostrum-derived phospholipid, or some unknown lipid component, blocked the attachment of *H. pylori* to its lipid receptors on antral epithelial cells (see Parodi, 2004).

2.4 Butyric acid

Milk fat is the sole dietary source of butyric acid (BA). BA is a potent anticancer agent that inhibits cell proliferation and induces differentiation and apoptosis. Substantial evidence suggests that BA, produced in the colon as a result of bacterial fermentation of fibre, protects against colon cancer. Experimental studies indicated a role for BA in treatment of and as adjuvant therapy for cancer, especially for malignancies associated with blood. Clinical experience, however, did not confirm this promise. The lack of response was attributed to the rapid metabolism of BA in the liver and its short half-life in the circulation. Subsequently, production of butyrate derivatives increased serum half-life in animals and humans. In milk fat, about one-third of all triacylglycerol molecules contain a butyrate residue. BA shows synergism with a number of common dietary items and pharmacological agents such as retinoic acid, $1{,}25(OH)_2D_3$, the plant antioxidant resveratrol, statin drugs (HMG-CoA reductase inhibitors) and aspirin, which can result in lower serum BA levels required to exert a physiological effect. Besides modulating cell growth, BA can inhibit angiogenesis and enhance the expression of glutathione *S*-transferase that is involved in the detoxification of dietary carcinogens (see the reviews by Parodi, 2004, 2006).

BA can also modulate previously discussed transcription factors; NF-κB is inhibited and PPARs are activated. In cultured endothelial cells, BA inhibited TNF-γ-induced intracellular adhesion molecule-1 and IL-1-induced expression of vascular cell adhesion molecule-1, which suggests both an anti-inflammatory and potentially an antiatherogenic role for BA (Zapolska-Downar *et al.*, 2004).

Two studies demonstrated that dietary butyrate inhibited chemically induced mammary tumour development in rats. In the first study, the addition of 6 g 100 g^{-1} sodium butyrate to a base diet containing 20 g 100 g^{-1} fat supplied as a safflower oil-based margarine significantly reduced the incidence of chemically induced mammary tumours (Yanagi *et al.*, 1993). In the second study, addition of tributyrin at a level of 1 g 100 g^{-1} (BA content equivalent to milk fat) or 3 g 100 g^{-1} to a sunflower oil-based diet reduced chemically induced tumour incidence by 20 and 52%, respectively. In this study, a milk fat diet produced fewer tumours than a sunflower oil-based diet (Belobrajdic & McIntosh, 2000).

2.5 Branched chain fatty acids

Milk fat contains a series of saturated branched chain fatty acids (BCFAs) (Jensen, 2002). Their origin is the structural lipids of certain rumen bacteria, which are absorbed from the cow's digestive tract to the circulation where they pass to adipose tissue and the lactating mammary gland for triglyceride synthesis. BCFAs have anticancer properties. Yang *et al.* (2000) reported that low concentrations of 13-methyltetradecanoic acid (13-MTDA, *iso*-15:0) inhibited the growth of a range of common human cancer cells through induction of apoptosis. In addition, dietary 13-MTDA reduced the growth of human lung and prostate cancer cells implanted orthotopically into mice. Wright *et al.* (2005) implanted squamous cell carcinoma fragments into the thigh muscle of rabbits. When a solution of 12-methyltetradecanoic acid was introduced to the artery supplying the tumour site, growth was inhibited in a dose-dependent manner.

Wongtangtintharn *et al.* (2004) tested the antitumour activity of a series of iso-BCFAs in two human breast cancer cell lines. The highest antitumour activity was found with *iso*-16:0 and the activity decreased with an increase or decrease in chain length from *iso*-16:0. *Anteiso*-BCFAs were also cytotoxic. The cytotoxicity of 13-MTDA in the breast cancer cells was comparable to RA. 13-MTDA, as well as RA, inhibited fatty acid synthase (FAS) activity. Increased FAS expression has been demonstrated in a number of human malignancies.

Milk fat also contains small amounts of the multi-BCFAs phytanic (3,7,11,15-tetramethylhexadecanoic) and pristanic (2,6,10,14-tetramethylpentadecanoic) acids. Phytanic acid is produced in the rumen from the phytol side chain of dietary chlorophyll and some of this phytanic acid can be metabolised in the liver to produce pristanic acid. Both these multi-BCFAs are agonists for PPARα (Zomer *et al.*, 2000). PPARα agonists are known to inhibit development of chemically induced tumours in animals (Parodi, 2004).

Rumen bacteria also produce small quantities of hydroxy fatty acids, which are transferred to milk fat (Chance *et al.*, 1998; Jensen, 2002). Hydroxy fatty acids can inhibit growth and induce apoptosis in human cancer cell lines. Acids with an 18-carbon chain length and the hydroxyl group at one end of the carbon chain and the carboxyl group at the other end appear to be the most effective (Abe & Sugiyama, 2005).

2.6 Fat-soluble components

2.6.1 *The vitamins*

Milk fat is a rich source of vitamin A and β-carotene, but its content of vitamins D, E and K are not high. Because these vitamins are common, not unique to milk fat, and their properties are covered in most nutrition texts, they will not be discussed here. Nevertheless, because of its growing importance, vitamin D is discussed briefly.

Vitamin D_3 is produced in the skin through UV irradiation of 7-dehydrocholesterol and is metabolised to 25-hydroxyvitamin D_3 (the storage form) in the liver and to the biologically active hormone $1,25(OH)_2D_3$ in the kidneys. Sunlight was considered to contribute about 90% of vitamin D requirements (Holick, 2004). However, there is accumulating evidence that a significant proportion of the population worldwide, even in sunny environments, are vitamin D insufficient. This situation stems from several factors that include decreased exposure to sunlight to reduce the risk of skin cancer and prevent skin wrinkling, increased smog, migration of people with pigmented skin from the tropics to temperate climates, an aging population that is house bound and the demonisation of fat that carries the dietary vitamin.

The pivotal role of vitamin D in calcium absorption and metabolism and the skeletal disorders of rickets, osteomalacia and osteoporosis are well appreciated, but evidence is accumulating that vitamin D plays an important role in the prevention of a variety of diseases including cancer, especially of the colon, breast and prostate, tuberculosis, inflammatory bowel disease, multiple sclerosis, rheumatoid arthritis, type 1 diabetes, hypertension and the metabolic syndrome (Holick, 2004; Peterlik, & Cross, 2005).

In some countries, dairy products are fortified with vitamin D. Because of the increasing incidence of vitamin D insufficiency and deficiency and the realisation of its importance in multiple chronic diseases, the non-fortification countries should consider its introduction.

2.6.2 *Cholesterol*

The role of cholesterol in carcinogenesis has been studied infrequently with conflicting results (Parodi, 2004). In serum cholesterol–sensitive rats, dietary cholesterol was shown to inhibit the formation of chemically induced colonic ACF and development of mammary tumours in two different models (El-Sohemy & Archer, 2000; Duncan *et al.*, 2004). It is believed that cholesterol from serum enters cells via the LDL receptor and acts as a negative feedback for HMG-CoA reductase, which converts HMG-CoA to mevalonate, the major step in cellular cholesterol synthesis (El-Sohemy & Archer, 2000). Mevalonate is also required for DNA synthesis and cell proliferation. Elevated mevalonate synthesis occurs in a range of cancer cells. Addition of mevalonate to a cultured human breast cancer cell line resulted in increased proliferation, and when these cells were injected into the mammary fat pads of mice tumour growth increased after administration of mevalonate (Duncan *et al.*, 2004).

2.6.3 *Other interesting components*

The cow has the special ability to extract bioactive components from its feed, mostly unsuitable for human consumption, and transfer them to its milk, or to extract components that are converted to bioactives in the rumen and other tissues. Examples are β-carotene from pasture, a portion of which is converted to vitamin A *in vivo*, with the provitamin and vitamin having both common and diverse nutritional properties. The bioproduction of RA and conversion of the phytol side chain of chlorophyll to phytanic and pristanic acids were cited previously. Cows that were fed cottonseed meal transfer the polyphenol gossypol to milk, and lucerne or alfalfa provides β-ionone. Both these compounds are anticancer agents and the latter is a HMG-CoA reductase inhibitor, which may exert hypocholesterolemic properties (Parodi, 2004, 2006).

2.7 Further nutritional benefits

A number of other nutritional benefits have been ascribed to milk fat and have been reviewed elsewhere by the author (Parodi, 2004). Epidemiological evidence suggests that children who drank low fat or skim milk had a several-fold higher risk of gastrointestinal illness than children who drank predominantly whole milk. This benefit may be related to the presence in milk fat of short- and medium-chain length fatty acids that were shown in *in vitro* and animal studies to inhibit the growth of a range of enteric pathogens that included *E. coli*, *S. enteritidis*, *L. monocytogenes* and *H. pylori*. As discussed previously, sphingolipids and their digestion products can protect against certain intestinal bacterial and viral infections.

There is also a growing body of epidemiological evidence that shows that whole milk compared to skimmed milk, and butter compared to margarine are associated with less allergic airway disorders such as asthma and allergic rhinitis (hay fever).

The long-chain ω-3 polyunsaturated fatty acids eicosapentaenoic acid (EPA, 20:5) and docosahexaenoic acid (DHA, 22:6) are now considered essential for normal growth and development and also for the prevention and treatment of CAD, hypertension, diabetes, arthritis, other inflammatory and autoimmune diseases and cancer (Simopoulos, 2002).

Dietary EPA and DHA are supplied in small amounts mainly from animal tissue lipids, but most are produced in the body from dietary α-linolenic acid by a series of elongations and desaturations by Δ^5 and Δ^6 desaturases. It was calculated that in early times the ratio of dietary ω-6 (exemplified by linoleic acid) to ω-3 acids was about 1:1. With the advent of modern agriculture and dietary advice to increase linoleic acid intake to prevent CAD, the ratio has changed to about 30:1 or more. Because the desaturases needed to convert α-linolenic acid to EPA and DHA are common to those required to metabolise linoleic acid, it is believed that the metabolic competition imposed by a high ω-6: ω-3 ratio can adversely affect the production of EPA and DHA in tissues.

Milk fat is not a rich source of linoleic and α-linolenic acids, but their ratio of about 2:1 is optimal. Several studies with rats have demonstrated that compared to various vegetable oils, milk fat produces a more favourable EPA and DHA profile in tissues and there is limited evidence that milk fat can at least increase EPA in humans. Milk fat may help bone growth and reduce plaque formation on teeth, but the evidence is limited. Human randomised clinical trials are required to substantiate all the benefits outlined in this section.

2.8 Perceived nutritional negatives for milk

Fat, including milk fat, has long been associated with CAD and more recently with cancer at a number of sites and the current obesity epidemic. A detailed review of these subjects is beyond the scope of this chapter but is covered in Parodi (2004, 2006). A brief summary of some key points is included here.

2.8.1 *Milk fat and coronary artery disease*

Milk fat consumption is associated with CAD risk because it contains saturated fatty acids and cholesterol, which may increase the level of serum cholesterol. Prospective epidemiological studies show that the risk of CAD increases as the level of serum cholesterol increases. However, data from the often-cited Seven Countries Study showed that while the level of serum cholesterol was lineally related to CAD mortality in all participating countries, the absolute levels of mortality were strikingly different. For instance, at a serum cholesterol level of 5.2 mmol L^{-1}, there was a 5-fold greater mortality in Northern Europe than in Japan after adjusting for age, smoking and systolic blood pressure. This suggests that other powerful factors influence CAD risk in Western countries.

Cholesterol is insoluble in aqueous solution and is carried in blood attached to proteins called apoproteins (apo). The resulting lipoproteins are heterogeneous and consist of distinct families based largely on density and have different pathophysiological properties. Initially total serum cholesterol was used as a biomarker for CAD; later it was found that LDL cholesterol was atherogenic, whereas HDL was antiatherogenic. The LDL particles were found to be heterogeneous with regard to size, density and composition, and the small dense LDL particles are associated with a several-fold higher risk of CAD than large buoyant particles.

Public health messages relating to diet, serum cholesterol and CAD are based largely on studies that measured serum total cholesterol levels. The question arises as to whether the association between serum cholesterol and CHD is causal, a correlate or a consequence of the disease. For instance, elevated serum homocysteine levels are a major risk factor for CAD. In hypercholesterolemic men, the content of homosysteine was highest in the LDL fraction and was fourfold higher than in normocholesterolemic men (Olszewski and McCully, 1991). Administration of homocysteine to animals produced atherosclerotic plaques, endothelial injury and an increase in serum cholesterol level. In a human cohort, serum homocysteine was significantly correlated with serum cholesterol levels. In those with CAD, treatment with a mixture of nutrients that included the homocysteine catabolising agents folate, pyridoxine and riboflavin significantly lowered not only serum homocysteine levels but also serum cholesterol and LDL apoB levels. Treatment of human LDL with homocysteine produced increased amounts of homocysteine containing small dense particles (McCully, 1993).

ApoB is the predominant protein of LDL and acts as a ligand for the removal of LDL from serum by cell surface LDL receptors. Deficiency and activity of LDL receptors influences cholesterol homeostasis. HDL contains mainly apoA. Results from many studies show that apoB, which represents total atherogenic particle number, is a better marker of CAD risk than total or LDL cholesterol and that the ratio of apoB/apoA is superior to total cholesterol/HDL cholesterol as an overall index of risk (Sniderman *et al*., 2003). These facts may help explain why more than half the patients with CAD have normal serum cholesterol levels.

2.8.2 *Saturated fatty acids*

Saturated fatty acids have been demonised because of the belief that they are hypercholesterolemic. However, not all fatty acids elevate serum cholesterol to the same extent. The short-chain fatty acids, butyric, caproic, caprylic and the medium-chain capric and stearic acid, have no discernable effect on cholesterol levels. Lauric, myristic and palmitic acids, on the other hand, do raise cholesterol levels, with myristic acid being the most potent. Nevertheless, these three acids concomitantly increase levels of antiatherogenic HDL cholesterol. Further, serum levels of the atherogenic lipoprotein[a] were lower on a diet high in cholesterol-raising saturated acids than on an oleic acid diet (see Parodi, 2004, 2006).

A diet high in saturated fat (Dreon *et al*., 1998) and milk-derived saturated acids (Sjogren *et al*., 2004) was associated with a reduction in atherogenic small dense LDL particles. Compared to a number of other saturated and unsaturated fatty acids, myristic acid stimulated the generation of endothelial NOS, generally considered to be antiatherogenic (Zhu & Smart, 2005). PPARγ ligands, of which milk fat RA is one, can also stimulate endothelial NO production (Polikandriotis *et al*., 2005). Thus, the antiatherogenic properties of saturated fatty acids may compensate (or more) for their cholesterol-raising action, and the isolated use of total and LDL cholesterol levels as markers to condemn them is inappropriate.

For more than three decades, national public health bodies and others have recommended a reduction in total and saturated fat intake to decrease the risk of CAD. How strong is the evidence that supports these recommendations? Prospective (cohort) epidemiological studies provide the most reliable evidence for determining association. This format was used to examine the association between saturated fatty acid intake and the risk of CAD in 18 cohorts. Follow-up results were presented at more than one time point for five cohorts that resulted in 23 published reports. Some cohorts stratified data by sex and some by age, which resulted in 31 data sets overall. Six of these sets showed a marginally statistically significant association between saturated far intake and risk of CAD; four associations were positive and two associations were negative. Similarly, the association between total fat intake and CAD risk was weak.

Even so, individuals do not consume saturated fat as a single entity, rather it is part of a total diet, and in the case of milk fat-derived saturated fatty acids, their origin is milk. Elwood *et al.* (2004) identified 10 prospective epidemiological studies that calculated the risk between cardiovascular disease and milk consumption. Only one small study noted a positive association, but a pooled estimate of risk found that drinking milk was associated with a small but worthwhile reduction in CAD and stroke risk.

Epidemiological associations, however, cannot be used to ascribe causality. For this a series of well-conducted randomised clinical trials are required. Oliver (1997) who reviewed the trials in this field pointed out that there are only two clinical trials on the effects of total and saturated fats alone on the risk of CAD. Both were small, conducted on patients who had already suffered a heart attack, and both failed to show any change in the incidence of subsequent CAD events.

Many trials investigated simultaneous changes in more than one dietary component and modification of other risk factors such as cigarette smoking, physical activity, weight and treatment of hypertension on the risk of CAD. Some of these were successful; however, the Cochrane Collaboration produced a meta-analysis of 27 such studies that included 40 intervention arms. There was no significant effect on total mortality, but there was a trend towards protection for cardiovascular mortality. A significant protection for cardiovascular events was noted but became non-significant on sensitivity analysis (Hooper *et al.*, 2000). Overall, the contention that saturated fat is implicated in CAD risk lacks adequate proof.

2.8.3 *Fat intake and cancer*

For decades there has been a strong belief that a high fat intake was linked to cancer of the colon, breast and prostate (the major non-smoking related cancers). The belief resulted from early studies with animals and from international comparison (ecological) studies that showed strong positive correlations between assumed per capita consumption of fat and cancer death rates. Ecological studies are a poor format for determining causality because they tell nothing about the diets of individuals who develop cancer and those who do not. Many other dietary items and environmental factors associated with cancer differ between cultures and countries and cannot be accounted for in these studies.

Later evidence from animal studies and epidemiology showed that total energy intake and physical activity were more likely to influence cancer development than fat intake. Evidence for a relationship between fat intake and cancer from case–control studies is inconsistent. On the other hand, evidence from the more reliable prospective studies and pooled analyses of these studies have not supported an association between fat intake and the risk of colon, breast and prostate cancer (Willett, 2001; Kushi & Giovannucci, 2002; Parodi, 2005, 2006).

The Women's Health Initiative Dietary Modification Trial, a large randomised clinical trial of 48 835 postmenopausal women is the first to directly assess the health benefits of promoting a low-fat dietary pattern. The goal for the intervention group was to reduce total fat intake to 20% of energy and increase consumption of fruit, vegetables and grains. After 8.1 years of follow-up, the low-fat dietary pattern did not result in a statistically significant reduction in invasive breast cancer (Prentice *et al.*, 2006) or colorectal cancer (Beresford *et al.*, 2006) compared to women who consumed a diet with less fruit, vegetables and grains and a fat intake that represented about 35% of energy.

2.8.4 *Dietary fat and obesity*

In affluent countries, the prevalence of overweight and obesity during the past three decades has increased dramatically to epidemic proportions and is now a major public health concern. Overweight adults are at increased risk for such chronic diseases as insulin resistance, type 2 diabetes mellitus, hypertension, cardiovascular disease, some types of cancer including colon and breast, sleep apnea and osteoarthritis of weight-bearing joints (Willett, 2002; Parodi, 2006).

Dietary fat is often implicated in the high incidence of obesity. The rationale is that ecological studies show that the incidence of obesity is high in affluent countries with a high fat intake and low in developing countries with a low fat intake. Also, fat is energy dense and supplies 9.0 kcal g^{-1} compared to 4.0 kcal g^{-1} for protein and carbohydrate. However, between populations, comparisons are seriously confounded by differences in lifestyle factors such as type and availability of food and in physical activity. Moreover, the drastic increase in the incidence of obesity occurred during the period when the fat intake as a percentage of total energy intakes had declined (Willett, 2002). Rather, the obesity epidemic relates mainly to an imbalance between energy intake and energy expenditure owing to the combination of an abundant supply of cheap, palatable, energy-dense foods and an increasingly sedentary lifestyle.

A Cochrane Collaboration (Pirozzo *et al.*, 2002) reviewed the results of randomised controlled trials of low-fat diets versus other weight-reducing diets. They found that low-fat diets were no better than other types of weight-reducing diets in achieving and maintaining weight loss in overweight or obese people over a 12- to 18-month period.

The energy value of a fatty acid is dependent on the ratio of its carbon atoms to its oxygen atoms. Thus, the heat of combustion, which is used to calculate energy value for BA (4:0), which has a low carbon-to-oxygen ratio, is 5.92 kcal g^{-1}. The value for stearic acid (18:0) with a high carbon-to-oxygen ratio, is 9.48 kcal g^{-1}. When milk fat is digested, fatty acids with chain length less than 12 carbons are absorbed from the intestine, pass to the portal circulation and then to the liver where they are oxidised. In contrast, the longer-chain acids are re-synthesised into triglycerides, packed into chylomicrons, pass to the venous circulation and are carried to peripheral tissues for utilisation or stored in adipose tissue, which can contribute to weight gain and obesity (Parodi, 2004, 2006). Although it has not been established in a clinical trial, the above facts suggest that milk fat would contribute less to adiposity than an equal weight of other dietary fats.

2.9 Conclusions

Health messages that suggest fat, especially milk fat, is associated with the major cancers of the colon, breast and prostate lack credible evidence. Likewise, there is no compelling evidence that fat is responsible for the current obesity epidemic or is implicated in weight gain independent of its caloric density. Saturated fatty acids have been implicated with

CAD through their capacity to elevate serum cholesterol levels. However, their other antiatherogenic properties may more than compensate for this and there is no good evidence from prospective epidemiological studies and the few randomised clinical trials that saturated fatty acids are associated with heart disease.

In restricting the consumption of milk fat, the diet is deprived of a number of zoochemicals with potential health benefits. Examples are RA and VA with antiatherogenic and anticancer properties. Other anticancer agents include sphingolipids, BA and the branched chain fatty acids. There is evidence that milk fat can prevent certain intestinal infections particularly in children, prevent some allergic disorders like asthma and hay fever and improve the serum and tissue levels of ω-3 polyunsaturated fatty acids. Fat is an essential dietary component and milk fat is noted for its ability to add flavour to food, and its inclusion in a balanced diet in most cases should not be feared.

References

Abe, A. & Sugiyama, K. (2005) Growth inhibition and apoptosis induction of human melanoma cells by ω-hydroxy fatty acids. *Anti-cancer Drugs*, **16**, 543–549.

Aro, A., Mannisto, S., Salminen, I., Ovaskainen, M.-l., Kataja, V. & Uusitupa, M. (2000) Inverse association between dietary and serum conjugated linoleic acid and risk of breast cancer in postmenopausal women. *Nutrition and Cancer*, **38**, 151–157.

Ballou, L. R., Laulederkind, S. J. F., Rosloniec, E. F. & Raghow, R. (1996) Ceramide signalling and the immune response. *Biochimica et Biophysica Acta*, **1301**, 273–287.

Banni, S., Angioni, E., Murru, E., Carta, G., Melis, M. P., Bauman, D., Dong, Y. & Ip, C. (2001) Vaccenic acid feeding increases tissue levels of conjugated linoleic acid and suppresses development of premalignant lesions in rat mammary gland. *Nutrition and Cancer*, **41**, 91–97.

Belobrajdic, D. P. & McIntosh, G. H. (2000) Dietary butyrate inhibits NMU-induced mammary cancer in rats. *Nutrition and Cancer*, **36**, 217–223.

Belury, M. A. (2002a) Dietary conjugated linoleic acid in health: physiological effects and mechanisms of action. *Annual Reviews in Nutrition*, **22**, 505–531.

Belury, M. (2002b) Inhibition of carcinogenesis by conjugated linoleic acid: potential mechanisms of action. *Journal of Nutrition*, **132**, 2995–2998.

Beresford, S. A. A., Johnson, K. C., Ritenbaugh, C., Lasser, N. L., Snetselaar, L. G., Black, H. R., Anderson, G. L., Assaf, A. R., Bassord, T., Bowen, D., Brunner, R. L., Brzyski, R. G., Caan, B., Chlebowski, R. T., Gass, M., Harrigan, R. C., Hays, J., Heber, D., Heiss, G., Hendrix, S. L., Howard, B. V., Hsai, J., Hubbell, F. A., Jackson, R. D., Kotchen, J. M., Kuller, L. H., LaCroix, A. Z., Lane, D. S., Langer, R. D., Lewis, C. E., Manson, J. E., Margolis, K. L., Mossavar-Rahmani, Y., Ockene, J. K., Parker, L. M., Rerri, M. G., Phillips, L., Prentice, R. L., Robbins, J., Rossouw, J. E., Sarto, G. E., Stefanick, M. L., van Horn, L., Vitolins, M. Z., Wactawski-Wende, J., Wallace, R. B. & Whitlock, E. (2006) Low-fat dietary pattern and risk of colorectal cancer. *JAMA*, **295**, 643–654.

Brownbill, R. A., Petrosian, M. & Llich, J. Z. (2005) Association between dietary linoleic acid and bone mineral density in postmenopausal women. *Journal of the American College of Nutrition*, **24**, 177–181.

Chance, D. L., Gerhardt, K. O. & Mawhinney, T. P. (1998) Gas-liquid chromatography – mass spectrometry of hydroxy fatty acids as their methyl esters *tert.*-butyldimethylsilyl ethers. *Journal of Chromatography A*, **793**, 91–98.

Chen, B.-Q., Xue, Y.-B., Liu, J.-R., Yang, Y.-M., Zheng, Y.-M., Wang, X.-L. & Liu, R.-H. (2003) Inhibition of conjugated linoleic acid on mouse forestomach neoplasia induced by benzo(a)pyrene and chemopreventive mechamisms. *World Journal of Gastroenterology*, **9**, 44–49.

Chujo, H., Yamasaki, M., Nou, S., Koyanagi, N., Tachibana, H. & Yamada, K. (2003) Effect of conjugated linoleic acid isomers on growth factor-induced proliferation of human breast cancer cells. *Cancer Letters*, **202**, 81–87.

Chung, S., Brown, J. M., Provo, J. N., Hopkins, R. & McIntosh, M. K. (2005) Conjugated linoleic acid promotes human adipocyte insulin resistance through NF-κB-dependent cytokine production. *Journal of Biological Chemistry*, **280**, 38445–38456.

Cinque, B., Di Marzio, L., Centi, C., Di Rocco, C., Riccardi, C. & Cifone, M. G. (2003) Sphingolipids and the immune system. *Pharmacological Research*, **47**, 421–437.

Corl, B. A., Barbano, D. M., Bauman, D. E. & Ip, C. (2003) *cis*-9, *trans*-11 CLA derived endogenously from *trans*-11 18:1 reduces cancer risk in rats. *Journal of Nutrition*, **133**, 2893–2900.

Desroches, S., Chouinard, P. Y., Galibois, I., Corneau, L., Delisle, J., Lamarche, B., Couture, P. & Bergeron, N. (2005) Lack of effect of dietary conjugated linoleic acids naturally incorporated into butter on the lipid profile and body composition of overweight and obese men. *American Journal of Clinical Nutrition*, **82**, 309–319.

Dreon, D. M., Fernstrom, H. A., Campos, H., Blanche, P., Williams, P. T. & Krauss, R. M. (1998) Change in saturated fat intake is correlated with change in mass of large low-density-lipoprotein particles in men. *American Journal of Clinical Nutrition*, **67**, 828–836.

Duan, R.-D. (2005) Anticancer compounds and sphingolipid metabolism in the colon. *In vivo*, **19**, 293–300.

Duncan, R. E., El-Sohemy, E. & Archer, M. (2004) Mevalonate promotes the growth of tumors derived from human cancer cells *in vivo* and stimulates proliferation *in vitro* with enhanced cyclin-dependent kinase-2 activity. *Journal of Biological Chemistry*, **279**, 33079–33084.

Eckhardt, E. R. M., Wang, D. Q.-H., Donovan, J. M. & Carey, M. C. (2002) Dietary sphingomyelin suppresses intestinal cholesterol absorption by decreasing thermodynamic activity of cholesterol monomers. *Gastroenterology*, **122**, 948–956.

El-Sohemy, A. & Archer, M. C. (2000) Inhibition of *N*-methyl-*N*-nitrosourea- and 7,12-dimethylbenz[a]anthracene-induced rat mammary tumourigenesis by dietary cholesterol is independent of Ha- *ras* mutations. *Carcinogenesis*, **21**, 827–831.

Elwood, P. C., Pickering, J. E., Hughes, J., Fehily, A. M. & Ness, A. R. (2004) Milk drinking, ischaemic heart disease and ischaemic stroke II. Evidence from cohort studies. *European Journal of Clinical Nutrition*, **58**, 718–724.

Fantini, J., Hammache, D., Delezay, O., Yahi, N., Andre-Barres, C., Rico-Lattes, I. & Lattes, A. (1997) Synthetic soluble analogs of galactosylceramide (GalCer) bind to the V3 domain of HIV-1 gp120 and inhibit HIV1-induced fusion and entry. *Journal of Biological Chemistry*, **272**, 7245–7252.

Gurr, M. I. (1998) Milk, nutrition and health. *Milk and Health*, pp. 9–22, Proceedings of 25th International Dairy Congress, Aarhus.

Hanada, K. (2005) Sphingolipids in infectious diseases. *Japanese Journal of Infectious Diseases*, **58**, 131–148.

Holick, M. F. (2004) Sunlight and vitamin D for bone health and prevention of autoimmune diseases, cancers and cardiovascular disease. *American Journal of Clinical Nutrition*, **80**, 1678S–1688S.

Hooper, L., Summerbell, C. D., Higgins, J. P. T., Thompson, R. L., Clements, G., Capps, N., Davy Smith, G., Riemersma, G. & Ebrahim, S. (2000) Reduced or modified dietary fat for preventing cardiovascular disease (Review). *The Cochrane Database of Systematic Reviews*, **2**, CD002137.

Ip, C., Banni, S., Angioni, E., Carta, G., McGinley, J., Thompson, H. J., Barbano, D. & Bauman, D. E. (1999) Conjugated linoleic acid-enriched butter fat alters mammary gland morphogenesis and reduces cancer risk in rats. *Journal of Nutrition*, **129**, 2135–2142.

Ip, C., Chin, S. F., Scimeca, J. A. & Pariza, M. W. (1991) Mammary cancer prevention by conjugated dienoic derivative of linoleic acid. *Cancer Research*, **51**, 6118–6124.

Ip, C., Dong, Y., Ip, M. M., Banni, S., Carta, G., Angioni, E., Murru, E., Spada, S., Melis, M. P. & Saebo, A. (2002) Conjugated linoleic acid isomers and mammary cancer prevention. *Nutrition and Cancer*, **43**, 52–58.

Ip, M. M., Masso-Welch, P. A. & Ip, C. (2003) Prevention of mammary cancer with conjugated linoleic acid: role of stroma and the epithelium. *Journal of Mammary Gland Biology and Neoplasia*, **8**, 103–118.

Jackson, S. M. & Demer, L. L. (2000) Peroxisome proliferator-activated receptor activators modulate the osteoblastic maturation of MC3T3-E1 preosteoblasts. *FEBS Letters*, **471**, 119–124.

Jensen, R. G. (1995) *Handbook of Milk Composition*, Academic Press, San Diego.

Jensen, R. G. (2002) The composition of bovine milk lipids: January 1995 to December 2000. *Journal of Dairy Science*, **85**, 295–350.

Kobayashi, T., Shimizugawa, T., Osakabe, T., Watanabe, S. & Okuyama, H. (1997) A long-term feeding of sphingolipids affected the level of plasma cholesterol and hepatic triacylglycerol but not tissue phospholipids and sphingolipids. *Nutrition Research*, **17**, 111–114.

Kritchevsky, D., Tepper, S. A., Wright, S., Czarnecki, S. K., Wilson, T. A. & Nicolosi, R. J. (2004) Conjugated linoleic acid isomer effects in atherosclerosis: growth and regression of lesions. *Lipids*, **39**, 611–616.

Kushi, L. & Giovannucci, E. (2002) Dietary fat and cancer. *American Journal of Medicine*, **113**, 63S–70S.

Larsson, S. C., Bergkvist, L. & Wolk, A. (2005) High-fat dairy food and conjugated linoleic acid intakes in relation to colorectal cancer incidence in the Swedish Mammography Cohort. *American Journal of Clinical Nutrition*, **82**, 894–900.

Li, Y. & Watkins, B. A. (1998) Conjugated linoleic acid alters bone fatty acid composition and reduce *ex vivo* prostaglandin E_2 biosynthesis in rats fed n-6 or n-3 fatty acids. *Lipids*, **33**, 417–425.

Libby, P., Ridker, P. M. & Maseri, A. (2002) Inflammation and atherosclerosis. *Circulation*, **105**, 1135–1143.

Lock, A. L. & Bauman, D. E. (2004) Modifying milk fat composition of dairy cows to enhance fatty acids beneficial to human health. *Lipids*, **39**, 1197–1206.

Lock, A. L., Corl, B. A., Barbano, D. M., Bauman, D. E. & Ip, C. (2005a) The anticarcinogenic effect of *trans*-11 18:1 is dependent on its conversion to *cis*-9, *trans*-11 CLA by δ9-desaturase in rats. *Journal of Nutrition*, **134**, 2698–2704.

Lock, A. L., Horne, C. A. M., Bauman, D. E. & Salter, A. M. (2005b) Butter naturally enriched in conjugated linoleic acid and vaccenic acid alters tissue fatty acids and improves the plasma lipoprotein profile in cholesterol fed hamsters. *Journal of Nutrition*, **135**, 1934–1939.

Lock, A. L., Parodi, P. W. & Bauman, D. E. (2005c) The biology of trans fatty acids: implications for human health and the dairy industry. *Australian Journal of Dairy Technology*, **60**, 134–142.

Loscher, C. E., Draper, E., Leavy, O., Kelleher, D., Mills, K. H. G. & Roche, H. M. (2005) Conjugated linoleic acid suppresses NF-κB activation and IL-12 production in dendritic cells through ERK-mediated IL-10 induction. *Journal of Immunology*, **175**, 4990–4998.

McCann, S. E., Ip, C., Ip, M. M., McGuire, M. K., Muti, P., Edge, S. B., Trevisan, M. & Freudenheim, J. L. (2004) Dietary intake of conjugated linoleic acid and risk of premenopausal and postmenopausal breast cancer, Western New York Exposures and Breast Cancer Study (WEB Study). *Cancer Epidemiology Biomarkers and Prevention*, **13**, 1480–1484.

McCully, K. S. (1993) Chemical pathology of homocysteine – 1. Atherogenesis. *Annals of Clinical and Laboratory Science*, **23**, 447–493.

Mitchell, P. L., Langille, M. A., Currie, D. L. & McLeod, R. S. (2005) Effect of conjugated linoleic acid isomers on lipoproteins and atherosclerosis in the Syrian golden hamster. *Biochimica et Biophysica Acta*, **1734**, 269–276.

Modrak, D. E., Lew, W., Goldenberg, D. M. & Blumenthal, R. (2000) Sphingomyelin potentates chemotherapy of human cancer xenografts. *Biochemical and Biophysical Research Communications*, **268**, 603–606.

Motouri, M., Matsuyama, H., Yamamura, J.-I., Tanaka, M., Aoe, S., Iwanaga, T. & Kawakami, H. (2003) Milk sphingomyelin accelerates enzymatic and morphological maturation of the intestine in artificially reared rats. *Journal of Pediatric Gastroenterology and Nutrition*, **36**, 241–247.

Noh, S. K & Koo, S. I. (2004) Milk sphingomyelin is more effective than egg sphingomyelin in inhibiting intestinal absorption of cholesterol and fat in rats. *Journal of Nutrition*, **134**, 2611–2616.

Noone, E. J., Roche, H. M., Nugent, A. P. & Gibney, M. J. (2002) The effect of dietary supplementation using isomeric blends of conjugated linoleic acid on lipid metabolism in healthy human subjects. *British Journal of Nutrition*, **88**, 243–251.

Oliver, M. F. (1997) It is more important to increase the intake of unsaturated fats than to decrease the intake of saturated fats: evidence from clinical trials relating to ischemic heart disease. *American Journal of Clinical Nutrition*, **66**, 980S–986S.

Olszewski, A. J. & McCully, K. S. (1991) Homocysteine content of lipoproteins in hypercholesterolemia. *Atherosclerosis*, **88**, 61–68.

O'Shea, M., Bassaganya-Riera, J. & Mohede, I. C. M. (2004) Immunomodulatory properties of conjugated linoleic acid. *American Journal of Clinical Nutrition*, **79**, 1199S–1206S.

Palmquist, D. L., Lock, A. L., Shingfield, K. J. & Bauman, D. E. (2005) Biosynthesis of conjugated linoleic acid in ruminants and humans. *Advances in Food and Nutrition Research*, **50**, 179–217.

Pariza, M. W. & Hargraves, W. A. (1985) A beef-derived mutagenesis modulator inhibits initiation of mouse epidermal tumours by 7.12-dimethylbenz[a]anthracene. *Carcinogenesis*, **6**, 591–593.

Pariza, M. W., Park, Y. & Cook, M. E. (2001) The biologically active isomers of conjugated linoleic acid. *Progress in Lipid Research*, **40**, 283–298.

Parodi, P. W. (2002) Health benefits of conjugated linoleic acid. *Food Industry Journal*, **5**, 222–259.

Parodi, P. W. (2003) Conjugated linoleic acid in food. *Advances in Conjugated Linoleic Acid Research*, Volume 2, (eds. J.-L. Sebedio, W. W. Christie & R. O. Adlof), pp. 101–122, AOCS Press, Champaign.

Parodi, P. W. (2004) Milk fat in human nutrition. *Australian Journal of Dairy Technology*, **59**, 3–59.

Parodi, P. W. (2005) Dairy product consumption and the risk of breast cancer. *Journal of the American College of Nutrition*, **24**, 556S–568S.

Parodi, P. W. (2006) Nutritional significance of milk lipids. *Advanced Dairy Chemistry*, Volume 2, 3rd edn., (eds. P. F. Fox & P. L. H. McSweeney), pp. 601–639, Springer, New York.

Patton, S. (2004) *Milk*, Transaction Publishers, New Brunswick.

Peterlik, M. & Cross, H. S. (2005) Vitamin D and calcium deficits predisposes for multiple chronic diseases. *European Journal of Clinical Investigation*, **35**, 290–304.

Pirozzo, S., Summerbell, C., Cameron, C. & Glasziou, P. (2002) Advice on low-fat diets for obesity (Review). *The Cochrane Database of Systematic Reviews*, 2, CD003640.

Polikandriotis, J. A., Mazzella, L. J., Rupnow, H. L. & Hart, C. M. (2005) Peroxisome proliferator-activated receptor γ ligands stimulate endothelial nitric oxide production through distinct peroxisome proliferator-activated receptor γ -dependent mechanisms. *Atherosclerosis, Thrombosis and Vascular Biology*, **25**, 1810–1816.

Prentice, R. L., Caan, B., Chleboski, R. T., Patterson, R., Kuller, L. H., Ockene, J. K., Margolis, K. L., Limacher, M. C., Manson, J. E., Parker, L. M., Paskett, E., Phillips, L., Robbins, J., Rossouw, J. E., Sarto, G. E., Shikany, M., Stefanick, M. L., Thomson, C. A., van Horn, L., Vitolins, M. Z., Wactawski-Wende, J., Wallace, R. B., Wassertheil-Smoller, S., Whitlock, E., Yano, K., Adans-Campbell, L., Anderson, G. L., Assaf, A. R., Beresford, S. A. A., Black, H. R., Brunner, R. L., Brzyski, R. G., Ford, L., Gass, M., Hays, J., Heber, D., Heiss, G., Hendrix, S. L., Hsai, J., Hubbell, F. A., Jackson, R. D., Johnson, K. C., Kotchen, J. M., LaCroix, A. Z., Lane, D. S., Langer, R. D., Lasser, N. L. & Henderson, M. M. (2006) Low-fat dietary pattern and risk of invasive breast cancer. *JAMA*, **295**, 629–642.

Riserus, U., Arner, P., Brismar, K. & Vessby, B. (2002a) Treatment with dietary *trans*-10 *cis*-12 conjugated linoleic acid causes isomer-specific insulin resistance in obese men with the metabolic syndrome. *Diabetes Care*, **25**, 1516–1521.

Riserus, U., Basu, S., Jovinge, S., Fredrikson, G. N., Arnlov, J. & Vessby, B. (2002b) Supplementation with conjugated linoleic acid causes isomer-dependent oxidative stress and elevated C-reactive protein. *Circulation*, **106**, 1925–1929.

Riserus, U., Vessby, B., Arnlov, J. & Basu, S. (2004) Effects of *cis*-9, *trans*-11 conjugated linoleic acid supplementation on insulin sensitivity, lipid peroxidation, and proinflammatory markers in obese men. *American Journal of Clinical Nutrition*, **80**, 279–283.

Roche, H. M., Noone, E., Nugent, A. & Gibney, M. J. (2001) Conjugated linoleic acid: a novel therapeutic nutrient? *Nutrition Research Reviews*, **14**, 173–187.

Roche, H. M., Noone, E., Sewter, C., Mc Bennett, S., Savage, D., Gibney, M. J., O'Rahilly, S. & Vidal-Puig, A. J. (2002) Isomer-dependent metabolic effects of conjugated linoleic acid. *Diabetes*, **51**, 2037–2044.

Rombaut, R. & Dewettinck, K. (2006) Properties, analysis and purification of milk polar lipids. *International Dairy Journal*, **16**, 1362–1373.

de Roos, B., Rucklidge, G., Reid, M., Ross, K., Duncan, G., Navarro, M. A., Arbones-Mainar, J. M., Guzman-Garcia, M. A., Osada, J., Browne, J., Loscher, C. E. & Roche, H. M. (2005) Divergent mechanisms of *cis*-9, *trans*-11- and *trans*-10, *cis*-12-conjugated linoleic acid affecting insulin resistance and inflammation in apoprotein E knockout mice: a proteomics approach. *FASEB Journal*, **19**, 1746–1748.

Rueda, R., Maldonado, J., Narbona, E. & Gil, A. (1998) Neonatal dietary gangliosides. *Early Human Development*, **53**, S135–S147.

Schmelz, E. M. (2004) Sphingolipids in the chemoprevention of colon cancer. *Fronteers in Bioscience*, **9**, 2632–2639.

Schmelz, E. M., Mottillo, E. P., Baxa, A. C., Doyon-Reale, N. & Roberts, P. C. (2006) Effect of sphingosine and enigmol on mouse ovarian surface epithelial cells representing early, intermediate, and late stages of ovarian cancer. *Journal of Nutrition*, **135**, 3045S.

Scimeca, J. A. (1999) Cancer inhibition in animals. *Advances in Conjugated Linoleic Acid Research*, Volume 1, (eds. M. P. Yurawecz, M. M. Mossoba, J. K. G. Kramer, M. W. Pariza & G. J. Nelson), pp. 420–443, AOCS Press, Champaign.

Simopoulos, A. P. (2002) Omega-3 fatty acids in inflammation and autoimmune diseases. *Journal of the American College of Nutrition*, **21**, 495–505.

Sniderman, A. D., Furberg, C. D., Keech, A., Roeters van Lennep, J. E., Frohlich, J., Jungner, J. & Walldius, G. (2003) Apolipoproteins versus lipids as indices of coronary risk and as targets for statin treatment. *Lancet*, **361**, 777–780.

Sprong, R. C., Hulstein, M. F. & van der Meer, R. (2001) Bactericidal activities of milk lipids. *Antimicrobial Agents Chemotherapy*, **45**, 1298–1301.

Sjogren, P., Rosell, M., Skoglund-Andersson, C., Zdravkovic, S., Vessby, B., de Faire, U., Hamsten, A., Hellenius, M-L. & Fisher, R. M. (2004) Milk-derived fatty acids are associated with a more favourable LDL particle size distribution in healthy men. *Journal of Nutrition*, **134**, 1729–1735.

Toomey, S., Harhen, B., Roche, H. M., Fitzgerald, D. & Belton, O. (2006) Profound resolution of early atherosclerosis with conjugated linoleic acid. *Atherosclerosis*, **187**, 40–49.

Terpstra, A. H. M. (2004) Effect of conjugated linoleic acid on body composition and plasma lipids in humans: an overview of the literature. *American Journal of Clinical Nutrition*, **79**, 352–361.

Tricon, S., Burdge, G. C., Kew, S., Banerjee, T., Russell, J. J., Jones, E. L., Grimble, R. F., Williams, C. M., Yaqoob, P. & Calder, P. C. (2004) Opposing effects of *cis*-9, *trans*-11 and *trans*-10, *cis*-12 conjugated linoleic acid on blood lipids in healthy humans. *American Journal of Clinical Nutrition*, **80**, 614–620.

Valeille, K., Ferezou, J., Amsier, G., Quignard-Boulange, A., Parquet, M., Gripois, D., Dorovska-Taran, V. & Martin, J.-C. (2005) A *cis*-9, *trans*-11-conjugated linoleic acid-rich oil reduces the outcome of atherogenic process in hyperlipidemic hamster. *American Journal of Physiology*, **289**, H652–H659.

Wahle, K. W. J., Heys, S. D. & Rotondo, D. (2004) Conjugated linoleic acids: are they beneficial or detrimental to health? *Progress in Lipid Research*, **43**, 553–587.

Watkins, B. A., Shen, C.-L., McMurtry, J. P., Xu, H., Bain, S. D., Allen, K. G. D. & Seifert, M. F. (1997) Dietary lipids modulate bone prostaglandin E_2 production, insulin-like growth factor-1 concentration and formation rate in chicks. *Journal of Nutrition*, **127**, 1084–1091.

Willett, W. C. (2001) Diet and cancer: one view at the start of the millennium. Cancer epidemiology. *Biomarkers and Prevention*, **10**, 3–8.

Willett, W. C. (2002) Dietary fat is not a major determinant of body fat. *American Journal of Medicine*, **113**, 47S–59S.

Wongtangtintharn, S., Oku, H., Iwasaki, H. & Toda, T. (2004) Effect of branched-chain fatty acid biosynthesis of human breast cancer cells. *Journal of Nutritional Science and Vitaminology*, **50**, 137–143.

Wright, K. C., Yang, P., Van Pelt, C. S., Hicks, M. E., Collin, P. & Newman, R. A. (2005) Evaluation of targeted arterial delivery of the branched chain fatty acid 12-methyltetradecanoic acid as a novel therapy for solid tumors. *Journal of Experimental Therapeutics and Oncology*, **5**, 55–68.

Yanagi, S., Yamashita, M. & Imai, S. (1993) Sodium butyrate inhibits the enhancing effect of high fat diet on mammary tumorigenesis. *Oncology*, **50**, 201–204.

Yang, Z., Liu, S., Chen, X., Chen, H., Huang, M. & Zheng, J. (2000) Induction of apoptotic cell death and *in vivo* growth inhibition of human cancer cells by a saturated branched-chain fatty, 13–methyltetradecanoic acid. *Cancer Research*, **60**, 505–509.

Yu, Y., Correll, P. H. & van den Heuvel, J. P. (2002) Conjugated linoleic acid decreases the production of pro-inflammatory products in macrophages: evidence for a PPARα-dependent mechanism. *Biochimica et Biophysica Acta*, **1581**, 89–99.

Zapolska-Downar, D., Siennicka, A., Kaczmarczyk, M., Zolodziej, B. & Naruszewicz, M. (2004) Butyrate inhibits cytokine-induced VCAM-1 and ICAM-1 expression in cultured endothelial cells: the role of NF-κB and PPARα. *Journal of Nutritional Biochemistry*, **15**, 220–228.

Zhu, W. & Smart, E. J. (2005) Myristic acid stimulates endothelial nitric-oxide synthase in a CD36- and AMP kinase-dependent manner. *Journal of Biological Chemistry*, **280**, 29543–29550.

Zomer, A. W. M., van der Burg, B., Jansen, G. A., Wanders, R. J. A., Poll-The, B. T. & van der Saag, P. T. (2000) Pristanic acid and phytanic acid: naturally occurring ligands for the nuclear receptor peroxisome proliferator-activated receptor α. *Journal of Lipid Research*, **41**, 1801–1807.

3 Separation and Standardisation of the Fat Content

M. Gunsing, H.C. van der Horst, D. Allersma and P. de Jong

3.1 Introduction

This chapter gives an overview of the methods applied to standardise the milk fat content in dairy products. Fat is a valuable component that has a large effect on flavour and texture of products. Standardisation of milk is needed because

- the fat content of raw milk varies with the season (see Figure 3.1);
- the fat content of raw milk varies with the breed of cow;
- different products need milks with different fat contents and
- in whey products fat is unwanted.

In the process of preparing dairy products, the standardisation of the fat content (consumption milk), the fat/protein ratio (cheese milk), dry solid content (for milk powder) or fat level in different types of creams are important steps for the quality of these products and for optimisation of the economics of these processes. Hence, the fat content of milk can be decreased or increased depending on the product that is produced. (Walstra *et al.*, 2006) For the production, for example, of Gouda cheese at 48+ g 100 g^{-1} fat-in-dry matter (FDM), a fat content of 3.5 g 100 g^{-1} milk is needed, while for the production of cream cheeses, 5 g fat 100 g^{-1} milk is required.

The process of adjusting the fat content of a milk stream to a pre-defined level is called standardisation. This process involves the decrease of the fat content in milk for 48+ FDM in cheese, milk powders and consumption milk, which also results in an additional stream with an increased fat content (Figure 3.2). This stream is called cream. The process can also involve an increase in fat content for the production of cream, resulting in a surplus of skimmed milk.

A widely applied method of standardisation of consumption milk and cheese milk is to de-fat part of the whole milk. The milk is heated in a plate heat exchanger to approximately 57°C, and subsequently separated into a high-fat fraction and a low-fat fraction. The low-fat product, called skimmed milk, is then mixed with a whole milk stream to obtain the specified fat content. A schematic overview of this method is shown in Figure 3.3. Furthermore, the fat content in whole milk can be increased by adding cream to it.

The process of de-fating has not really changed since the first invention of the centrifuge at the end of the nineteenth century. Only, the reliability of the equipment and the scale at which these processes are run are far beyond the imagination of the inventors. Owing to up-to-date process control, standardisation processes are very reliable, and applying advances in process control has opened up the possibility to operate close to the optimal performance.

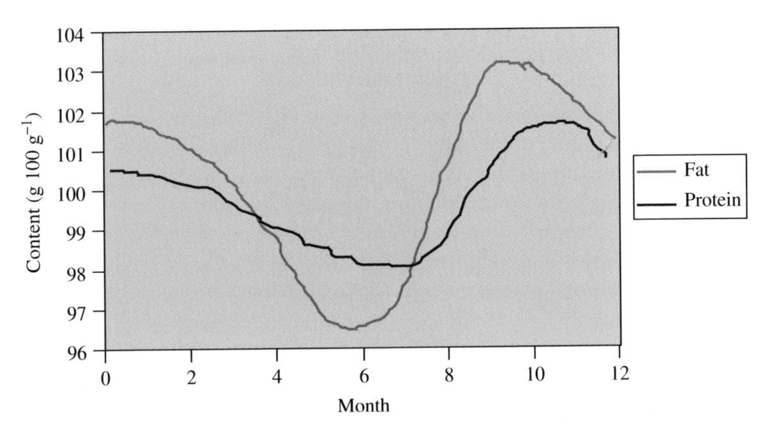

Fig. 3.1 Seasonal variations in fat and protein content of raw milk. Note that 100% is the yearly average of fat or protein content.

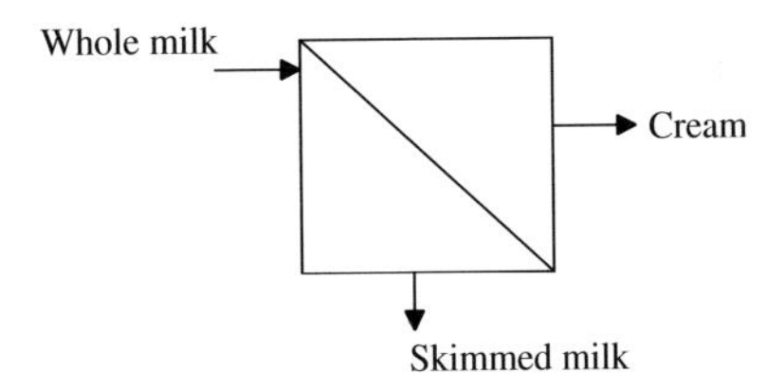

Fig. 3.2 Schematic representation of separation process of milk.

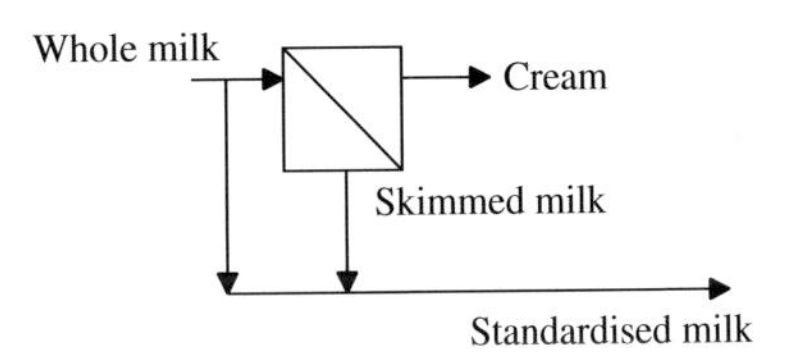

Fig. 3.3 Schematic representation of continuous standardisation of the fat content in milk.

In the sections below, a brief overview of the history of milk fat separation will be given. This is followed by an explanation of the physics behind centrifuges, an overview of standardisation methods and a number of practical considerations.

3.2 Overview of the history of milk fat separation

On 18 April 1877, the German trade journal *Milch-Zeitung* reported that a new device for enhancing the separation of cream from milk had been invented (Anonymous, 2003). This apparatus consisted of a drum that was rotated for some time. After stopping the rotation, the cream floated on top of the skimmed milk. After reading this article, Gustav de Laval (1845–1913) started to work on what became the first continuously working

Fig. 3.4 A view of one of the first separators the Alfa 1 made in 1882. Reproduced with permission of Tetra Pak AB, Lund, Sweden.

centrifugal separator (The Hindu, 14 December 2004) (see Figure 3.4). His invention was first demonstrated on 15 January 1879.

In 1888, Freiherr von Bechtolsheim patented the use of conical disks in centrifuges to improve the performance of centrifuges. This patent was acquired by a Swedish company partly owned by Gustav de Laval. In 1968, Westfalia Separators was among the first to introduce a solid ejection system into their centrifuges. This allowed an increase in the run time of a centrifuge from 10 to over 20 hours without cleaning, and also guaranteed a continuous capacity over 20 hours.

Since about 50 years ago, the general design of centrifuges has not changed, although the capacity of the equipment has grown to around 60 000 L h^{-1} of milk. This will surely increase even more in the years to come. However, minor changes made over the same period are (a) the positioning of the engine, minimising the number of mechanical parts and simplifying maintenance, (b) installation of more disk area to improve the separation quality and reduce ejection of solids, resulting in a higher yield; product losses due to this ejection of solids (i.e. mainly proteins) are now typically in the range of 0.05 g 100 g^{-1} and (c) improvement of the hygienic design. A major drawback of the separators, their tremendous noise, has not been solved.

Only a limited number of alternatives for centrifuging are known. The most promising is perhaps a micro-filtration technique that was patented by Goud'ranche *et al.* (1998). A ceramic micro-filtration membrane with an average pore size of 2 μm was used to separate fat globules that exceed a 2-μm diameter at a temperature of 50°C. They also succeeded in decreasing the fat content of the whole milk from 39 to 17 g kg^{-1}. The products from this milk did not differ in taste from products that were made from normal centrifuged milk, but their texture and mouth feel did differ (Goud'ranche *et al.*, 2000). The advantage of membranes over centrifuges is that they contain no moving parts and are, therefore, less sensitive to malfunctioning. Nowadays, the dairy industry is trying to re-use as much protein as possible and, as a consequence, the proteins contained in the ejected solids are processed and used as a protein source for other products, such as desserts.

Another interesting development is the use of centrifuges at low temperatures. This is applied for the production of non-pasteurised cheese milk. The drawback is that the increased viscosity of the milk reduces the capacity of the equipment, even using specially designed tanks, by 30% of the capacity at 57°C.

3.3 Physical models

The technique used to separate milk fat from the milk is based on two criteria:

- The fact that milk fat is lumped together in small globules
- The fact that the density of fat differs from the density of the surrounding serum

Because of their lower density, the fat globules experience a buoyancy force. Once they start to move, they also experience a drag force that works in the opposite direction. Combining these two forces results in a small upward motion of the fat globules and, as a consequence, the fat globules slowly move to the surface of the milk.

The buoyancy force (F) can be computed by

$$F = \Delta\rho g \frac{1}{6}\pi d_g^3 \tag{3.1}$$

where $\Delta\rho$ is the density difference between the fat and the surrounding serum, typically around 48 (kg m^{-3}), g the gravitational acceleration (m s^{-2}) and d_g is the diameter of the fat globule, typically around 3×10^{-6} (m).

The drag force (F_d), which acts in the opposite direction, reduces the acceleration of the fat globule, and can be computed from

$$F_d = -\frac{1}{8}C_d \rho_l \pi d_g^2 v^2 \tag{3.2}$$

where C_d is the drag coefficient (–), ρ_l is the density of the surrounding serum (kg m^{-3}), d_g is the diameter of the fat globule, typically around 3×10^{-6} m, and v is the velocity of the rising droplet (m s^{-1}).

At low velocities the drag coefficient (C_d) can be estimated using Stokes' law (Krijgsman, 1993):

$$C_d = \frac{24}{Re} = \frac{24}{\frac{\rho v d}{\eta}} \tag{3.3}$$

where Re is the Reynolds number (–), $\rho v d$ is the product of density, velocity and characteristic diameter, and η the dynamic viscosity (kg m^{-1} s^{-1}). In the Reynolds number, the ratio between the initial forces and the viscous forces is given.

By combining equations 3.1, 3.2 and 3.3, the rise velocity of a fat globule through the surrounding liquid can be estimated.

$$v = g\frac{(\rho_p - \rho_l) \cdot d^2}{18 \cdot \eta_p} \tag{3.4}$$

Where d is the diameter of the globules, ρ and η are the density and viscosity, respectively, the subscripts p and l refer to the density of the plasma and lipid phase of milk, respectively, and g is the acceleration, which is 9.8 m s^{-2} for gravitational separation.

From equation 3.4, it can be concluded that the time needed to separate the fat from the milk can only be changed by changing the diameter of the fat globules, by changing the gravitational constant or by changing the distance over which the fat globules should move.

Changing the diameter of the fat globules is difficult, but increasing the gravitational acceleration constant can be achieved by replacing it by a centrifugal acceleration.

$$a = r\omega^2 \tag{3.5}$$

where a is the centrifugal acceleration constant (m s^{-2}), r is the radius of the centrifuge (m) and ω is the angular velocity (rad s^{-1}).

This is exactly what is done in a centrifugal separator. In Figure 3.5, a schematic overview of a centrifuge is given.

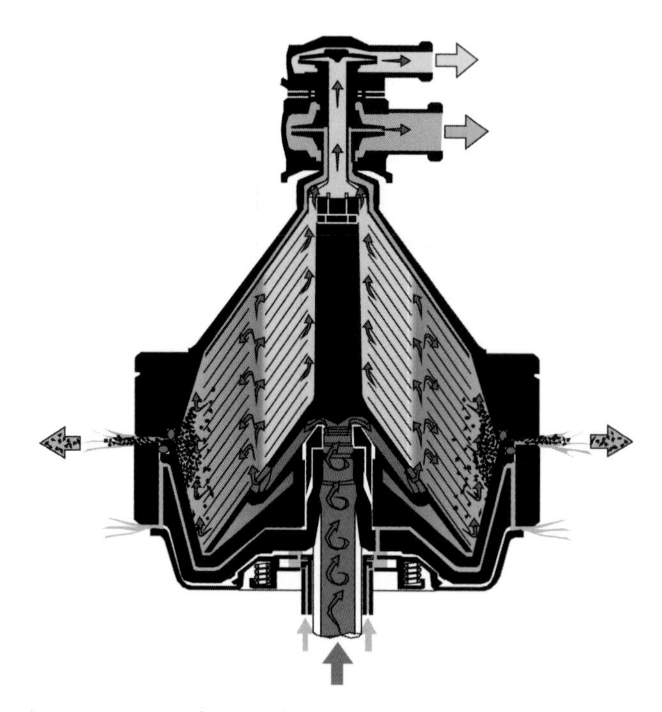

Fig. 3.5 Schematic representation of a centrifuge. Note that whole milk is coming in from the bottom, while cream (yellow) and skimmed milk (blue) are released through the top; the solid material is ejected through the outer walls of the centrifuge. After Anonymous (2003). Reproduced with permission of Tetra Pak AB, Lund, Sweden.

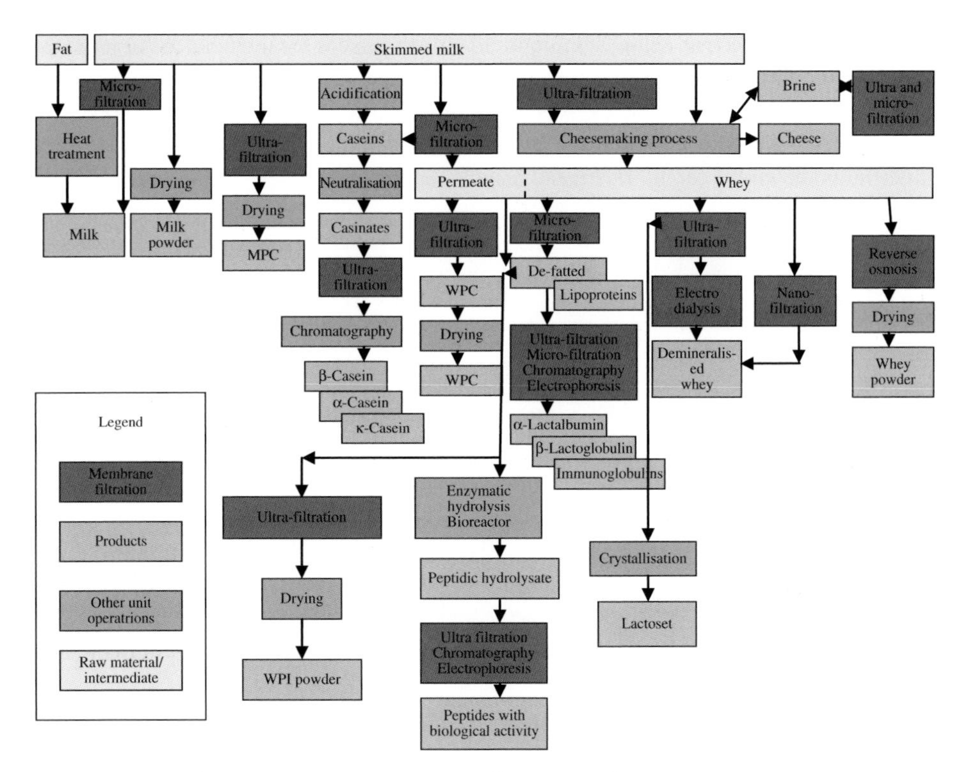

Fig. 3.6 Overview of dairy production processes of different speciality products.
MPC = milk protein concentrate; WPC = whey protein concentrate; WPI = whey protein isolate.

From equation 3.4, it also becomes clear that the viscosity of the milk has a large influence on the terminal velocity of the fat globules and, therefore, a large impact on separation efficiency. Viscosity is highly temperature dependent and, consequently, working at increased temperatures is advantageous to both the separation efficiency and the capacity of the equipment. The optimal temperature is around 57°C. Above this temperature the milk proteins start to degrade, and a further increase of the temperature is, therefore, not desirable. Since 57°C is also in the range of the thermisation temperature, these two process steps are often combined.

To increase the (economic) yield of the dairy industry, more and more by-product streams that were considered as waste in the past are now processed to high-value products; new techniques are applied to achieve this. A well-known example is the downstream processing of whey. Special whey centrifuges strip off the last traces of milk fat, and several membrane processes and chromatographic steps are applied to obtain products with specific functionalities. Figure 3.6 shows an overview of a wide range of these products.

3.4 Standardisation of the fat content of milk

In this section, different ways of standardisation will be described. Once the part of the raw milk has been split into skimmed milk and cream, the two streams can be mixed again in a ratio such that the resulting mixture will fulfil the product specifications.

Before coming to the actual techniques that can be used, the methods to compute the final fat content will be discussed. Davis (1966) reported the following example: if A is the cream fat content (40 g 100 g^{-1}), B is the skimmed milk fat content (0.05 g 100 g^{-1}), and C is the fat content of the end product (3 g 100 g^{-1}); then the ratio of cream and skimmed milk that should be added can be calculated using equations 3.6 and 3.7.

$$M_a = \frac{X(C - B)}{(C - B) + (A - C)} \tag{3.6}$$

$$M_b = \frac{X(A - C)}{(C - B) + (A - C)} \tag{3.7}$$

where M_a is the mass of product a (kg), M_b is the mass of product b (kg) and X is the mass of the desired product. This can also be represented graphically as shown in Figure 3.7. This method only takes the fat content into account. For many products, the ratio between the fat and the protein content is important.

As already shown in Figure 3.1, the average fat and protein contents of raw milk changes during the year. It can be seen that the position of the maximum fat content differs from that of the protein content. Since the fat/protein ratio is important to many dairy products, both should be adjusted to obtain a high-quality end product. A standardisation approach as described above will, therefore, not satisfy this need, and more advanced models are needed.

The introduction of computers into the process control of dairy plants has opened up the possibility of applying those more sophisticated mathematical models. These models have two distinct advantages over the above-described method: First, more than two process streams with different composition can be taken into account, and second, standardisation of the protein and milk fat can be accomplished at the same time.

Figure 3.8 shows a screen shot of such a system, which has been developed by NIZO Food Research. It is capable of selecting the optimal combination to produce the required product. This system is based on mass balances over all process streams, and can be applied to both protein and milk fat standardisation. The tool can be used off-line using the NIZO Premia simulation environment as well as in-line using the NIZO Premic environment (van den Hark, 2006). Use of these more advanced tools can lead to a more economic application of all process streams and a higher degree of control over the end product.

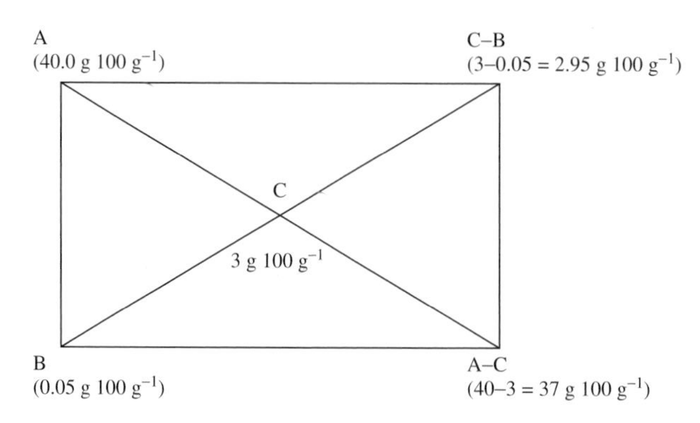

Fig. 3.7 Graphical representation of the calculation of the fat content in product C.

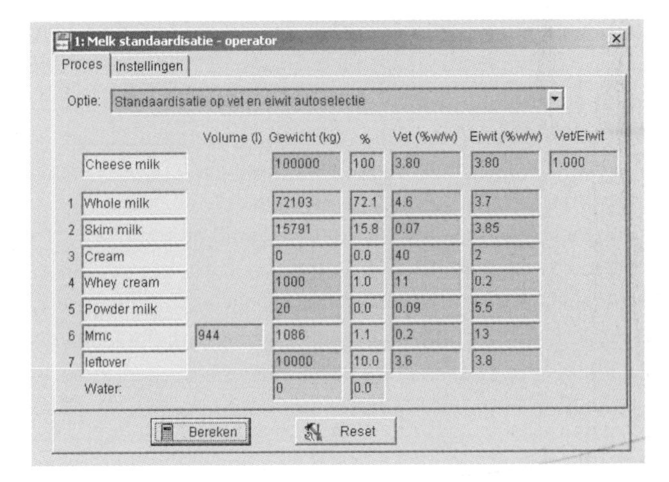

Fig. 3.8 Computer screen of NIZO Premia standardisation module.

In practice, mixing of milk can be done in several ways. The oldest approach is probably the batch standardisation method (Anonymous, 1989) where the fat content of an amount of whole milk is analysed, and it is calculated how much skimmed milk should be added to reach the required fat content. This amount is added to the tank and the total content is mixed.

In modern dairy plants, continuous standardisation is used. This means that the whole milk and the skimmed milk are mixed continuously in the process lines. In this process, several degrees of automation can be found. Here distinction is made between systems that advise the operator (on-line systems) and systems that act directly on the process (in-line systems). In the most advanced case, in-line measurement of the density is used to estimate the fat content of the milk, and the mixing ratio is continuously adjusted to this measured value. In modern systems, within 3 minutes of a disturbance in the feed, the composition is regulated as close as +0.0.1 g fat 100 g^{-1} to the required equilibrium value (Figure 3.9).

Advantages of the in-line standardisation are as follows (Alderlieste, 1992):

- Less intermediate storage required
- Shorter residence times of the raw material in the plant and
- Faster start up of the downstream processes

Most dairy plants use density measurement to estimate the fat content of the streams.

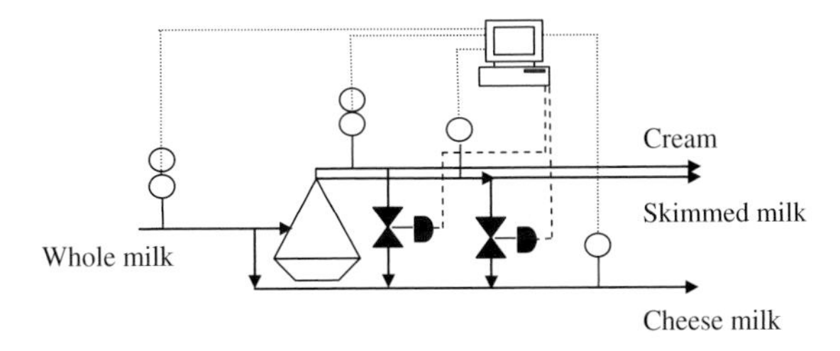

Fig. 3.9 Schematic representation of an in-line cheese milk standardisation system.

3.5 Conclusion

Since the first introduction of centrifuges in the dairy industry, centrifugation has become one of the standard unit operations. The technique has improved, but the physical principle has never changed. Although other techniques continue to be investigated, the reliability of this equipment is of crucial importance to the industry. Therefore, the centrifuge is still the equipment of choice for the separation of fat from milk or whey.

With developments in analytical techniques and the rise of using automated process control, ways of standardising raw milk have changed. Computer models have opened up the possibility to combine more streams to get the right product, and developments will no doubt continue to improve standardisation processes.

References

Alderlieste, P. J. (1992) In-line Standaardiseren van het Vetgehalte van Melk, *VMT*, 1992 **16/17**, pp. 9–11.

Anonymous (1989) In-line Standaardiseren van Melkproducten in Nieuw Daglicht, *Food Management*, January 1989.

Anonymous (2003) *Dairy Processing Handbook,* (ed. G. Bylund), 2nd and revised edn., pp. 99–122, Tetra Pak Processing Systems AB, Lund.

Davis J. G. (1966) Standardisation of the fat content of cream (or milk). *Dictionary of Dairying*, 2nd edn, pp. 976–977, Leonard Hill, London.

Goud'ranche, H., Fauquant, J. & Maubois J.-L. (1998) *Separation of Different Sizes of Fat Globules in Milk*, **French Patent Application,** 2 776 208.

Goud'ranche, H., Fauquant, J. & Maubois J.-L. (2000) Fractionation of globular milk fat by membrane filtration. *Lait*, **80**, 93–98.

van den Hark, M. (2006) De fabriek van 2012. *VMT* 2006 **20**, pp. 14–15.

Krijgsman, J. (1993) *Sedimentation and Centrifugation*. Course notes to advanced course on downstream processing TU-Delft 2–6 June 2003.

Walstra P., Wouters J. T. M. & Geurts T. J. (2006) Milk components. *Dairy Science and Technology*, 2nd edn, pp. 17–108, CRC Press (Taylor & Francis Group), Boca Raton.

4 Cream and Related Products

M.A. Smiddy, A.L. Kelly and T. Huppertz

4.1 Introduction

Dairy cream was traditionally considered a luxury product, but is now readily used in many forms and for a variety of purposes. Although originally associated with desserts and fresh fruit, it is also used as an ingredient in sweet and savoury dishes, e.g. ice cream, soup, custard bases and cakes, and is the primary raw material for the manufacture of butter and butter oil (Hoffmann, 2002a). Cream is a concentrated emulsion of milk lipid globules in skimmed milk and is prepared commercially by centrifugal separation of the less-dense lipid phase from skimmed milk. Different types of cream are primarily classified according to their fat content (g 100 g^{-1}):

- Light coffee cream (<10)
- Coffee cream (15–18)
- Single or half cream (15–25)
- Cream or full cream (30–40)
- Double cream (45–50)

Some current regulations for the fat content of cream products in various countries are outlined in Table 4.1, but because not all countries market all classes of cream, and countries use different names to describe similar products, a uniform international definition or classification system for cream products cannot be prepared. Cream products are also classified according to their function, e.g. whipping cream, coffee cream or cream liqueur, or by the processing received, e.g. pasteurised cream, ultra-high-temperature (UHT)-treated cream, frozen cream, dried cream and cultured or sour cream. The various cream products are discussed in detail in Sections 4.3 to 4.8.

Medium-fat cream (25 g fat 100 g^{-1}) contains ~68.5 g moisture, ~2.5 g protein, ~3.0 g carbohydrate and ~0.5 g ash 100 g^{-1}; these concentrations decrease with increasing fat content (Bassette & Acosta, 1988). The lipids in cream exist, as in milk, in the form of milk lipid globules ranging in diameter from ~0.5 to 10 μm, which are surrounded by the milk lipid globule membrane (MLGM). The physico-chemical properties of milk lipid globules are described in detail in Chapter 1. The physico-chemical properties of cream depend on several factors, such as the state of the lipid globules and the MLGM, the concentration of lipid globules, the type and concentration of non-fat milk solids in cream, for example, proteins and salts and added emulsifiers and stabilisers, the temperature of the cream and the physical handling of the cream, for example, pumping, aeration and agitation, which can cause disruption or agglomeration of the globules (Towler, 1986).

Table 4.1 Regulations in several countries with regard to lipid content in cream.

Country	Cream type	Lipid content (g 100 g^{-1})
Australia and New Zealand	Cream	18–40
France	Light cream	12–30
	Cream	30–40
Germany	Coffee cream	10–30
	Whipping cream	30–40
The Netherlands	Cream	10–30
	Whipping cream	30–40
United Kingdom	Half cream	12–18
	Single cream	18–35
	Whipping cream	35–48
	Double cream	48–55
United States	Half and half cream	10–18
	Light cream	18–30
	Light whipping cream	30–36
	Heavy cream	36–45

Data collated from Hoffmann (2002b).

In this chapter, the manufacture, properties and uses of common cream products are discussed. The common processing treatments applied in the preparation of cream products, that is, separation, standardisation, heat treatment and homogenisation, are briefly discussed in Section 4.2. In subsequent sections, different cream products, i.e. whipping cream, aerosol whipping cream, cream liqueur, cultured cream, coffee cream, frozen cream and dried cream, are discussed.

4.2 Cream processing

The basic technology for the preparation of cream products, i.e. separating the cream phase from the milk, standardising the cream to the desired fat content and heating the cream to increase its shelf-life, is universal to (almost) all industrially processed cream products. Another post-separation process, homogenisation, is applied only to certain cream products, either with the aim of improving product properties or increasing the physical shelf-life of the product (Varnam & Sutherland, 1994).

4.2.1 *Separation*

Cream was originally separated from whole milk by gravity. Whole milk was allowed to stand for a period during which time, lipid globules, being less dense than the skim phase of milk, rose to the surface, causing the formation of a cream layer, which was subsequently skimmed off by hand (Anonymous, 2003). As the lipid globules rise through the skim milk phase under the influence of gravity, they will eventually attain a constant velocity (v), which can be calculated using Stokes' Law:

$$v = a\frac{(\rho_p - \rho_l) \cdot d^2}{18 \cdot \eta_p}$$

where d is the diameter of the globules, ρ and η are the density and viscosity, respectively, the subscripts p and *l* refer to the plasma and lipid phases of milk, respectively, and a is the acceleration, which is 9.8 m s^{-2} for gravitational separation (Anonymous, 2003). For gravitational creaming, a equals the gravitational attraction of the earth, i.e. 9.8 m s^{-2}. However, by modern standards, gravitational separation is obviously insufficient and centrifugal separation is therefore used, whereby a centrifugal force is created by rotation. This creates a centrifugal acceleration, a, which increases with distance from the axis of rotation, the radius, r, and with the speed of rotation, expressed as angular velocity, ω, according to:

$$a = r\omega^2$$

Centrifugal separators contain stacks of conical discs with vertically aligned distribution holes through which whole raw milk is introduced. Under the influence of a centrifugal force, the sediment and lipid globules in the milk settle radially outwards or inwards respectively, in the separation channels, according to their densities relative to that of the continuous medium (skimmed milk; Anonymous, 2003). Solid impurities of high density in the milk rapidly settle out to the outside of the separator and are collected in the sediment space. Since the cream is less dense than skimmed milk, it moves inwards towards the axis of rotation and is removed through an axial outlet. Skimmed milk moves outwards beyond the disc stack and is removed through a channel between the top of the disc stack and the conical hood of the separator bowl to a skimmed milk outlet (Anonymous, 2003). Separation of cream from whole milk, therefore, involves concentration of lipid globules, followed by removal of the cream from the skim phase (Huppertz & Kelly, 2006) to produce two streams, skimmed milk and cream, the latter of which commonly amounts to ~10% of the total throughput.

Separation by centrifugal separators allows rapid and efficient separation and, although very high centrifugal forces are applied, not all lipid globules are removed. Skim milk usually has a residual fat content of ~0.1 g 100 g^{-1}, primarily containing lipid globules <1 μm in diameter (Anonymous, 2003). Separation is discussed in greater detail in Chapter 3 and by Faulks (1989).

4.2.2 *Standardisation*

Since traditional cream separators could not be accurately operated to produce a cream of specific fat content, i.e. within 0.5% of that required, it was necessary to firstly produce cream with a higher fat content, for example, 50 g 100 g^{-1}. This cream was collected and mixed in cream tanks from which cream was sampled for fat content. The Gerber test was normally used to determine the fat content of cream, which was then reduced via addition of skimmed milk, in a process called standardisation (Faulks, 1989). The purpose of standardisation is to produce cream of a defined, guaranteed, fat content. Nowadays, direct in-line standardisation is usually combined with separation (Anonymous, 2003), and equipment that automatically monitors and controls the fat content of cream is used for standardisation. On discharge from the separator, the skim and cream streams are mixed and the proportion of cream included determines the fat content of the cream product prepared. A system of control valves, flow and density meters and a computerised control loop is used to adjust the fat content of milk and cream to the desired value (see Anonymous, 2003 for a full description of this system). It is important that the fat content is as close to the stipulated value as possible; too high a fat content results in economic losses for the processor, whereas too low a fat content may result in cream products not meeting legal requirements. During standardisation, the temperature of cream can be over 40°C, so bacterial growth may occur. Hence, it is essential that standardisation is carried out rapidly, to prevent cream being held at these temperatures for unnecessarily long periods. Subsequent pasteurisation and cooling steps must also follow rapidly (Faulks, 1989). Standardisation is discussed in greater detail in Chapter 3.

4.2.3 *Heat treatment*

All cream products are heat treated, principally to inactivate spoilage and pathogenic micro-organisms, as well as enzymes. Heat-induced inactivation of micro-organisms and enzymes ensures the safety of cream products for the consumer, while also extending the shelf-life of the product. The effects of heating depend mainly on the intensity of the treatment, that is, the temperature and duration of heating (Walstra *et al.*, 2006). For heat treatment of cream, either high-temperature short-time (HTST) pasteurisation or UHT sterilisation is commonly applied. The high fat content of cream protects microbes during heating, necessitating a more severe heat treatment than that required for pasteurisation of milk (Early, 1998). For HTST pasteurisation of cream, heating at 75°C for 15 seconds for cream containing <20 g fat 100 g^{-1} or to >80°C for 15 seconds for cream containing >20 g fat 100 g^{-1} is recommended by the International Dairy Federation (IDF, 1996). Most vegetative cells, including pathogens and yeasts and moulds, are inactivated by pasteurisation, but some thermoduric bacteria, including spore-forming thermoduric *Bacillus* spp., survive pasteurisation (Early, 1998).

Where cream is UHT-treated, the quality of the raw material is particularly important owing to the longer shelf-life of this product. UHT processing of cream involves heating the product to 135 to 150°C for a few seconds, whereby minimal chemical, physical and organoleptic changes to cream occur, but spoilage and pathogenic micro-organisms are inactivated (Early, 1998). UHT treatment of cream does not necessarily produce a sterile product, as some bacterial spores, especially those of *Bacillus* spp., can survive UHT treatment (Early, 1998). The indigenous milk enzymes, lipoprotein lipase and the proteinase plasmin, cause lipolysis and proteolysis, respectively, during storage. Although lipase is largely inactivated by pasteurisation (Kosinski, 1996), considerable plasmin activity remains even after UHT treatment. Furthermore, bacterial lipases and proteinases are particularly heat-stable and can survive UHT treatment (Muir & Kjaerbye, 1996). Production of bacterial enzymes must therefore be prevented in UHT cream. In UHT-treated whipping cream, the formation of undesirable cream plugs (small lumps of partially solidified lipid material), which cannot be re-dispersed by gentle shaking, may also occur after prolonged storage

(Streuper & van Hooydonk, 1986; Huppertz & Kelly, 2006). An extensive review of UHT cream was published by IDF (1996).

4.2.4 *Homogenisation*

The main objective of homogenisation is to prevent, or at least minimise, the creaming phenomenon. However, alteration of the characteristics of cream products or the production of new product structures is also possible by selecting the appropriate homogenizing parameters (Hinrichs & Kessler, 1996). While homogenisation is essential for all types of single and half cream and may be applied to double cream, it can also have negative effects on cream functionality, for example, in whipping creams, as is further discussed in Section 4.3.4.

4.2.5 *Quality of cream*

The use of poor-quality raw milk contributes to most of the difficulties experienced during preparation of cream. Also, the physical, chemical and microbiological properties and yield of cream depend on the quality of the raw milk used, as well as on the processing and packaging efficiency, and on the conditions of distribution and storage of products before consumption (Rothwell, 1989). Good-quality raw milk combines a satisfactory composition with production under hygienic conditions to prevent contamination by undesirable spoilage and pathogenic micro-organisms (Kosinski, 1996). Furthermore, raw milk should also be unadulterated, free of taints, antibiotics, blood and visible sediments, be refrigerated before cream preparation and be processed as soon as possible to minimise growth of undesirable micro-organisms and production of microbial proteinases and lipases (Kosinski, 1996). Raw milk must also be handled in a way that minimises damage to the MLGM. Excessive agitation and/or pumping can cause air to be drawn in to the milk, which can damage the MLGM resulting in free fat, which may coalesce or 'churn', making separation difficult, and which may be hydrolysed to fatty acids imparting rancid off-flavours (Kosinski, 1996).

One of the important factors determining consumer acceptability of cream is visual assessment of its 'body' or 'viscosity' (Rothwell *et al*., 1989). Many factors, including every aspect of plant configuration, processing and handling, affect, although sometimes slightly, the viscosity of cream. Factors associated with the milk (e.g. the triacylglycerol content), its processing (e.g. the separation temperature, fat content, heat treatment and homogenisation pressure) and storage (e.g. the temperature and length of storage) all affect the viscosity of the resulting cream (Varnam & Sutherland, 1994).

Since cream is still regarded as something of a luxury, it must have an excellent flavour, and as it is a high-fat product, any off-flavour of the milk lipids will be concentrated. For example, rancidity will not be detected by the majority of consumers in milk that contains 1 mmol of free fatty acids 100 g^{-1}, but a cream made from such milk will taste rancid. Milk used for preparation of cream must therefore be of high quality, especially with regard to lipolysis (Walstra *et al*., 2006). During storage, cream must be protected from the environment to minimise chemical, organoleptic and bacteriological degradation (Bull, 1989). Control of exposure to both the atmosphere and light are very important as the synergistic effects of oxygen and UV light cause oxidative degradation of lipids, producing off-flavours and rancidity (IDF, 1982).

4.3 Whipping cream

Whipped cream is valued by consumers for its taste and texture and is often considered a luxury product. The whipped product is created by beating air into cream until a stiff-foamed product is produced, which finds many applications, including those in desserts and cakes. This section will outline the production process for whipping cream and the process of whipping of cream, as well as the influence of a variety of factors thereon.

4.3.1 *Production of whipping cream*

A general outline of the process for the manufacture of pasteurised and UHT whipping cream is outlined in Figure 4.1. Cream is first standardised to the desired fat content, which is generally between 30 and 40 g 100 g^{-1}. Stabilisers, which are discussed in more detail in Section 4.3.6, may be added to the standardised cream before heat treatment. Pasteurisation of cream is commonly carried out at ~80°C and, following pasteurisation, cream is cooled, packaged and distributed; pasteurised cream has a shelf-life of <3 weeks at refrigeration temperature. The need for a cream with an extended shelf-life has led to the production of UHT whipping cream, where the cream, after the optional addition of stabilisers, is heated at >135°C for a few seconds. As a result, a shelf-life of several months can be attained. However, the physical shelf-life of such cream is limited, owing to the fact that creaming of lipid globules occurs, so a homogenisation step needs to be applied to reduce lipid globule size and minimise separation of the globules during storage. However, as discussed

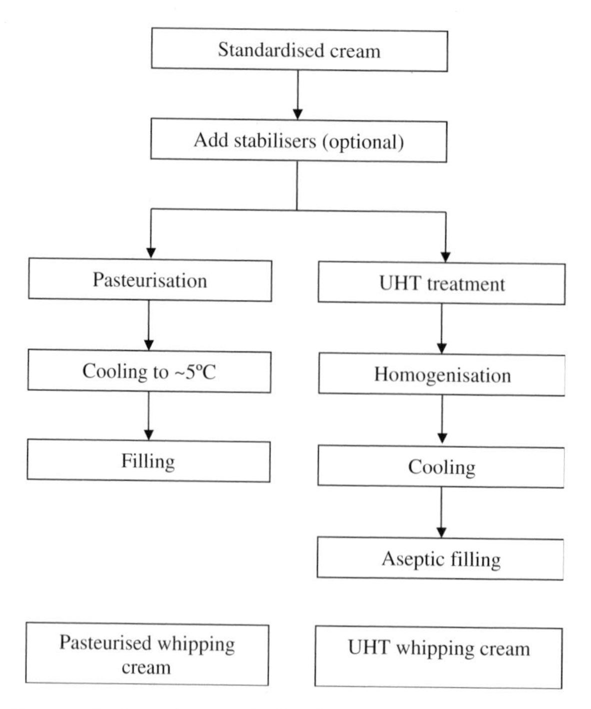

Fig. 4.1 Process for the manufacture of pasteurised and UHT whipping cream.

in more detail in Section 4.3.4, homogenisation impairs the whipping properties of cream, so choosing optimal conditions for homogenisation of UHT whipping cream represents a trade-off between maintaining optimal whipping characteristics and minimising separation of lipid globules during storage. Following homogenisation, UHT-treated whipping cream is cooled, packaged aseptically and distributed (Hoffmann, 2002a).

4.3.2 *Whipping of the cream*

During the first stage of the whipping process, air is beaten into the cream, resulting in the formation of a coarse foam, which contains air bubbles with an average diameter of ~150 μm (Noda & Shiinoki, 1986). The air bubbles in the foamed cream rapidly become covered by milk proteins, which stabilise them against collapse (Brooker, 1993). A large proportion of the protein absorbed on the bubble surface is formed by the surface-active β-casein, which is present in large quantities in non-micellar form at the low temperatures at which cream is commonly whipped, but other caseins and whey proteins are also found at the bubble interface (Brooker *et al.*, 1986; Anderson *et al.*, 1987). On further whipping, air bubble size is reduced approximately three-fold (Noda & Shiinoki, 1986) and milk lipid globules displace some of the proteins from the bubble interface; on adsorption at the bubble interface, the globules shed their globule membrane material from the contact area with the air bubble in the process, thereby creating an air–lipid interface (Buchheim, 1978; Brooker *et al.*, 1986; Brooker, 1993). Prolonged whipping leads to stiffening of the cream, as a result of the creation of a network of partially coalesced lipid globules. Partial coalescence may be induced as a result of mechanical damage to the milk lipid globules during whipping; this process is enhanced by the presence of large lipid crystals in the globules. Partial coalescence of the lipid globules can also result from the collapse of air bubbles in the foam. Since the globules have shed their original membrane material on the area that interacts with the bubble interface, the collapse of air bubbles will lead to partially uncovered globules, which are extremely susceptible to partial coalescence. Eventually, a stiff foam will result, in which the air bubbles are surrounded and stabilised by a network of coalesced lipid globules, whereas a network of coalesced lipid globules in the serum phase of the foam adds structure and stability and prevents collapse of the whipped cream (Anderson & Brooker, 1988).

4.3.3 *Characterisation of whipped cream*

Three principal parameters are commonly used to characterise the whippability of cream and the properties of the obtained whip, that is, the overrun, the stiffness of the whipped cream and the whipping time. The overrun is a measure of the amount of air incorporated into the whipped cream and can be calculated from the density of the cream before and after whipping according to the equation,

$$\text{Overrun} = \frac{(\text{density of unwhipped cream} - \text{density of whipped cream})}{\text{density of whipped cream}}$$

In practice, the density values are commonly replaced by the weight of a given volume of cream before and after whipping. The stiffness of the whipped cream is a measure of the degree of textural development in the product and can be determined by common methods for texture analysis. From a consumer perspective, a stiff whipped cream is usually preferable. The whipping time of the cream is defined as the time required to reach a set end point in the whipping process. The most commonly applied end points are the point of maximum overrun and the point of maximum stiffness. These are not equivalent, as maximum stiffness occurs after maximum overrun has been reached and has started to decline. In the following sections, the influence of cream composition, additives and processing conditions on overrun, whipping time and stiffness of the whipped cream are described.

4.3.4 *Influence of processing conditions on whipping characteristics of cream*

The various processing steps outlined in Section 4.3.1 can significantly affect the whipping characteristics of cream. Whipping characteristics of cream obtained from the same milk can vary considerably, just by use of a different separator for separating the cream (Anderson *et al.*, 1987), and overrun of whipped cream increases with increasing temperature of separation in the range 25 to 50°C (Kieseker & Zadow, 1973). Homogenisation, which is required to prevent creaming of lipid globules in UHT whipping creams, increases the whipping time of the cream and reduces the overrun and stiffness of the whipped cream. Such changes are related to the fact that, as outlined in Chapter 1, the membrane of homogenised lipid globules is composed primarily of caseins, which makes them less susceptible to absorption onto the air bubble interface (Anderson & Brooker, 1988) and shear-induced aggregation (Melsen & Walstra, 1989), compared to the globules in unhomogenised cream. To improve whipping characteristics of homogenised cream, emulsifiers may be added to the cream, as described further in Section 4.3.6. Heat treatment *per se* does not drastically affect the whipping properties of cream (Smith *et al.*, 2000), but it is commonly applied in conjunction with homogenisation, which is why highly heat-treated creams generally have poor whipping properties.

4.3.5 *Compositional factors affecting whipped cream characteristics*

Of the constituents of cream, the protein and fat content have the largest influence on the properties of the whipped cream. Reducing the protein content of cream has little effect on the overrun of whipped cream, but reduces the whipping time and stiffness of the cream considerably, whereas increasing the protein content of the cream has the opposite effect (Anderson & Brooker, 1988; Needs & Huitson, 1991). As only a small amount (<5 g 100 g^{-1}) of the total protein present in cream is required to stabilise the air bubbles in cream during the initial stage of whipping, the influence of protein content on the whipping time and stiffness may be a result of the influence of protein concentration on the viscosity of the serum phase of the cream (Anderson *et al.*, 1987); in this respect, it is interesting to note that, as described in Section 4.3.6, the addition of stabilisers, which also increase serum phase viscosity, has an effect on whipping properties similar to that on increasing the protein content. Whipping time may increase with increasing serum phase viscosity, owing to the fact that shear-induced partial coalescence of lipid globules decreases with increasing serum phase viscosity, thereby increasing the time to build up the required network of partially coalesced lipid globules.

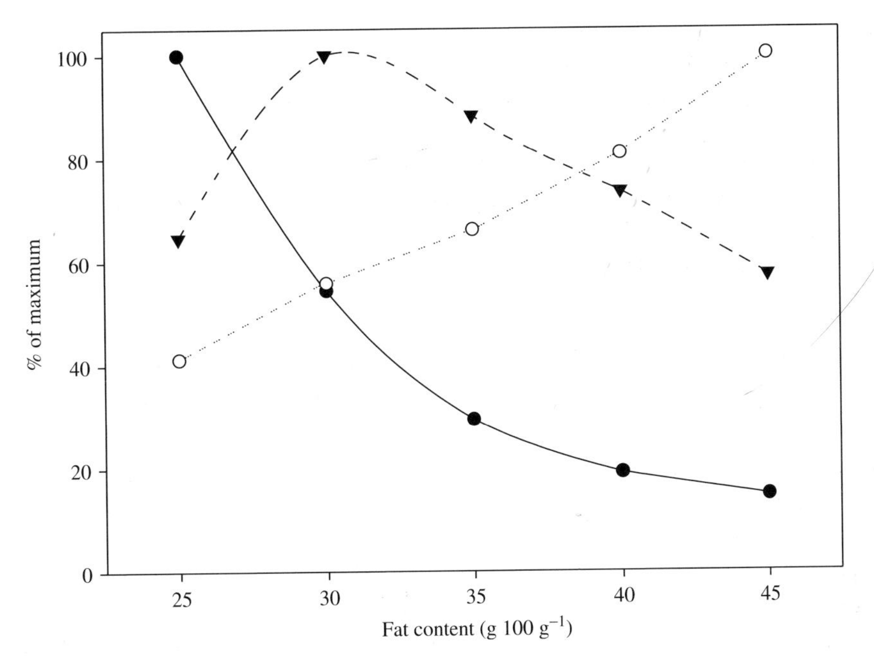

Fig. 4.2 Influence of lipid content of the cream on the whipping time (•), stiffness (○) and overrun (▼) of the whipped cream. Note: Values are expressed as a percentage of the maximum value attained. After Anderson & Brooker (1988).

Fat content also has a strong influence on the whipping properties of cream, as clearly illustrated in Figure 4.2. The whipping time and stiffness of the whipped cream progressively decreases or increases, respectively, with increasing fat content in the range 25 to 45 g 100 g^{-1}. Such trends are readily explainable, considering that the stiffness of whipped cream is derived primarily from a network of partially coalesced lipid globules and that a higher concentration of lipid globules facilitates the more rapid formation thereof. The overrun has a clear maximum at ~30 g fat 100 g^{-1} (Figure 4.2); this may be the result of increased fat content, obviously facilitating the formation of a network of partially coalesced lipid globules, but also reducing the amount of air whipped into the cream as a result of increased cream viscosity. Natural variations in the concentration of free fatty acids in cream appear to have little influence on its whipping properties (Needs *et al.*, 1988). The influence of other cream constituents, for example, lactose or salts, on the whipping properties of cream has not been studied in detail to date.

Besides the concentration of fat in the cream, the crystallisation state of the lipids is also of great importance in obtaining a whipped cream. The presence of some lipid crystals inside the globules is required to induce partial coalescence and prevent destabilisation of foam bubbles through spreading of excessive amounts of liquid lipid over the interface. Cream is, therefore, commonly whipped at ~5°C, and the whipping properties of cream can be further influenced by manipulating the size of the lipid crystals. Warming the whipping cream to 25 to 30°C causes melting of some, but not all, of the crystalline lipid in cream. Subsequently, cooling the cream back to 5°C causes rapid growth of the existing crystals. The whipping properties of cream are improved significantly by such tempering of the cream, and can even lead to a whipped cream, which can be stored for several weeks without any noticeable structural change (Drelon *et al.*, 2006).

4.3.6 *Influence of stabilisers and emulsifiers on whipping characteristics of cream*

To improve the whipping properties of cream, both stabilisers and emulsifiers may be added to cream in jurisdictions that permit such additives. Emulsifiers are principally added to improve the destabilisation and partial coalescence of the lipid globules and are of particular relevance for improving the whipping properties, particularly reducing whipping time, of homogenised cream products (Anderson & Brooker, 1988). Emulsifiers commonly used for this purpose are mono- and diglycerides, and the use of the phospholipids from buttermilk has also been suggested; care should be taken with the buttermilk approach because adding powdered buttermilk also increases the protein content of the cream considerably, which, as discussed in Section 4.3.5, has adverse effects on the whipping properties of cream.

The main purpose of the addition of stabilisers to whipping cream is to reduce the extent of lipid globule separation in the cream during storage and improve the stiffness and stability of the whipped cream. As such, stabilisers are particularly useful additives for cream products with a long shelf-life, that is, sterilised or UHT whipping cream, and cream products that form a weak whip, for example, reduced-fat cream products. The most commonly used stabilisers are polysaccharides, for example, carrageenans, alginates or starches (De Moor & Rapaille, 1982), whereas polyamides, for example, gelatine, can also be used. Stabilisers exert their positive effect on the storage stability of the whipping cream and stiffness of the whipped cream by increasing the viscosity of the serum phase of milk, analogous to the manner in which increased protein content is believed to increase the stiffness of whipped cream. Similar to the increase in protein content, the addition of stabilisers to whipping cream also has the less desirable effects of increasing whipping time and decreasing overrun (Camacho *et al.*, 1998). Such effects may be overcome by the combined inclusion of stabilisers and emulsifiers in whipping cream.

4.4 Aerosol-whipped cream

The term aerosol-whipped cream refers to a foamed cream that is produced by an aerosol can. The aerosol can contains a cream under pressure, which is supersaturated with a gas, often nitrous oxide, better known as laughing gas, which functions as the propellant. Pressing the top of the nozzle opens the aerosol can, and cream is forced through the nozzle and expands into a foam as a result of the decrease in pressure. Aerosol-whipped cream may be produced from either dairy or non-dairy cream, but only the former is considered here. Non-dairy aerosol creams generally contain vegetable fat, and are often referred to as toppings.

4.4.1 *Production of aerosol-whipped cream*

Before being filled into the aerosol can, standardised cream is heat-treated to extend its shelf-life. UHT treatment is most commonly applied in this respect as it provides a long shelf-life, while causing minimum heat damage to the constituents and organoleptic properties of the cream (Wijnen, 1997). The microbial shelf-life of aerosol-whipped cream is further extended by being stored under a nitrous oxide atmosphere in the aerosol can (Juffs *et al.*, 1980). After heating, the cream is homogenised at a low pressure to reduce lipid globule size and thus prevent creaming and partial coalescence of the cream on storage. Since the cream does not contain sufficient surface-active material to cover the increased milk lipid globule surface area after homogenisation, an emulsifier, for example, monoglycerides, is added to improve homogenisation efficiency. Furthermore, a stabiliser, for example, carrageenan, is added to increase the viscosity of the product and thereby decrease the rate of creaming. Finally, sugar and flavourings may be added to create a desirable flavour (Wijnen, 1997).

Following the aforementioned processing steps, the cream is aseptically filled into tinplate or aluminium cans, which are then closed and sealed with a valve. The food-grade propellant, usually nitrous oxide, is added to the cans via the valve, while the cans are shaken to speed up dissolution of the gas in the cream. Nitrous oxide is commonly used as the propellant, because its solubility in cream is $\sim$50 times higher than the solubility of air in cream. The pressure in the aerosol can is $\sim$0.5 to 1 MPa, and therefore, most of the nitrous oxide is dissolved in the cream (Wijnen, 1997).

The aerosol can is closed by a valve (Figure 4.3a) where the inside gasket (3) prevents the pressurised cream from leaving the can. The aerosol can is opened by pressing the valve in the direction of the arrow (Figure 4.3b), which results in a local pressure drop as a result of the direct contact between the inside and outside of the can. This causes the flow of cream from the can and leads to the formation and growth of bubbles, because the nitrous oxide comes out of solution as a result of the reduced solubility at the lower, atmospheric pressure. The instant foam formation is greatly affected by the amount of gas dissolved in the cream, which needs to be sufficiently high to obtain satisfactory foaming properties, but is not allowed to exceed 1.5–2.0 MPa, particularly for household use, to avoid the risk of explosion (Wijnen, 1997).

4.4.2 *Properties of aerosol-whipped cream*

The obvious advantage of aerosol-whipped cream over traditionally whipped cream is the speed and ease of production of foam in controllable portions. The foam is characterised by a very high overrun, that is, $\sim$400 to 600%, which is $\sim$4 times higher than that of traditional whipped cream. Although aerosol-whipped creams are commercially available and give a satisfactory mouth-feel and firmness, they are not regarded as being as good as traditional whipped creams. This is due to the fact that the lipid globule network, which stabilises traditionally whipped cream, does not form in aerosol-whipped cream (Wijnen, 1997). The lack of a lipid globule network in aerosol-whipped cream also has implications for the stability of the foamed product. As illustrated in Figure 4.4, the aerosol-whipped cream collapses rapidly, a process which is aided by the high solubility of nitrous oxide in the cream, thus making the foamed product very susceptible to disproportionation (Wijnen & Prins, 1995). The use of a gas with a lower solubility in cream may overcome this problem, but would require considerably higher pressures inside the aerosol can to obtain a product of comparable overrun and foam bubble size distribution; however, such pressures are unsuitable for household use (Wijnen, 1997).

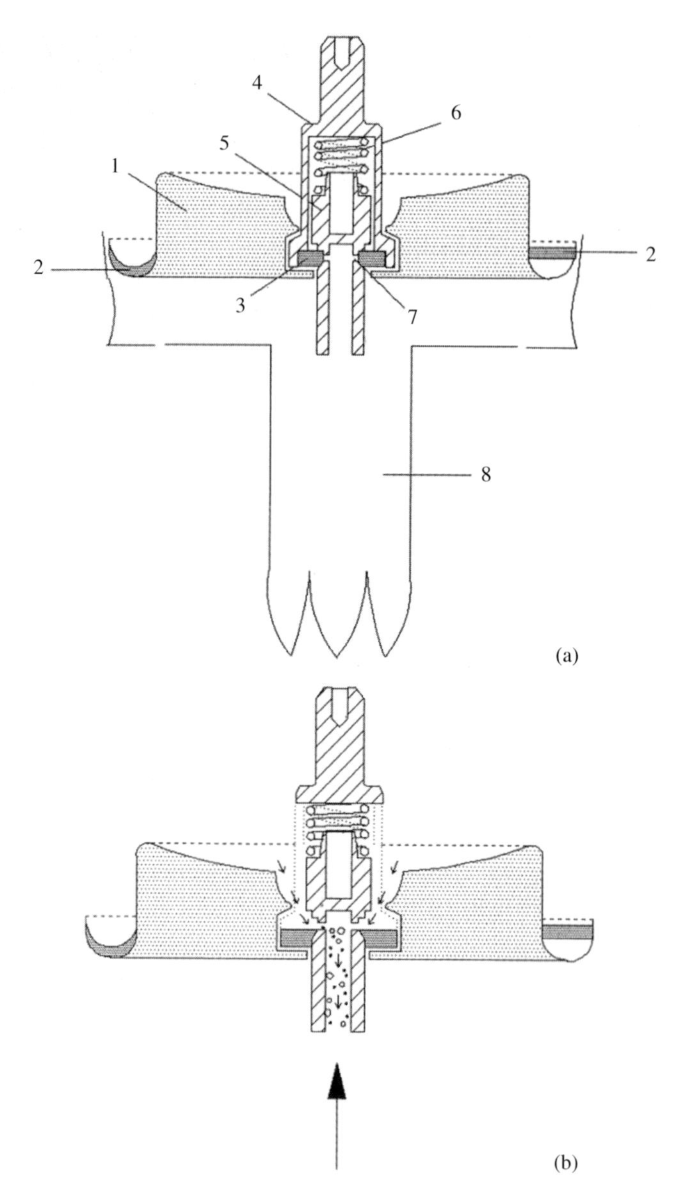

Fig. 4.3 Schematic drawing of an aerosol valve in (a) closed or (b) open position. Note: (1) valve holder; (2) outside gasket; (3) inside gasket; (4) valve casing; (5) valve cone; (6) spring; (7) hole and (8) spout. The arrow indicates the direction of opening of the aerosol can. Reproduced from Wijnen (1997).

4.5 Cream liqueur

The term cream liqueur refers to a group of liqueurs that contain dairy cream as an ingredient. The roots of cream liqueur lie in home-made products that trace back for centuries, such as the Scottish product Atholl Brose, which consists of dairy cream and honey, to which whisky, in which oatmeal was steeped to impart a pleasant nutty flavour, is

(a) (b)

Fig. 4.4 Aerosol-whipped cream (a) immediately after foam formation and (b) 15 minutes after foam formation. Reproduced from Wijnen (1997).

added (Banks & Muir, 1988). Attempts in the second half of the twentieth century to produce Atholl Brose and other home-made cream liqueurs on a commercial scale largely failed, until 1974, when Baileys Irish Cream Liqueur became the first commercialised cream liqueur. The most famous cream liqueurs are the Irish product, Baileys Irish Cream and the South African product, Amarula.

4.5.1 *Composition of cream liqueur*

In terms of commercial practice, a standard cream liqueur does not exist; each manufacturer decides on the preferred components, and concentrations thereof, which enable a product of desired organoleptic properties, combined with a sufficiently long shelf-life, to be produced. The composition of a standard cream liqueur, which appears to represent most commercially available products, is shown in Table 4.2. The 14 g 100 g^{-1} concentration of ethanol in Table 4.2 equates to ~17 g per 100 mL ethanol. Of the ingredients in Table 4.2, sodium caseinate is used as an emulsifier, whereas citrate is added to prevent serum separation during storage, as outlined in Section 4.5.3.

Table 4.2 The composition of a standard cream liqueur.

Component	Concentration (g kg^{-1})
Milk fat	160
Sucrose	195
Sodium caseinate	30
Milk solids-not-fat (SNF)	14
Total solids (TS)	399
Ethanol	140
Water	461

Data collated from Banks & Muir (1988).

4.5.2 *Processing of cream liqueur*

Figure 4.5 shows general single- and two-stage processes for the manufacture of cream liqueur. Although there is no universal manufacturing process for cream liqueur, the principal steps of mixing of the ingredients and homogenising the cream base are universal. The sodium caseinate can be dissolved in hot water (~85°C), after which the other ingredients can be added to produce the cream base. As illustrated in Figure 4.5, ethanol may be added to the cream base either before or after homogenisation. The function of homogenisation is to reduce lipid globule size, thereby prolonging the physical stability of the product by preventing creaming and/or the formation of the cream plug on storage, as outlined in Section 4.5.3. A general rule-of-thumb for homogenisation efficiency is that >98% of the milk lipid globules should have a diameter <0.8 μm (Banks & Muir, 1988).

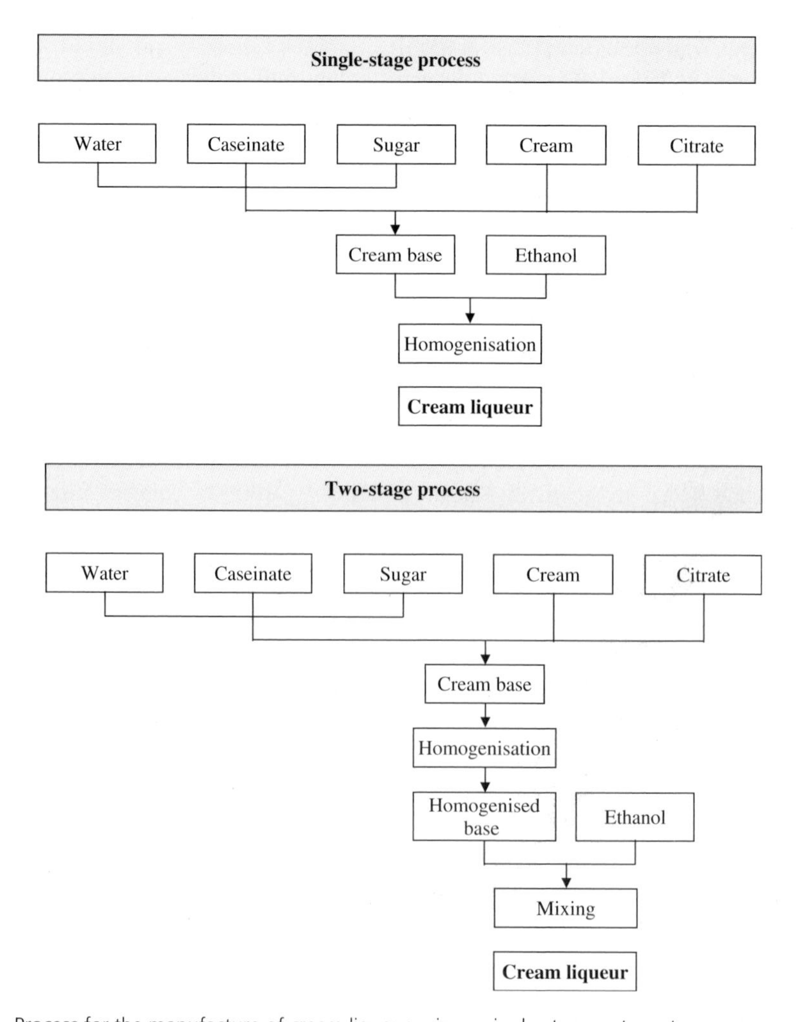

Fig. 4.5 Process for the manufacture of cream liqueur using a single-stage or two-stage process.

4.5.3 *Shelf-life of cream liqueur*

Like other cream products, the shelf-life of cream liqueurs consists of two aspects, the microbiological shelf-life and the physico-chemical shelf-life. The microbial shelf-life of cream liqueurs causes little concern, because human pathogens are unable to grow in this medium of high ethanol and sugar content. The physico-chemical shelf-life of cream liqueurs is of far greater concern and three different types of physico-chemical defects have been noted in cream liqueurs: (a) cream plug formation, (b) serum separation and (c) precipitate formation.

Cream plug formation

Cream plug formation of fat in the neck of the bottle during storage, which cannot be re-dispersed into the product on shaking or agitation, is the result of insufficient homogenisation. As a result, lipid globules rise to the top, which may ultimately lead to a cream plug. Since the plug cannot be re-dispersed on shaking or agitation, it is likely that the creamed lipid globules have aggregated or coalesced; the belief that coalescence occurs is supported by the fact that almost all of the lipids in the cream plug are freely extractable (Dickinson *et al.*, 1989a). The extent of cream plug formation correlates positively with the proportion of large lipid globules in the product and can be prevented by increasing the homogenisation pressure (Banks *et al.*, 1982) or increasing the number of passes through the homogeniser (Muir & Banks, 1986). Homogenisation efficiency may also be increased, and hence creaming reduced, by the inclusion of low levels ($\sim$0.5 g 100 g^{-1}) of low-molecular-weight surfactants in the cream liqueur (Dickinson *et al.*, 1989a). Cream plug formation is likely to be enhanced at a low pH ($<$6.0), at a high concentration of ionic calcium, at a low concentration of emulsifier and by temperature fluctuations during storage (Dickinson *et al.*, 1989b).

Serum separation

Gelation of cream liqueur results in syneresis and hence separation of a serum layer from the product. The shelf-life of cream liqueur, determined as the number of days the product can be stored at 45°C before noticeable serum separation, increases in a sigmodial manner with increasing pH (Banks *et al.*, 1981). Using washed cream (i.e. cream devoid of most of the skimmed milk solids) or anhydrous milk fat, as the source of milk fat, improves the shelf-life of cream liqueurs considerably, as does the addition of the calcium-chelating agent, trisodium citrate (Banks *et al.*, 1981). Such observations led to the conclusion that serum separation is governed predominantly by the concentration of ionic calcium, which destabilises the casein-covered lipid globules. The addition of citrate reduces the concentration of ionic calcium and hence increases the shelf-life of cream liqueurs by up to 2 orders of magnitude (Banks *et al.*, 1981). Replacing caseinate with whey protein concentrate also increased product stability (Kaustinen & Bradley, 1987), presumably because whey proteins are less sensitive to calcium-induced aggregation than caseins. Furthermore, the use of sorbitol, rather than sucrose, as the carbohydrate component significantly increases the shelf-life of cream liqueurs (Banks *et al.*, 1982).

Precipitation

While the addition of citrate to cream liqueur enhances its stability against serum separation, it also causes a defect in the form of a slightly granular precipitate, consisting primarily of calcium and citrate, at the bottom of the bottle. The precipitate is particularly noticeable on storage at above ambient temperature, because the solubility of calcium citrate decreases with increasing temperature. As a result, the amount of citrate added to cream liqueur becomes a trade-off between preventing serum separation and minimising the risk of precipitate formation.

From the preceding text, it is clear that the successful production of a shelf-stable cream liqueur is an intricate process, which involves a constant trade-off between desired organoleptic properties and physico-chemical stability. For instance, a stable cream liqueur with increased alcohol content can be prepared when reducing the total solids content of the product, which will invariably affect the organoleptic perception of the product (Banks & Muir, 1985). Furthermore, increasing the lipid content of cream liqueur generally improves the body and mouth-feel of the product, but also reduces its physico-chemical shelf-life. Likewise, from model studies, it has been predicted that a lower caseinate concentration reduces serum separation, but the necessity of sufficiently small lipid globules constrains the reductions in caseinate achievable without inducing creaming and cream plug formation (Banks & Muir, 1988).

4.6 Cultured, fermented or sour cream

4.6.1 *Background*

Cultured, fermented or sour cream products are prepared by fermenting fresh cream and have many uses, including, as condiments, as dip bases for snacks and vegetables and as ingredients in sauces and dressings. The fat content of sour cream can vary from 10 to 40 g 100 g^{-1} (Hoffmann, 2002a; see also Tamime & Marshall, 1997).

4.6.2 *Production of cultured, fermented or sour cream*

The process for the production of cultured cream is outlined in Figure 4.6. Following separation and standardisation of the cream, dry matter may be enriched, and stabilisers, such as caseinates or hydrocolloids, may be added to improve product texture and prevent syneresis. The cream is then heated for 15 seconds to 30 minutes at 85 to 95°C or for a few seconds at 120 to 130°C and homogenised; homogenisation following heating allows for better texture than a homogenisation step later in the process. For cream of a high fat content, homogenisation pressures should be kept low, to avoid the formation of homogenisation clusters. Homogenised lipid globules partake directly in the acid-coagulation process and in the network structure of the cultured product (Hoffmann, 2002a; see also Lyck *et al*., 2006). After homogenisation, cream is cooled to the inoculation temperature of 20 to 24°C, and fermentation is initiated by the addition of mesophilic lactic acid bacteria. Mesophilic cultures commonly used in the production of cultured creams include *Lactococcus lactis* subsp. *lactis, Lactococcus lactis* subsp. *cremoris, Lactococcus lactis* subsp. *lactis* var. *diacetylactis* and *Leuconostoc mesenteroides* subsp. *cremoris*, with *Lactobacillus acidophilus* also being increasingly used, owing to its probiotic status (see also Tamime *et al*., 2005).

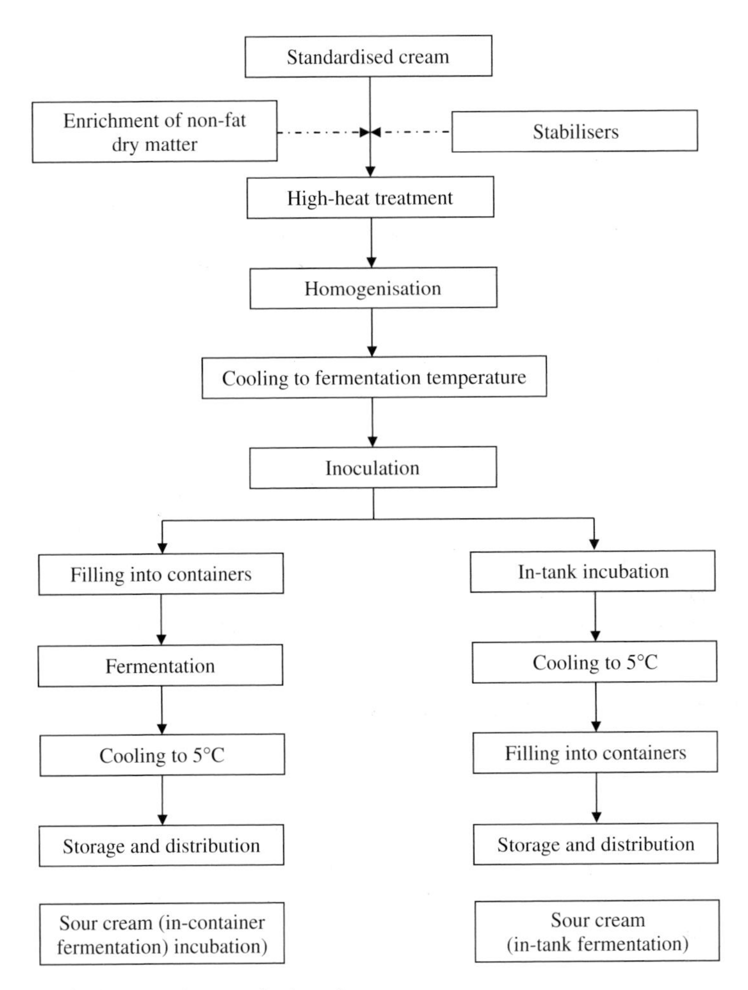

Fig. 4.6 Process for the manufacture of cultured cream.

Fermentation, which takes 14 to 24 hours at 20 to 24°C, may occur in filled cream containers to produce set-style cultured cream, or in a fermentation tank from which cream containers are subsequently filled. Disadvantages exist with both types of fermentation. Following in-tank fermentation, the cream has to come in contact with the production equipment again as it has to be further processed, thereby increasing the possibility of re-infection and reduction in the viscosity of the cream and, consequently, increasing the need for addition of hydrocolloids. Although set-style cultured cream is thicker, it tends to become inhomogeneous during the long fermentation at ambient temperatures (Hoffmann, 2002b). When the desired pH is reached, further fermentation is inhibited by rapid cooling of cream to 5°C. A freshly produced cultured cream should have a pH of ~4.5, resulting in a slightly acid, mild 'cheesy' or 'buttery' flavour, and the texture should be uniform, viscous and creamy. The flavour and texture of cultured cream are greatly influenced by the fermentation culture and the product storage time (Folkenberg & Skriver, 2001), and a lack of flavour and a grainy texture are the most common defects associated with cultured cream

(Mann, 1980). Cultured creams attaining a near-plastic consistency (by using appropriate processing conditions), may be used as low-fat spreads (Hoffmann, 2002a).

Cream may also be chemically acidified, using glucono-δ-lactone (GDL) or food-grade acids. This method of direct acidification has several advantages over conventional culturing practices, including elimination of culture-handling problems, improved production efficiency and quality control. Chemically acidified sour creams have a similar appearance and texture as those of cultured sour creams (Kwan *et al.*, 1982), but the latter have a superior flavour (Freeman & Bucy, 1969; Kwan *et al.*, 1982).

4.7 Coffee cream

Coffee cream is a popular product that is mainly used for whitening of coffee, as well as for imparting a pleasant flavour to coffee (Towler, 1986), in the preparation of food and drinks, and for direct consumption (Spreer, 1998). Coffee cream has a minimum shelf-life of 4 months at room temperature and normally contains 10 to 12 g fat 100 g^{-1}, and less often, 15 to 20 g fat 100 g^{-1}.

4.7.1 *Processing of coffee cream*

An outline of the production process for coffee cream is given in Figure 4.7. Following separation and standardisation, the cream is heat treated at 90 to 95°C and cooled to 6°C (Hoffmann, 2002a). Subsequently, stabilising salts, e.g. phosphates or citrates, may be added to increase the pH and to reduce the concentration of ionic calcium, thereby reducing the susceptibility of casein micelles to aggregation during the following sterilisation process or on addition to hot coffee (Hoffmann, 2002a). Cream is then subjected to the first two-stage homogenisation step, followed by sterilisation. Traditionally, coffee cream was sterilised in cans or bottles but, in the last 20 years, continuous-flow sterilisation in a UHT plant, followed by aseptic packaging, has largely replaced the former process (Hoffmann & Buchheim, 2006). A second homogenisation step follows after the flow sterilisation. Both homogenisation steps operate at a total pressure of 20 MPa and a temperature of ~70°C. Flow-sterilised coffee cream is, subsequently, cooled to 25°C and aseptically filled, usually into plastic cups (7.5–15 g capacity) or preformed cups or can (100–200 g capacity; Hoffmann, 2002a).

4.7.2 *Properties of coffee cream*

In addition to good sensory properties, the main criteria for assessing the quality of coffee creams are as follows (Buchheim *et al.*, 1986; Hoffmann *et al.*, 1996; Hoffmann & Buchheim, 2006):

- Emulsion stability, that is, the ability of the particulate constituents (primarily lipid and protein particles) to remain evenly dispersed throughout the coffee cream during its long shelf-life.

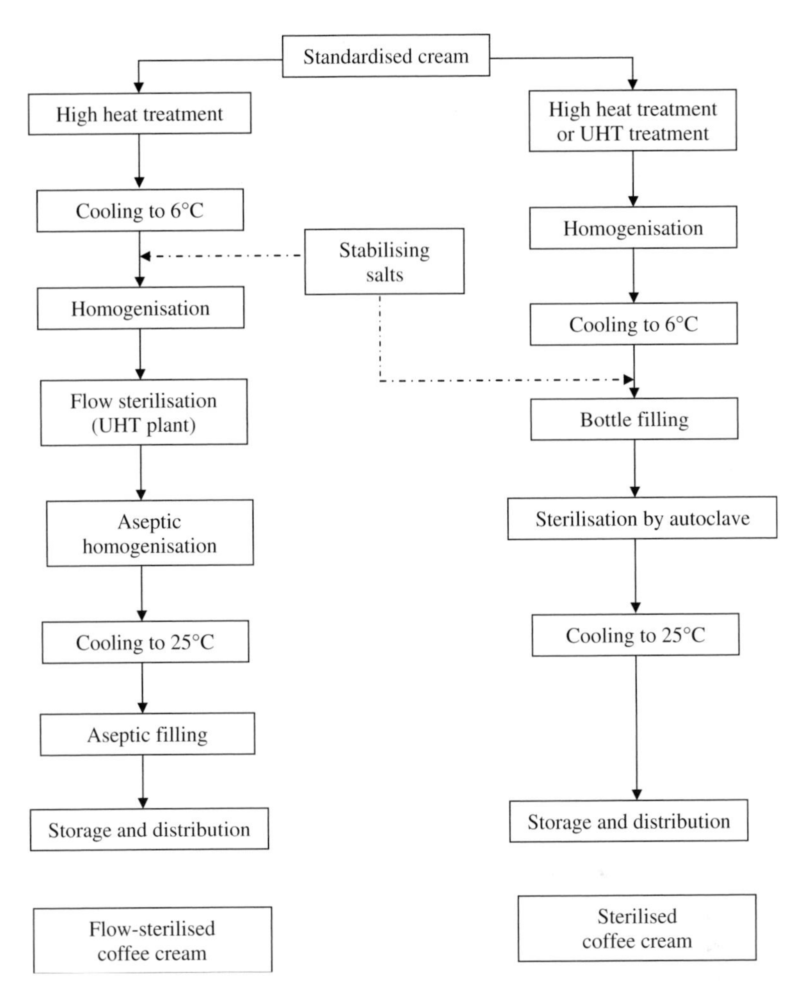

Fig. 4.7 Process for the manufacture of sterilised and flow-sterilised coffee cream.

- The whitening effect and the ability to form a milky, homogenous mixture on addition to coffee, irrespective of the composition, brewing conditions and temperature of the coffee.

One particularly important defect with respect to coffee creams is the phenomenon known as 'feathering', i.e. the formation of macroscopic aggregates containing proteins and lipids through acid-induced flocculation of partially protein-covered lipid globules in hot coffee drinks (Lewis, 1989). Although aggregates always form in hot coffee, it is those that are visible (i.e. $\geq$100 μm) that render the coffee cream defective (Abrahamsson *et al.*, 1988), because the consumer commonly perceives such aggregates as an indication of sourness of the product. The susceptibility of coffee cream to feathering is often referred to as its coffee stability. Feathering is affected by several factors, for example, compositional factors (fat content, fat:protein ratio and mineral balance), processing conditions (homogenisation and heating of the cream) as well as the properties of the coffee (the composition of the coffee and the tap water and the brewing conditions used) (Koops, 1967a,b; Glazier, 1967; Hoffmann, 2002b), and also the acidity of the cream or the coffee (Buchheim *et al.*, 1986).

Although no correlation was established between the average lipid globule size and the degree of feathering (Abrahamsson *et al.*, 1988), it is widely believed that homogenisation has an important influence on feathering of coffee cream (Burgwald, 1923; Abrahamsson *et al.*, 1988; Geyer & Kessler, 1989a; Varnam & Sutherland 1994; Spreer, 1998; Hoffmann, 2002b). Hoffmann & Buchheim (2006) suggest that using low fat contents and optimising heating and homogenisation conditions allow greatest control of the physical properties of flow-sterilised cream. Unhomogenised cream does not feather visibly in hot coffee, as only the relatively small protein particles are involved in feathering in this instance (Geyer & Kessler, 1989b). Single-stage homogenisation is less effective in preventing feathering than two-stage homogenisation (Burgwald, 1923; Abrahamsson *et al.*, 1988), but application of excessive homogenisation pressure also increases the susceptibility to feathering (Burgwald, 1923). Homogenisation of cream results in the alteration of the composition of the MLGM, with caseins and denatured whey protein becoming a predominant part of the membrane material in homogenised cream. The formation of large thermally induced protein aggregates and especially lipid–protein complexes must be avoided to obtain good quality, which is why coffee cream is frequently subjected to the second two-stage homogenisation step after heating, to disrupt such heat-induced lipid/protein aggregates (Hoffmann & Buchheim, 2006). Similarly, Buchheim *et al.* (1986) observed that the composition and structure of the interfacial layer of lipid globules in homogenised cream are mainly responsible for feathering in hot coffee solutions.

Flow sterilisation is another important factor that affects the susceptibility of coffee cream to feathering (Hoffmann, 2002b; Hoffmann & Bucheim, 2006). The temperature at which the sterilisation is carried out has a greater effect on the number and size of the aggregates than the duration of the sterilisation process; the number of aggregates is reduced by treatment at $\leq 130^{\circ}C$ rather than at temperatures of $\geq 135^{\circ}C$ (Hoffmann & Buchheim, 2006). Preheating coffee cream, in a manner similar to pre-treatment of evaporated milk to improve its stability, did not alter the stability of coffee cream to feathering (Abrahamsson *et al.*, 1988).

The method of mixing the coffee and the cream affects the stability of coffee cream. Cream was more susceptible to feathering when coffee was added to cream and sugar, than when cream was added to coffee (Burgwald, 1923). The use of different grades of coffee or different methods of coffee making, i.e. boiling, percolating or by French drip, had no effect on the coffee stability of coffee cream (Burgwald, 1923). Ageing of cream for 7 to 10 days at 1 to $2^{\circ}C$, freezing of cream or increasing the fat content from 18–20 to 30–35 (g 100 g^{-1}), has little effect on the coffee stability of coffee cream (Burgwald, 1923).

Since coffee cream has a long shelf-life and is stored at ambient temperatures, creaming and sedimentation, which result in larger, non-dispersible aggregates in stored cream, should be avoided or minimised (Hoffmann & Buchheim, 2006). Sediment formation, that is, the formation of a rather compact sediment layer, occurs slowly and is only visible in coffee cream after 1 to 2 months of storage. Pieces of the sediment layer, which originally may be up to 1-mm thick and comprise mainly of large protein aggregates, may remain when coffee cream is added to hot coffee (Buchheim *et al.*, 1986). Fortunately, sedimentation rarely occurs in UHT-treated coffee creams.

Like UHT milk, homogenised UHT creams exhibit gelation phenomenon, generally attributed to the activity of heat-resistant proteolytic enzymes, after prolonged storage (Buchheim *et al*., 1986).

4.8 Other cream products

4.8.1 *Frozen cream*

The development of frozen cream has allowed consumers to store cream for a considerable period before use and have it available when needed. If properly processed, frozen cream provides the consumer with a cream similar in quality to fresh cream, since freezing of cream inhibits bacterial growth and spoilage (Pearce, 1989) and can be used to obtain a good-quality whipped cream (Cooper, 1978). Frozen cream may also be used in soups, where imparting flavour is its main function, or in recombined milk or ice cream (Towler, 1986).

The preparation of frozen cream, as outlined in Figure 4.8, involves separation, standardisation and, typically, heat treatment at ~82°C. This relatively severe heat treatment step is essential to inactivate indigenous milk enzymes such as lipase, catalase and peroxidase. Since enzymatic oxidation of milk lipids continues even at freezing temperatures, such activity results in cream that has a rancid odour on thawing (Pearce, 1989). Following pasteurisation, cream is cooled rapidly to 4°C and is frozen in moulds. Rapid freezing is essential to ensure

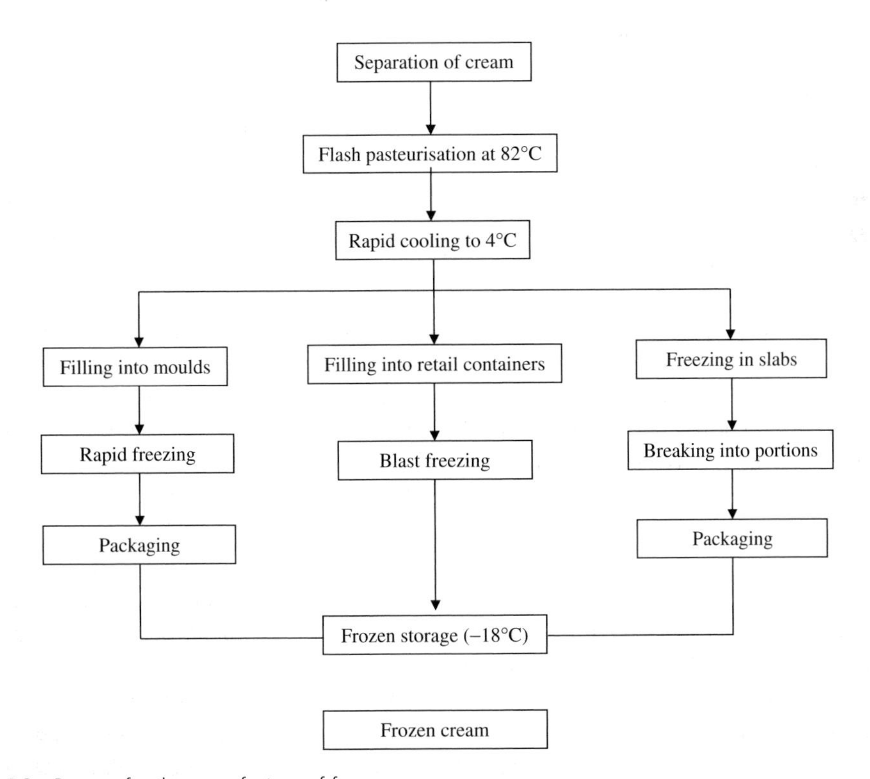

Fig. 4.8 Process for the manufacture of frozen cream.

formation of small ice crystals, which greatly assists in maintaining good physical properties of thawed cream (Pearce, 1989). When rapid freezing of cream is not possible, the addition of emulsifiers and stabilisers aids in its freeze–thaw stability (Towler, 1986). Once frozen at a temperature of approximately -15°C, cream is packaged and frozen further to -18°C before distribution (Pearce, 1989).

4.8.2 *Dried cream*

Cream can also be dried to a powdered form for shipment to distant markets. Dried cream has a low moisture (<2 g 100 g^{-1}) high-fat (40–70 g 100 g^{-1}) product, prepared by the removal of water from cream (Varnam & Sutherland, 1994). Dried cream may be used as an ingredient in dried soups, ice creams, desserts or packet cake mixes, although it has limited functionality when reconstituted, as lipid globules do not reform to natural globules unless specific emulsification and homogenisation procedures are used (Towler, 1986). Dried cream has a long shelf-life and has been produced commercially, in small quantities, for some years (Towler, 1986).

Spray drying is commonly used for drying cream, although some difficulties may be encountered, especially with the high proportion of lipids in the total solids in the end product, since the lipids are liquid at the temperature at which it exits the spray-drying chamber. The lipid globules must, therefore, be encapsulated and protected by solids-non-fat components, which necessitates the addition of protein and a suitable carbohydrate carrier (Towler, 1986). Furthermore, the lipids must be solidified by cooling, to prevent caking, which would occur if the protective MLGM is disrupted (Towler, 1986). The high-fat content of cream makes it susceptible to lipid oxidation and the addition of an antioxidant may be necessary. It is essential that lipases are destroyed during heating of cream to prevent lipolysis of cream. The addition of an anti-caking agent to help prevent caking and the storage of cream powder at low ambient temperatures (to minimise liquid fat content and consequently, minimise caking and deterioration of flavour) are recommended (Towler, 1986).

4.9 Conclusion

From the preceding text, it is apparent that cream is an extremely versatile product with applications that reach far outside the dairy sector. Cream products have been traditionally regarded as luxury products and their popularity may be affected in the current climate by consumer concerns about high-fat intake. Such developments may resurrect interest in scientific research relating to cream products, which has declined considerably in the last two decades. With recent advances in understanding of milk and cream constituents and interactions thereof with other ingredients, the development of reduced fat cream products with properties equivalent to their traditional high-fat counterparts should be possible.

References

Abrahamsson, K., Frennborn, P., Dejmek, P. & Buchheim, W. (1988) Effects of homogenization and heating conditions on physico-chemical properties of coffee cream. *Milchwissenschaft*, **43**, 762–765.

Anderson, M. & Brooker, B. E. (1988) Dairy foams. *Advances in Food Emulsions and Foams* (eds. E. Dickinson & G. Stainsby), pp. 221–255, Elsevier Applied Science Publishers, London.

Anderson, M., Brooker, B. E. & Needs, E. C. (1987) The role of proteins in the stabilization/destabilization of dairy foams. *Food Emulsions and Foams* (ed. E. Dickinson), pp. 100–109, The Royal Society of Chemistry, Cambridge.

Anonymous (2003) Centrifugal separators and milk standardization. *Dairy Processing Handbook*, 2nd edn., pp. 99–122, Tetra Pak Processing Systems AB, Lund.

Banks, W. & Muir, D. D. (1985) Effect of ethanol content on emulsion stability of cream liqueurs. *Food Chemistry*, **18**, 139–152.

Banks W. & Muir, D. D. (1988) Stability of alcohol-containing emulsions. *Advances in Food Emulsions and Foams* (eds. E. Dickinson & G. Stainsby), Elsevier Applied Science, London.

Banks, W., Muir, D. D. & Wilson, A. G. (1981) Extension of the shelf-life of cream-based liqueurs at high ambient temperatures. *Journal of Food Technology*, **16**, 587–595.

Banks, W., Muir, D. D. & Wilson, A. G. (1982) Formulation of cream-based liqueurs: a comparison of sucrose and sorbitol as the carbohydrate component. *Journal of the Society of Dairy Technology*, **35**, 41–43.

Bassette, R. & Acosta, J. S. (1988) Composition of milk products. *Fundamentals of Dairy Chemistry* (eds. N. P. Wong, R. Jenness, M. Keeney & E. H. Marth), 3rd edn., pp. 39–79, Van Nostrand Reihold, New York.

Brooker, B. E. (1993) The stabilization of air in foods containing fat–review. *Food Structure*, **12**, 115–122.

Brooker, B. E., Anderson, M. & Andrews, A. T. (1986) The development of structure in whipped cream. *Food Microstructure*, **5**, 277–285.

Buchheim, W. (1978) Mikrostruktur von geschlagenem Rahm. *Gordian*, **78**, 184–188.

Buchheim, W., Falk, G. & Hinz, A. (1986) Ultrastructural aspects of physico-chemical properties of ultra-high-temperature (UHT)-treated coffee cream. *Food Microstructure*, **5**, 181–192.

Bull, R. M. (1989) Packaging requirements for fresh cream. *Cream Processing Manual* (ed. J. Rothwell), 2nd edn., pp. 34–41, The Society of Dairy Technology, Cambridge.

Burgwald, L. H. (1923) Some factors which influence the feathering of cream in coffee. *Journal of Agricultural Research*, **26**, 541–546.

Camacho, M. M., Martinez-Navarette, N. & Chiralt, A. (1998) Influence of locust bean gum/λ-carrageenan mixtures on whipping and mechanical properties and stability of dairy creams. *Food Research International*, **31**, 653–658.

Cooper, H. R. (1978) Preparation of whipping cream from frozen cream. *New Zealand Journal of Dairy Science and Technology*, **13**, 202–208.

De Moor, H. & Rapaille, A. (1982) Evaluation of starches and gums in pasteurised whipping cream. *Progress in Food and Nutritional Sciences*, **6**, 199–207.

Dickinson, E., Narhan, S. K. & Stainsby, G. (1989a) Stability of cream liqueurs containing low-molecular-weight surfactants. *Journal of Food Science*, **54**, 77–81.

Dickinson, E., Narhan, S. K. & Stainsby, G. (1989b) Factors affecting the properties of cohesive creams formed from cream liqueurs. *Journal of the Science of Food and Agriculture*, **48**, 225–234.

Drelon, N. Gravier, E., Daheron, L., Boisserie, L., Omari, A. & Leal-Calderon, F. (2006) Influence of tempering in the mechanical properties of whipped dairy creams. *International Dairy Journal*, **16**, 1454–1463.

Early, R. (1998) Liquid milk and cream. *The Technology of Dairy Products* (ed. R. Early), 2nd edn., pp. 1–49, Blackie Academic and Professional, Melbourne.

Faulks, B. (1989) Separation and standardization. *Cream Processing Manual* (ed. J. Rothwell), 2nd edn., pp. 12–25, The Society of Dairy Technology, Cambridge.

Folkenberg, D. M. & Skriver, A. (2001) Sensory properties of sour cream as affected by fermentation culture and storage time. *Milchwissenschaft*, **56**, 261–264.

Freeman, T. R. & Bucy, J. L. (1969) Flavour characteristics of nonculture sour cream. *Journal of Dairy Science*, **52**, 341–344.

Geyer S. & Kessler, H. G. (1989a) Effect of manufacturing methods on the stability to feathering of homogenized UHT coffee cream. *Milchwissenschaft*, **44**, 423–427.

Geyer, S. & Kessler, H. G. (1989b) Influence of individual milk constituents on coffee cream feathering in hot coffee. *Milchwissenschaft*, **44**, 284–288.

Glazier, L. R. (1967) Some observations on cream feathering. *American Dairy Review*, **29**, 70–76.

Hinrichs, J. & Kessler, H. G. (1996) Processing of UHT cream. *UHT Cream*, Document No. 315, pp. 17–22, International Dairy Federation, Brussels.

Hoffmann, W. (2002a) Cream. *Encyclopaedia of Dairy Sciences* (eds. H. Roginski, P. F. Fox & J. W. Fuquay), pp. 545–551, Academic Press, London.

Hoffmann, W. (2002b) Cream products. *Encyclopaedia of Dairy Sciences* (eds. H. Roginski, P. F. Fox & J. W. Fuquay), pp. 551–557, Academic Press, London.

Hoffmann W. & Buchheim, W. (2006) Significance of milk fat in cream products. *Advanced Dairy Chemistry* (eds. P. F. Fox & P. L. H. McSweeney) Vol. 2, Lipids, 3rd edn., pp. 365–375, Springer, New York.

Hoffmann, W., Moltzen, B. & Buchheim, W. (1996) Photometric measurement of coffee cream stability in hot coffee solutions. *Milchwissenschaft*, **51**, 191–194.

Huppertz, T. & Kelly, A. L. (2006) Physical chemistry of milk fat globules. *Advanced Dairy Chemistry* (eds. P. F. Fox & P. L. H. McSweeney) Vol. 2, Lipids, 3rd edn., pp. 173–212, Springer, New York.

IDF (1982) *Technical Guide for the Packaging of Milk and Milk Products*, Document No. 143, International Dairy Federation, Brussels.

IDF (1996) *UHT Cream*, Document No. 315, pp. 4–34, International Dairy Federation, Brussels.

Juffs, H. S., Smith, S. R. J. & Moss, D. C. (1980) Keeping quality of whipping cream stored in dispensers pressurized with nitrous oxide. *Australian Journal of Dairy Technology*, **35**, 132–136.

Kaustinen, E. M. & Bradley, R. L. (1987) Acceptance of cream liqueurs made with whey protein concentrate. *Journal of Dairy Science*, **70**, 2493–2498.

Kieseker, F. G. & Zadow, J. G. (1973) The whipping properties of homogenized and sterilized cream. *Australian Journal of Dairy Technology*, **28**, 108–113.

Koops, J. (1967a) Preparation and properties of sterilized coffee cream. 1. In-bottle sterilization. *Netherlands Milk and Dairy Journal*, **21**, 29–49.

Koops, J. (1967b) Preparation and properties of sterilized coffee cream. 2. Continuous flow-sterilization and aseptic packaging. *Netherlands Milk and Dairy Journal*, **21**, 22–49, 50–63.

Kosinski, E. (1996) Raw material quality. *UHT Cream*, Document No. 315, pp. 12–16, International Dairy Federation, Brussels.

Kwan, A. J., Kilara, A., Friend, B. A. & Shahani, K. M. (1982) Comparative B-vitamin content and organoleptic qualities of cultured and acidified sour cream. *Journal of Dairy Science*, **65**, 697–701.

Lewis, M. J. (1989) UHT cream processing. *Cream Processing Manual* (ed. J. Rothwell), 2nd edn., pp. 52–65, The Society of Dairy Technology, Cambridge.

Lyck, S., Nilsson, L.-E. & Tamime, A. Y. (2006) Miscellaneous fermented milk products. *Probiotic Dairy Products* (ed. A. Y. Tamime), pp. 217–236, Blackwell Publishing, Oxford.

Mann, E. J. (1980) Cultured cream. *Dairy Industries International*, **45**, 29, 39, 59.

Melsen, J. P. & Walstra, P. (1989) Stability of recombined milk-fat globules. *Netherlands Milk and Dairy Journal*, **43**, 63–78.

Muir, D. D. & Banks, W. (1986) Multiple homogenization of cream liqueurs. *Journal of Food Technology*, **21**, 229–232.

Muir, D. D. & Kjaerbye, H. (1996) Quality aspects of UHT cream. *UHT Cream*, Document No. 315, pp. 25–34, International Dairy Federation, Brussels.

Needs, E. C. & Huitson, A. (1991) The contribution of milk serum proteins to the development of whipped cream structure. *Food Structure*, **10**, 353–360.

Needs, E.C., Anderson, M. & Kirby, S. (1988) Influence of somatic cell count on the whipping properties of cream. *Journal of Dairy Research*, **55**, 89–95.

Noda, M. & Shiinoki, Y. (1986) Microstructure and rheological behaviour of whipping cream. *Journal of Texture Studies*, **17**, 189–204.

Pearce, G. (1989) Frozen cream. *Cream Processing Manual* (ed. J. Rothwell), 2nd edn., pp. 99–103, The Society of Dairy Technology, Cambridge.

Rothwell, J. (1989) Pasteurisation and homogenization. *Cream Processing Manual* (ed. J. Rothwell), 2nd edn., pp. 26–33, The Society of Dairy Technology, Cambridge.

Rothwell, J., Jackson, A.C. & Faulks, B. (1989) Modification and control of cream viscosity. *Cream Processing Manual* (ed. J. Rothwell), 2nd edn., pp. 83–87, The Society of Dairy Technology, Cambridge.

Smith, A.K., Goff, H.D. & Kakuda, Y. (2000) Microstructure and rheological properties of whipped cream as affected by heat treatment and addition of stabilizer. *International Dairy Journal*, **10**, 295–301.

Spreer, E. (1998) Market milk, milk drinks and cream products. *Milk and Dairy Product Technology*, pp. 155–201, Marcel Dekker, New York.

Streuper, A. & van Hooydonk, A.C.M. (1986) Heat treatment of whipping cream. II. Effect on cream formation. *Milchwissenschaft*, **41**, 547–552.

Tamime, A.Y. & Marshall, V.M.E. (1997) Microbiology and technology of fermented milks. *Microbiology and Biochemistry of Cheese and Fermented Milk* (ed. B.A. Law), 2nd edn., pp. 57–152, Blackie Academic & Professional, London.

Tamime, A.Y., Saarela, M., Korslund Søndergaard, A.A., Mistry, V.V. & Shah, N.P. (2005) Production and maintenance of viability of probiotic micro-organisms in dairy products. *Probiotic Dairy Products* (ed. A.Y. Tamime), pp. 39–72, Blackwell Publishing, Oxford.

Towler, C. (1986) Developments in cream separation and processing. *Modern Dairy Technology – Advances in Milk Processing* (ed. R.K. Robinson), Vol. 1, pp. 51–92, Elsevier Applied Science, London.

Varnam, A.H. & Sutherland, J.P. (1994) Cream and cream-based products. *Milk and Milk Products – Technology, Chemistry and Microbiology*, pp. 182–223, Chapman & Hall, Melbourne.

Walstra, P., Wouters, J.T.M. & Guerts, T.J. (2006) Cream products. *Dairy Science and Technology*, 2nd edn., pp. 447–466, CRC Press, Boca Raton.

Wijnen, M.E. (1997) *Instant Foam Physics: Formation and Stability of Aerosol Whipped Cream*, PhD Thesis, Wageningen Agricultural University, The Netherlands.

Wijnen, M.E. & Prins, A. (1995) Disproportionation in aerosol whipped cream. *Food Macromolecules and Colloids* (eds. E. Dickinson & D. Lorient), pp. 309–311, The Royal Society of Chemistry, Cambridge.

5 Butter

R.A. Wilbey

5.1 Introduction

Butter is a water-in-oil (W/O) emulsion, generally recognised as having a minimum of 80 g milk fat 100 g^{-1} and a maximum of 16 g moisture 100 g^{-1}. It is probably one of the oldest milk products, being produced by the concentration of milk fat following the destabilisation of the oil-in-water (O/W) milk or cream emulsion.

Prior to the invention of continuous mechanical separation, the fat content of raw milk would be concentrated by overnight flotation in basins, the cream being decanted into a hand-operated wooden churn and subjected to shear and mild aeration, either using a plunger or by rotating the vessel. Once the emulsion had been destabilised and the fat formed clumps, the serum was poured off and efforts were made to consolidate the fatty mass and remove surplus moisture. This was not a hygienic process; in most cases, the cream would have been souring by the time it was made into butter and the wooden equipment would be extremely difficult to keep in a clean state. Problems would have been compounded by a lack of refrigeration, and one of the few preservation methods was the addition of salt to the butter grains before working. The presence of significant quantities of lactic acid from the sour cream would have contributed to the subsequent preservation of the butter. Butter has also been stored in containers immersed in peat bogs, taking advantage of the lower temperature and virtually anaerobic conditions, though it is not clear whether this was a routine approach to providing a backup food source or a response to an emergency where food must be hidden.

The late nineteenth century saw the inventions of mechanical separation, mechanical refrigeration and the better realisation of the advantages of heat treatment to improve the shelf-life of dairy products. This led to the establishment of creameries where milk could be brought for separation, and the availability of larger quantities of cream led to the mechanisation of buttermaking. Initially, the churns were of wooden construction, essentially a scale-up of the barrels used for hand production, but then slowly replaced by aluminium and then stainless steel till the technology was overtaken in the second half of the twentieth century by the development of continuous processes. By the beginning of the twenty-first century, batch churning had been replaced in all but the smallest dairies by continuous churning processes. McDowall (1953) provides further information on the development of batch churning and the early development of continuous buttermaking.

5.2 Cream preparation

Cream is separated from milk by centrifugation. Normally, the raw milk will have been preheated to above 40°C to ensure that all of the fat is liquid so that the milk-fat globules are less susceptible to shear damage. The optimum temperature for separation is 63°C, higher temperatures causing denaturation of whey proteins which, though not critical for buttermaking, may adversely affect the properties of the skimmed milk. For batch churning the cream may be separated at 35 or up to 40 g fat 100 g^{-1}, while for continuous buttermakers the fat content is normally 40 to 48 g fat 100 g^{-1}, depending on the particular machine.

While a few very small manufacturers, for instance those operating on a farmhouse scale, may batch pasteurise the cream at 63 to 66°C for a minimum of 30 minutes, it is virtually common practice to use a high-temperature short time (HTST) treatment. The minimum treatment is at 72°C for 15 seconds though most manufacturers use a slightly more severe treatment such as at 74 to 76°C for 15 seconds. Bodyfelt *et al.* (1988) reported a minimum of 74°C for 30 minutes or 85°C for 15 seconds in the United States. However, severe heat treatments should be avoided, firstly to minimise the generation of cooked flavours (more easily detected in sweet cream butter) and secondly to minimise the uptake of copper onto the fat globule membrane from the serum. Copper is a very powerful pro-oxidant at levels of 10 ppb (parts per billion), compared to typical serum levels of 20 ppb (Walstra & Jenness, 1984).

Plate heat exchangers are most commonly used for heating and cooling duties. The heat exchangers should be sized to maintain good heat transfer without creating excessive shear on the fat globules. Centrifugal pumps can cause problems in this respect, and it is recommended that moving cavity, for example, lobe, pumps be used in order to minimise damage to the cream. All seals must be fat resisting, for instance, by the use of nitrile rubber sealing rings in pipework, or fat absorption and the subsequent breakdown of the seals will result in product contamination.

One of the problems associated with buttermaking is that it may be used as the ultimate process for milk that cannot be processed into more valuable products, this recovery role implying that sub-standard milks and surplus creams may need to be treated. Where there have been problems with taints in the milk, whether from pasture weeds consumed by the cattle or as a result of storage problems, further treatment may be needed. In some countries, for instance, Australia and New Zealand, a Vacreator was used for multi-stage vacuum treatment of cream. This equipment has now been replaced by spinning cone evaporators where the volatile compounds are removed from a thin film under vacuum (Hill, 2003). Where less severe flavour problems may occur, Cream Treatment Units (CTUs) may be used. Here the cream is heated to ~90°C, then flash cooled by spraying into a chamber where a pressure of ~20 kPa is maintained. The loss of water on cooling is accompanied by reduction in any other volatile component. The additional shear forces can result in an increase in small fat globules that might be carried over into the final butter and improve the texture, though there is also a risk of increased fat loss in the buttermilk on churning.

5.2.1 *Sweet cream*

Sweet cream for buttermaking is potentially the simplest to prepare. The hot cream should be cooled as quickly as possible, usually within the plate heat exchanger used for pasteurisation and, where possible, saving energy by the use of regeneration. By the end of the hold at pasteurisation temperature, all of the fat in the milk-fat globules will be liquid and, as the cream is cooled, the fat will first be supercooled then start to crystallise, releasing latent heat of crystallisation. This process is not completed during the cooling operation, and the cream will exit the heat exchanger with the fat globules containing a mixture of crystalline and supercooled, liquid milk fat, the proportions varying with the nature of that particular fat and the time–temperature profile of the heat exchanger.

Milk fat contains a very wide range of fatty acids, and hence triglycerides, crystallising as a mixture of predominantly α- and β′-crystals. The continuing crystallisation releases more heat, mainly within 2 hours from cooling, and causes the cream to warm by about 2°C. The extent of the crystallisation will depend on the temperature and on the composition of the fat. Ideally, the cream should be cooled from 4 to 5°C immediately after pasteurisation, so that even with the release of the remaining latent heat the temperature should remain below 7°C. Where this is not possible, additional cooling should be provided, either by cooling pads on the tank wall or by circulation through an external heat exchanger. The cooled cream should be held for at least 4 hours before buttermaking to permit adequate crystallisation – at least 50% of the milk fat should be crystalline. Overnight ageing is the preferred approach when buttermaking is carried out on a single shift.

5.2.2 *Ripened/fermented/cultured cream*

The mechanisation of buttermaking, particularly the introduction of pasteurisation and adequate refrigeration, prevented the development of acidity and associated fermented flavours in the cream. In many markets, these flavours were highly desired and steps were taken to reintroduce an appropriate microflora and carry out a controlled fermentation.

Pasteurisation conditions were usually more severe than for sweet cream, for example, 90 to 95°C for 15 seconds or 105 to 110°C with no hold (Boutonnier & Dunant, 1985). The increased protein denaturation reduces the redox potential, aiding growth of the culture. Cooling is limited to about 20°C, with a typical fermentation time of 12 to 18 hours depending on the activity of the starter. The culture normally consists of a mixture of mesophilic lactic acid bacteria, strains of *Lactococcus lactis* subsp. *lactis* and *Lactococcus lactis* subsp. *cremoris* providing lactic acid while citrate positive strains of *Lactococcus lactis* subsp. *lactis* biovar *diacetylactis* produce flavour compounds, predominantly diacetyl and its precursor, acetoin. The inclusion of *Leuconostoc mesenteroides* subsp. *cremoris* or *Leuconostoc mesenteroides* subsp. *citrovorum* increases diacetyl production whilst avoiding a yoghurt-like flavour by reducing acetaldehyde to ethanol. The fermentation must progress to a pH below 5.3 for the generation of diacetyl. For mild lactic flavours, the fermentation may be halted at this point by cooling, although in northern Europe and Scandinavia the fermentation is often allowed to progress further, for instance, to pH ranging from 4.5 to 4.6, to give a stronger lactic flavour. Cooling, whether by a cooling jacket on the incubation vessel or by pumping through an external heat exchanger, normally commences before the target pH in order to achieve the desired average value. Once cooled, the cream has to be aged as for sweet cream.

5.2.3 *Modifications of cream ageing*

The seasonal variation in fatty acid composition leads to shifts in the texture of the final butter, the effect being more marked with the slower cooling often associated with ripened creams. Samuelsson & Petersson (1937) introduced a tempering regime for ripened cream, commonly referred to as the Alnarp process, to reduce these effects.

The milk fat in cream produced during the winter normally contains relatively higher levels of short chain fatty acids and correspondingly lower levels of oleic acid, giving higher melting properties and hence potentially a harder butter. It was found that for winter cream this effect could be reduced by first cooling to 8°C for 1 to 2 hours immediately following the pasteurisation (to promote fat crystallisation), during which time the starter culture should be added. The cream should then be warmed to 19°C for 2 hours before cooling to 16°C for the duration of the fermentation (14–20 h), at the end of which the cream is cooled to 12°C and held for $\geqslant$ 4 hours before churning.

In summer, when the milk fat is less saturated and there is a tendency to produce a softer than average butter, the seasonal effect may be reduced by a different temperature–time regime: initially cooling the freshly pasteurised cream to 19°C and holding for 2 hours (thus achieving less crystallisation than with the winter cream profile). Starter culture may be added at this stage, but at a higher level to compensate for the relatively short fermentation time. The temperature is then reduced slightly to 16°C and held for 3 hours before cooling to 8°C and holding overnight.

Several other temperature–time combinations have been proposed, some based on the iodine value (IV) of the milk fat. Frede *et al.* (1983) adopted a more fundamental approach based on the melting and solidification curves of the milk fat as obtained by differential scanning calorimetry (DSC). Examples of the approach are as follows:

For winter cream

- Cool to 6°C and hold for 3 hours.
- Warm to 2 to 3°C above the melting point of the lower MP fraction (~16–21°C), add starter and hold for at least 2 hours till target pH is achieved.
- Cool to an intermediate temperature between the solidification points of the two main fractions and hold before buttermaking.

For summer cream

- Cool to a point above the upper solidification point of the milk fat, add culture and ferment to the desired pH.
- Cool to 6°C and hold for 3 hours.
- Warm to a point below the lower melting peak before buttermaking.

Although DSC is not usually available in creameries, simpler nuclear magnetic resonance (NMR) techniques could be used to identify shifts in the melting curves for the milk fat and used as a basis for seasonal shifts in temperature–time profiles. These shifts will be accompanied by large changes (e.g. 1–7%) in bulk starter addition to compensate for the variation in fermentation times. Although developed for lactic butter production, the principles may equally be applied to the treatment of sweet cream, for instance, a simplified heat treatment using a plate heat exchanger as reported by Dixon (1970). More recently, Danmark & Bagger (1989a, b) reported a series of production trials with sweet cream for continuous buttermaking, using the IV as an indicator of the degree of unsaturation of the milk fat. Variations with IV included the heating temperatures for the cream following a 2-h hold at 8°C with the traditional winter method, the winter method with stepwise cooling and with the introduction of extra heating and cooling steps. The summer method of cooling to 19°C, holding for 2 hours then cooling to 8.6°C for 19 hours plus the basic method of ageing for 19 hours at 8.6°C were also used. Creams were heated to 12°C before continuous buttermaking. An optimum consistency range was identified and consistency scores mapped against heating temperature and IV, the shapes of the optimum zones being slightly different, depending on whether the traditional winter method or stepwise cooling was being applied. Both fat loss and initial moisture values on buttermaking increased slightly with IV and were inversely proportional to the fat content of the cream.

5.3 Batch churning

Batch buttermaking is the traditional method of manufacture. The butter churn is part filled with aged cream at 8 to 10°C. If the cream is at lower temperatures, then the churning process will be extended, occasionally higher temperatures may be used. It is essential that the churn be less than half filled.

When the rotation of the churn is commenced, part of the cream will be lifted then tumble on top of the remainder. This tumbling action will introduce a coarse aeration and inflict some impact damage on the fat globules. At the relatively warm temperature of the aged cream in the churn, the fat globules will contain a mixture of fat crystals and liquid fat, ideally in a 50:50 ratio. The presence of the fat crystals reduces the flexibility of the fat globules so that on distortion by the impact forces the fat crystals may penetrate through the milk-fat globule membrane (MFGM), and allow some of the liquid fat to escape. This liquid fat will be attracted to the air–water interface of the air cells and form aggregates with damaged and undamaged fat globules. However, the air cells in the cream are poorly stabilised as β-casein is less soluble in the aqueous phase at this temperature; hence the air cells will tend to coalesce, reducing the surface:volume ratio such that the interfacial film becomes more concentrated. Larger air cells will desorb from the mass of the cream more rapidly, leaving behind an accumulation of hydrophobic free fat and fat globules, microscopic butter grains.

With continuing agitation, the degree of breakdown of the O/W emulsion of the cream continues and the resultant butter grains become visible. The accretion of fatty material into the butter grains is accompanied by a gradual reduction of free fat globules in the aqueous phase, the latter being known as buttermilk. When the butter grains reach the size of small rice grains, the churn is stopped and the buttermilk drained off. The buttermilk will typically contain 0.5 to 1 g milk fat 100 g^{-1}, and is normally put through a milk separator in order to recover most of the fat, though small fat globules ($<1\ \mu m$) and the MFGM will remain in the buttermilk.

The often poor microbiological quality of the cream used for buttermaking in the past has led to chilled potable water being added to the butter grains, the churn resealed and rotated for a few revolutions to wash most of the remaining buttermilk out of the butter grains. This process reduces both the residual microflora and the substrate to support its growth, as well as cooling the grains so that they become firmer. The washing process may be repeated. These rinsings are a significant contributor to the effluent from the buttermaking plant.

Once the rinse water has been drained, the dry salt may be scattered over the butter grains if a salted butter is to be prepared. The door of the churn is then resealed, and the churn restarted on a slow speed to ensure that the tumbling action is achieved. The objective of the tumbling action is to apply sufficient force firstly to push the butter grains together into a continuous fatty mass and secondly to disperse the moisture into fine droplets as a

W/O emulsion. The formation of a large mass of butter, often over 500 kg, results in high percussive loadings during this working operation and requires substantial foundations for the butter churns. Smaller, farm-scale butter churns are often too small to allow such shear forces to be developed and must rely on post-churn working by other means.

5.4 Continuous butter manufacture

In the past 50 years, the Fritz method of continuous buttermaking has become the dominant technology, at least in western Europe. The Fritz method is sought to carry out similar steps to that of the traditional batch buttermaking, but converting relatively small quantities (at any point of time) at a much higher rate, creating the potential for greater production capacity and process control. Outputs of 5 tonnes h^{-1} are common, with more than 10 tonnes h^{-1} possible.

5.4.1 *Cream feed to buttermaker*

The buttermaker must be supplied with a consistent feed if it is to operate optimally with a minimum of corrective action. This requires the cream to be fed with consistent composition (fat, pH), physical characteristics (viscosity, degree of fat crystallisation), temperature and feed rate throughout the working day. In part, these requirements can be met by bulking together the cream for a day's production in a single silo. Ideally, the cream would then be supplied within $\pm 0.5\%$ to the buttermaker by a variable speed pump controlled by a programmable flow-meter. The pipework and fittings should be designed and constructed to ensure that the pump is neither starved nor there be excessive shear exerted on the cream during transfer, pipeline flow rates of 0.2 to 0.4 ms^{-1} being satisfactory for sweet cream with slightly lower rates for cultured creams, depending on viscosity.

The ideal ageing temperature is often lower than the optimum temperature for destabilisation of the cream in the buttermaker, and it is expensive to introduce that energy by mechanical action. This shortfall can be corrected by passing the cream through a preheater, typically a plate heat exchanger using warm water as the heating medium. Small temperature differentials of 1 to 2°C should be employed to minimise the risk of overheating, whether in normal production or during any stoppages. The cream outlet temperature should be controlled within ± 0.25°C of the target temperature. This temperature will vary with the fatty acid profile in the cream, higher temperatures being needed in the winter to compensate for the greater proportion of saturated fat so that the temperature will approximate to that needed for 50% of the fat in the globules to be in the liquid state. There is also a tendency for cream feed temperature to be lowered with increasing fat content.

As a general rule, all handling of cream prior to buttermaking should avoid damage to the milk-fat globules, since damaged fat globules will tend to agglomerate and may block the pipework. However, controlled destabilisation has been used in the past as a pre-treatment immediately before the buttermaker, to increase its production capacity.

5.4.2 *Conversion to butter-grains*

Most continuous buttermakers employ a two-stage approach to the conversion of cream to butter-grains. The first stage employs a high-shear mixing of the cream with air; analogous to a continuous whipping machine, but at a temperature guaranteed to produce an unstable foam. Rotor speeds may be of the order of 1000 revolutions per minute (rpm) with a residence time of 1 to 2 seconds, the actual conditions varying with the make of the machine. The air inclusion creates bubbles that attract milk protein to the gas–liquid interface, though with less β-casein than might be expected with whipping cream as a result of the higher temperature. At the same time, the shear forces and turbulence result in collisions between fat globules that result in damage to the fat globule membrane and leakage of liquid fat into the serum. This liquid fat is hydrophobic, and will associate with the fat of other damaged globules and migrate to the gas–liquid interface of the air bubbles. This second effect leads to displacement of protein from the interface and an increased rate of coalescence of the bubbles. These larger bubbles have a smaller surface-to-volume ratio so the damaged fat globules on their surface are brought closer together. Larger bubbles are less stable in the suspension; so, when the large bubbles desorb from the cream, the damaged fat globules are released as fatty agglomerates and may trap undamaged fat globules in the matrix, as illustrated in Figure 5.1. Approximately half of the MFGM is lost into the serum. Note that while aeration commonly contributes to the destabilisation, it is not essential.

In many designs of continuous buttermaker, the mixture of small granules and serum falls into the second-stage cylinder, typically of much greater diameter and rotating relatively slowly at about 35 rpm. The first part of this cylinder acts as an aggregator, the slow rotation creating a tumbling action so that the small granules are pushed together and serum squeezed out by the collisions (over-rapid rotation in the first-stage cylinder can lead to too large a granule that loses serum less readily at this point, resulting in high-moisture butter). The serum then drains off through perforations in the latter part of the drum and is collected in a tray as buttermilk. The buttermilk contains a dispersion of fat droplets plus the MFGM. Some butter grains may be recovered by sieving, preferably using a gyratory sieve, and part of the fat may be recovered by centrifugation but about 0.5 g fat 100 g^{-1} will remain in the buttermilk as a colloidal dispersion. Some designs of buttermaker use a spinning disc clarifier as the first stage of fat recovery.

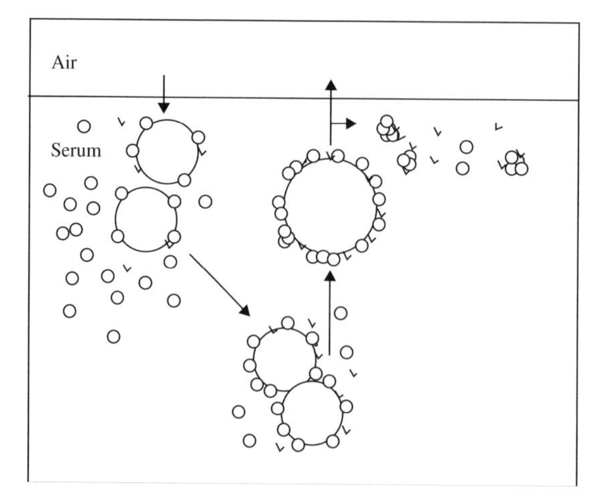

Fig. 5.1 Schematic illustration of role of air cells in destabilisation of the cream emulsion. *Note*: O = fat globules; L = milk-fat globule membrane (MFGM) fragment.

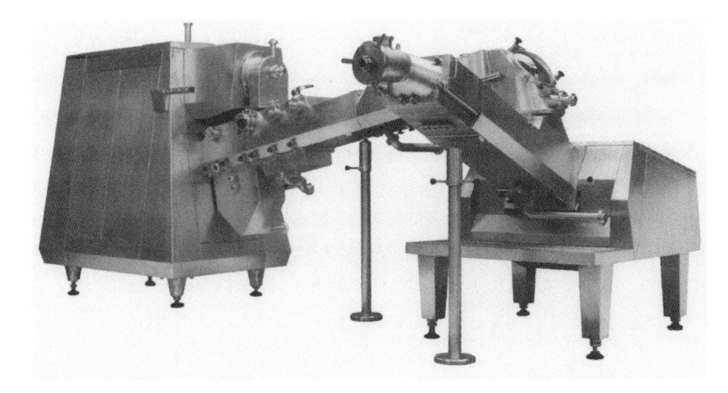

Fig. 5.2 Continuous buttermaker using a single-stage destabilisation of the cream. With permission of Simon SAS (Anonymous, 2008a), Cherbourg.

By this point the butter grains are of similar size to cereal grains, but are still relatively open aggregates of fatty particles with pools of serum and variable amounts of air trapped within them.

Not all buttermakers use the two-stage approach; for instance, the Contimab Simon range of machines employ a single-stage destabilisation unit running at 600 to 1000 rpm, as illustrated in Figure 5.2.

5.4.3 *Working*

Continuous consolidation of the butter grains is achieved by forcing the particles together using a pair of contra-rotating augers. These are normally mounted with a slight upward slope so that, as the particles are pushed together, surplus serum that exudes from the aggregates can flow back to a reservoir at the base of the worker unit and be drained away. The height of serum in the reservoir can be adjusted as a means of standardising the moisture in the butter.

The consolidation process is supplemented by forcing the butter through a series of orifice plates interspersed with mixing vanes driven by an extension of the augers. The shear rate can be varied by changing the diameter of the holes in the orifice plate and by the design of the mixing vanes; perpendicular mixing vanes increase shear while the more common pitched vanes increase the pumping effect. Passage through the orifice plates should complete the consolidation of the butter grains and disperse the moisture droplets within the continuous liquid fat phase. Ideally, the moisture droplets should be less than 10 μm in diameter in order to confer microbiological stability but this might not be achieved in practice. Butter containing 16 g 100 g^{-1} moisture with droplet diameters averaging $<$10 μm would contain 38 million droplets per millilitre so that even butters with a high residual microflora would average less than 0.1% of the droplets containing a viable organism (Wilbey, 2002). In these circumstances the small volume of the few contaminated droplets would rapidly limit both the substrate and space available for microbial growth.

However, in practice the distribution of moisture droplet sizes is not tightly clustered about the mean, and agglomeration of moisture droplets can also occur during subsequent handling. Free, that is, visible, moisture has been found in those butters produced with single working units. Although this might be similar to the situation with traditional batch-churned butters, the keeping quality of the butter is unnecessarily compromised and there was a move to running buttermakers with two working sections (also known as canons). It is an advantage to be able to control the canons independently so as to optimise the texture of the butter.

Also, the butter grains contain a small but variable amount of air that is not completely squeezed out during the consolidation step. Variation in the residual air leads to fluctuations in the density of the worked butter. Since most retail butter is sold by weight but dispensed by volume, these variations in density would lead to greater packing losses. With the installation of two working units, often referred to as canons, it became easier to install a degassing stage, inserting this between the canons, and thus reducing subsequent packing losses.

Butter extruded from the buttermaker is relatively plastic and pumpable. If not delivered directly to a butter silo, a lobe pump may be included in the discharge from the buttermaker. Subsequent pipework must be relatively large in diameter, for example, 100 or 150 mm, and any bends have a radius of 500 mm or more to minimise resistance to flow.

5.4.4 *Salting*

The continuous buttermaking process described earlier would be suitable for unsalted butters. Accurate dosing of crystalline solids into a viscous semi-solid posed severe problems that were circumvented by dosing brine into the butter at a point between the orifice plates in the first working section. A number of problems remained as follows:

- Salt has limited solubility, ~26 g 100 mL^{-1} (i.e. brine) at room temperature, so that adding 2 g salt 100 g^{-1} butter would also add about 6 g 100 g^{-1} to the moisture. This would require making the butter grains with $\leqslant$ 10 g 100 g^{-1} moisture prior to working.
- Incomplete mixing of the injected brine with the original moisture in the butter would lead to variations in the salt content of the moisture droplets with consequent risk of lower microbiological stability. This would be aggravated by the osmotic effect, where moisture would be drawn from the original, unsalted droplets to the salty droplets, creating large droplets and a greater chance of free moisture in the butter.

Moisture addition was reduced by preparing a 50 g 100 mL^{-1} slurry of fine salt crystals, for instance to British Standards (1969), and injecting this into the butter, thus reducing the water addition to approximately 2 g 100 g^{-1}, and requiring a less dramatic reduction in the moisture of the butter grains on discharge from the second cylinder. For Welsh extra-salted butter with 3 to 4 g salt 100 g^{-1}, the slurry would need to contain up to 70 g salt 100 mL^{-1}. Incomplete mixing and consequent osmotic effects were reduced by better and longer mixing during the two-stage working.

5.5 Alternative processes for cultured butters

Although the Fritz process was originally developed to handle ripened creams, there are some limitations to the manufacturing process:

- The use of ripened cream leads to the production of lactic buttermilk as a by-product. This is less heat stable than fresh buttermilk and, generally, has less economic value; it thus has to be often disposed of as animal feed.

- In producing the ripened cream, there is migration of copper from the serum to the MFGM. Since copper is a powerful oxidation catalyst, the shelf-life of cultured butter is potentially less than that of fresh cream butters.

One of the earliest methods of getting around these problems was the NIZO (Netherlands dairy research institute) process, whereby fresh cream was fed to the buttermaker to yield fresh buttermilk and butter grains. These butter grains were then modified in the first worker unit by injection of the typical starter organisms, a lactic acid preparation and a flavour concentrate rich in diacetyl, the latter two being prepared by fermentation. The resulting butter had a pH, microbiology and flavour similar to traditional cultured butters, but with a reduced copper content – for further details see NIZO (1976) and Wilbey (1994). Subsequently other producers have made claims to have simplified the additions to achieve a similar product.

5.6 Alternative technologies for continuous buttermaking

Though the World War II interrupted many non-military developments, the 1940s were notable for the range of innovations in continuous buttermaking. These techniques fall into three broad classes:

- The low-fat route, involving shearing of cold, crystallised cream ($\sim$40 g fat 100 g^{-1}), of which the Fritz process has become the most popular, particularly in western Europe.
- The shearing of a high-fat cream ($\sim$80 g fat 100 g^{-1}) followed by standardisation, cooling, crystallisation and shear.
- The shearing of a cream, followed by further separation and creation of the emulsion that was cooled, crystallised and sheared; this was the least popular approach.

5.6.1 *Low-fat route*

The methods under this route were reviewed by Wiechers & DeGoede (1950) and are summarised as follows:

- Senn method (Switzerland), based on the batch shearing of cream by high-speed stirring under 0.2 to 0.4 MPa of carbon dioxide, the butter being washed then kneaded and discharged by a pair of augers, not dissimilar to those used in the Fritz machine. Three 1 tonne h^{-1} machines were reported to have been built in Zurich, but they needed 0.2 kg of liquid carbon dioxide 40 kg^{-1} batch of cream.
- Rohrwasser (Germany) continuous process included a churning apparatus with the cream separator. Milk was introduced to the centre of the centrifuge, cream being ejected into the surrounding chamber where it was churned by the rotating beaters. Cream may also be introduced directly to the beating chamber. Cooling elements were included and in one variant a 'ripening compartment' was included – at that time it appeared that this process would have as good a chance of industrial application as the Fritz process.
- The Westphalia Separator (Germany) process used movable beating blades in the chamber, claiming increased capacity compared to the Fritz process. It should be

noted that Westphalia Separator AG (currently within the GEA group) was one of the manufacturers of the Fritz machines.

- The Wolf patent (Germany) claimed to overcome the cleaning problems with movable blades by using a closed drum with fixed blades giving an annular churning space.

5.6.2 *Shearing high-fat cream*

One of the drawbacks of buttermaking from 40 g fat 100 g^{-1} cream is the need to carry out a twofold increase in the level of milk fat in order to achieve the minimum 80 g milk fat 100 g^{-1} required in butter. As with batch churning, the continuous churning methods outlined earlier were not as efficient as would be desired, and some of the milk fat is lost in the serum removed from the cream during the concentration. Only part of the milk fat may be recovered from the buttermilk by re-separation, and arrangements have to be made to process or otherwise dispose of the buttermilk. The improvements in separator design in the 1930s introduced the possibility of taking the fat level in the cream to or above the final fat level for butter, thus minimising fat losses via buttermilk (Wiechers *et al.*, 1950).

- H.D. Wendt claimed patents in the United States for high-fat cream processes in 1931 and 1934; his patent, in 1937, gave the construction of a drum cooler for the cream.
- In 1939, the Sharples Process (USA) improved the Wendt process. Cream was separated to ~83 g fat 100 g^{-1}, and standardised by skimmed milk addition. Salt and other substances were added, and then the cream was cooled and kneaded to butter using an auger and a pair of orifice plates with paddles. This invention also combined aspects of the Fritz and the Alfa (i.e. known as the Swedish Alfa Process) processes described later.
- A US patent claimed by van der Meulen & Levowitz in 1939 suggested souring of 33 to 35 g fat 100 g^{-1} cream followed by partial neutralisation and dilution by hot water to give a mixture at 70°C, then re-separation at half the normal feed rate with the cream screw adjusted to give 80 g fat 100 g^{-1}, plus modification to the outlet to permit flow of the viscous product. Partial breakdown of the fat globules was claimed to give butter on cooling.
- The Alfa method, developed in Germany in 1942, used two separators to produce cream at 78 g fat 100 g^{-1} (or higher, depending on the moisture content of the butter to be made), which was then cooled and sheared to bring about phase inversion to give butter. The second separator was designed to minimise foaming and the cream was then passed directly to the cooler. The cooling (using brine at −4°C) and phase inversion took place in three tubular heat exchangers containing screw-like ribbed rotors to produce the shear. This type of cooler is comparable to the scraped-surface heat exchangers (SSHE) that have blades mounted on the rotating central shaft or mutator. During passage through the coolers the temperature was lowered from 60 to 20°C by brine cooling, then from 20 to 10°C, during which phase inversion took place and finally the butter produced in this second cylinder was warmed from 10 to 15°C by warm water. This method produced a sweet cream butter with fine moisture distribution (MD) and good keeping qualities, better than the Fritz butter at that time. The Alfa method had the advantage over the Fritz method that there was no loss of fat in buttermilk.
- An essentially similar process built by A.B. Separator was reported to be running in Sweden in 1946: 20 g fat 100 g^{-1} cream being re-separated to 15.8 g 100 g^{-1} moisture.

Lactic acid, diacetyl and colouring were added in aqueous solution by a dosing device while 1 g 100 g^{-1} dry salt was added. The dosed high-fat cream was cooled to 13°C then passed through the cooler, 'driven by a worm'.

- The Creamery Package process, introduced in the United States in 1946: First, pasteurised cream ~30 g fat 100 g^{-1} was heated at 75 to 90°C, then separated to produce 80 g fat 100 g^{-1} cream. A special separator was needed for sour cream. The high-fat cream was sheared by a high-pressure homogeniser to break the emulsion. Fat and/or serum may be removed at this point using either a centrifugal separator or a separator chamber with conical plates – any unwanted serum that had been separated going back into the 30 g fat 100 g^{-1} cream feed so that fat losses may be minimised. The fat phase from the separator (~98 g fat 100 g^{-1}) was then passed to a multi-piston metering pump where the aqueous portion containing salt solution and some souring agent was dosed in the correct ratio. Batch standardising tanks may also be used. Another modification used separation to 90 g fat 100 g^{-1} and addition of the salt solution. Cooling and working of the emulsion was by a two-stage SSHE, giving a product at 7–13°C.
- The Cherry-Burrell (USA) process in 1947 heated 30 to 40 g fat 100 g^{-1} cream in stages to 68°C by direct steam injection then indirectly to 88°C, after which the temperature was lowered to 56°C mainly by vacuum evaporation. The treated cream was then re-separated at about 52°C to give a cream containing 80–90 g fat 100 g^{-1}, part of which has been de-emulsified, cooled and stabilised using a form of SSHE, followed by a 'texturator' in commercial plants.
- In 1947, the Kraft process (USA) was based on acid cream, adding lactic acid if necessary to give a pH of 4.2. These batches of acid cream were heated for 30–40 minutes to ~99°C, and separated using a centrifugal separator to give a high-fat cream with 85 to 95 g fat 100 g^{-1}, which was re-separated to 99.5 g fat 100 g^{-1}, with an option for vacuum drying to give an anhydrous milk-fat intermediate, that may then be blended with cultured milk and the O/W emulsion cooled by passage through the SSHE.
- The New-way process used in Australia and New Zealand was again similar to the Alfa process, but used brine at −20°C to cool the butter to −2°C. Salt may be added in solution to the cream in the vacuum float chamber or directly into the balance tank during standardisation prior to cooling (see McDowall, 1953).

During the early 1950s, the Alfa-type processes were used in Europe while the Cherry-Burrell and Creamery Package methods were used in the United States, though by 1960 they had been largely replaced. However, the high-fat route was used extensively in the erstwhile USSR where the Meleshin process was based on the inversion of high-fat cream during scraped-surface cooling using greater shear than the Alfa-type process with working applied for a 150- to 200-second period during the crystallisation stage (Munro, 1986).

More recently, improvements in SSHE technology led to the development of the Ammix process in New Zealand. This process used a combination of the separation technology as applied to the production of anhydrous milk fat and the SSHE technology that had been developed for other yellow fats, as illustrated in Figure 5.3. As with the earlier high-fat cream-based processes, the recovery of the buttermilk was more easily controlled than with the Fritz process, and part of the process could easily be adapted to preparation of anhydrous milk fat. By blending cream with the milk fat, it was possible to incorporate both the serum and discrete fat globules from the cream, thus replicating the structural components expected

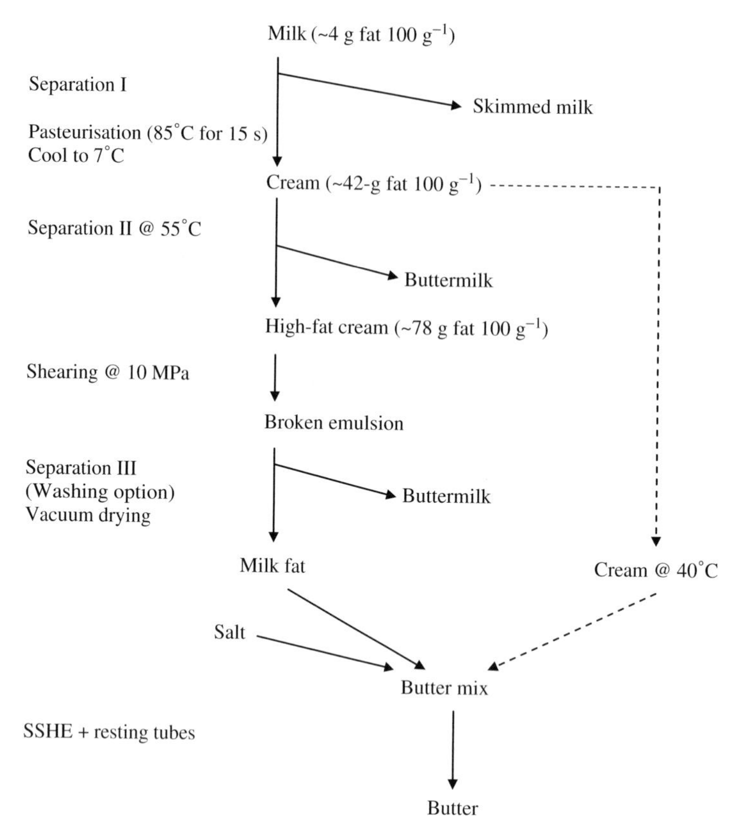

Fig. 5.3 Outline of the Ammix process.

for traditional butter. Crystallisation of the fat was achieved by rapid cooling on passage through the scraped-surface cooling units. This was augmented by low shear mixing in a pin working unit and extrusion via a pair of resting tubes that were used alternately to allow the butter to become further crystallised before extrusion to the packer.

Kawanari (1992) reported comparative work that had been carried out on commercial butters produced from high-fat cream using the Cherry-Burrell process and that by both batch and Fritz processes using Westfalia & Contimab machines, with the aim of producing a butter by the high-fat SSHE route and its properties for the making of French-style pastry. The production of lactic butter by SSHE technology was reported by Herrmann *et al.* (1995). An emulsion was prepared from milk fat, acidified skimmed milk and flavour enhancer. It was claimed that the spreadability of the butter could be improved by adjusting the cooling regime. Pedersen (1997) reported trials to produce a match to MD Foods' Danish Lurpak butter. Cream was re-separated to 76 g fat 100 g^{-1} and salt was added. This emulsion was then cooled by passage through an SSHE with three chilling tubes, inversion of the emulsion occurring spontaneously in the first chilling tube. The positioning of the pin working unit between the first and second cooling units was found to give the best result. Two such high-fat phase inversion plants (produced by Gerstenberg & Agger A/S) have been supplied for production of butter in Europe – one in Poland and one in Spain.

5.7 Recombined butter

Recombination has been practised primarily in countries where there has not been sufficient surplus cream to support a buttermaking industry, and where importation and repacking of bulk butter is neither practicable nor economic. The technology is essentially that used in the margarine industry with preparation of an emulsion of an aqueous phase, whether recombined skimmed milk (possibly fermented and/or salted), recombined whey or brine, dispersed in melted milk fat. The emulsion is then preferably heat treated before cooling using SSHE technology with a pin working unit and possibly resting tubes as described earlier. This technology does allow the use of milk fractions to modify the rheological properties of the finished butter.

5.8 Reduced-fat butters

Current European Union legislation permits the production of reduced- and low-fat butters (EU, 2007); standards are summarised in Table 5.1.

In making dairy or blended spreads, small reductions in fat content can be tolerated using continuous Fritz-type buttermakers, but as the fat content decreases there will be an increase in the moisture content, with greater likelihood of creating large moisture droplets and channels. This can be overcome by including monoglycerides, or other surface-active agents, into either all or part of the cream. Figure 5.4 illustrates a process proposed by Simon SAS (Anonymous, 2008a) for a three-quarter fat butter with 60–62 g fat 100 g^{-1}, where monoglyceride is incorporated into the hot (i.e. heat treated) cream before it is cooled and aged.

With further reduction in the fat content, scraped-surface technology becomes by far the best approach, particularly for the half-fat butters. At this point, when the volume of

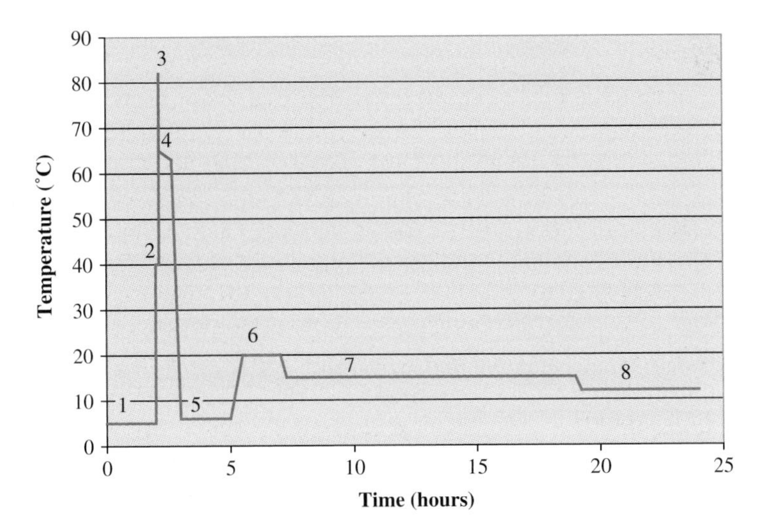

Fig. 5.4 Temperature–time plot for production of a three-quarter fat butter (adapted from Anonymous, 2008a). *Note*: 1 = raw milk; 2 = preheating and separation; 3 = high-temperature short time (HTST) treatment of cream; 4 = blending with monoglycerides; 5 = first ageing step; 6 = second ageing step; 7 = third ageing step; 8 = feed to buttermaker.

Table 5.1 Summary of European Union standards for dairy-based spreadable fats[a].

Fat group	Sales description	Additional description of the category with an indication of the % fat content by weight (i.e. g 100 g^{-1})
Milk fats[b]	Butter	The product with a milk-fat content of not less than 80 but less than 90, a maximum water content of 16 and a maximum non-fat milk-material content of 2.
	Three-quarter fat butter[c]	The product with a milk-fat content of not less than 60 but not more than 62.
	Half-fat butter[d]	The product with a milk-fat content of not less than 39 but not more than 42.
	Dairy spread X%	The product with the following milk-fat contents: • Less than 39 • More than 41 but less than 60 • More than 62 but less than 80
Fats composed of plant and/or animal products[e]	Blend	The product obtained from vegetable and/or animal fats with a fat content of not less than 80 but less than 90.
	Three-quarter fat blend[f]	The product obtained from vegetable and/or animal fats with a fat content of not less than 60 but less than 62.
	Half-fat blend[g]	The product obtained from vegetable and/or animal fats with a fat content of not less than 39 but less than 41 g.
	Blended spread X%	The product obtained from vegetable and/or animal fats with the following fat contents: • Less than 39 • More than 41 but less than 60 • More than 62 but less than 80

[a] Data compiled from EU (2007); this regulation came into force on 1 January 2008; see also Chapter 7.
[b] Products in the form of a solid, malleable emulsion, principally of the W/O type, derived exclusively from milk and/or certain milk products, for which the fat is the essential constituent of value. However, other substances necessary for their manufacture may be added, provided those substances are not used for the purpose of replacing, either in whole or in part, any milk constituents.
[c] Corresponding to 'smør 60' in Danish.
[d] Corresponding to 'smør 40' in Danish.
[e] Products in the form of a solid, malleable emulsion, principally of the W/O type, derived from solid and/or liquid vegetable and/or animal fats suitable for human consumption, with a milk-fat content of between 10 g 100 g^{-1} and 80 g 100 g^{-1} of the fat content.
[f] Corresponding to 'blandingsprodukt 60' in Danish.
[g] Corresponding to 'blandingsprodukt 40' in Danish.
Note: The milk-fat component of the products listed in this table may be modified only by physical processes, that is, by fractionation but not by hydrogenation or inter-esterification.

the aqueous phase is about one-third higher than that of the lipid phase, stabilisation of the aqueous phase is far more difficult. Monoglycerides play a greater role in modifying the crystallisation of the fats, but can promote O/W emulsions rather than the desired W/O emulsion. Thus, the mobility of the aqueous phase has often been reduced by including one or more thickeners such as milk proteins, gelatine, starches and gelling agents such as alginates.

5.9 Spreadable butters

The very wide range of fatty acids, and hence triglycerides, present in milk fat results in their displaying a gradual melting over a wide temperature range. Plasticity is normally achieved when there is less than 40 g 100 g^{-1} of the fat in the solid state, with ease-of-spreading between 20 and 30 g 100 g^{-1} solid fat. This level corresponds to about 15°C, midway between refrigerator and room temperature, hence not easy to achieve. Below 20 g 100 g^{-1} solid fat, the butter becomes too soft for most purposes.

Modification of the cream ageing temperatures has already been discussed as a means of minimising seasonal variations in the hardness of butters. This approach is not sufficient to overcome the high solid fat in the butter at refrigeration temperatures and the alternatives are to:

- Reduce the fat content as described earlier; not a very satisfactory approach in the context of spreadability as this does not alter the solid-to-liquid-fat ratio so that the product has a tendency to crumble rather than be plastic if too cool. A similar problem may be observed with whipped butter.
- Make major modifications to the cows' diet so that a less saturated milk fat is produced. The original work in this area used specially protected spray-dried emulsions containing unsaturated fats, in order to bypass the natural hydrogenation processes in the rumen. This approach was too expensive for commercial exploitation and subsequent work has used oilseeds to provide the protection (Frede *et al.*, 1992). Although this approach minimises the technological input, it does require virtually all of the cows in the milk-field supplying the butter factory to be fed with the special ration (over 20 000 L of milk are needed per tonne of butter), thus requiring a high degree of cooperation with the suppliers and the acceptance that the less saturated milk fat resulting from this approach may be less suitable in other products, for instance, whipping cream or spray-dried powders.
- Fractionate the fat and recombine selected fractions, preferably the highest and lowest melting fractions, in order to extend the temperature range where the milk fat is at optimum spreadability. This option is arguably not only the most effective method but also the most demanding in terms of technology, as illustrated in Figure 5.5. Milk-fat modifications are reviewed by Augustin & Versteeg (2006).
- Blend the cream or milk fat with a vegetable oil that is liquid at refrigeration temperatures to produce a blend (as defined in Table 5.1). This is the next simplest to the second option discussed earlier technologically, but probably the simplest and least expensive route, albeit not producing a butter but a blend. Methods have been developed using batch buttermaking techniques and updated to continuous buttermakers as well as use of scraped-surface technology (Wilbey, 1994).

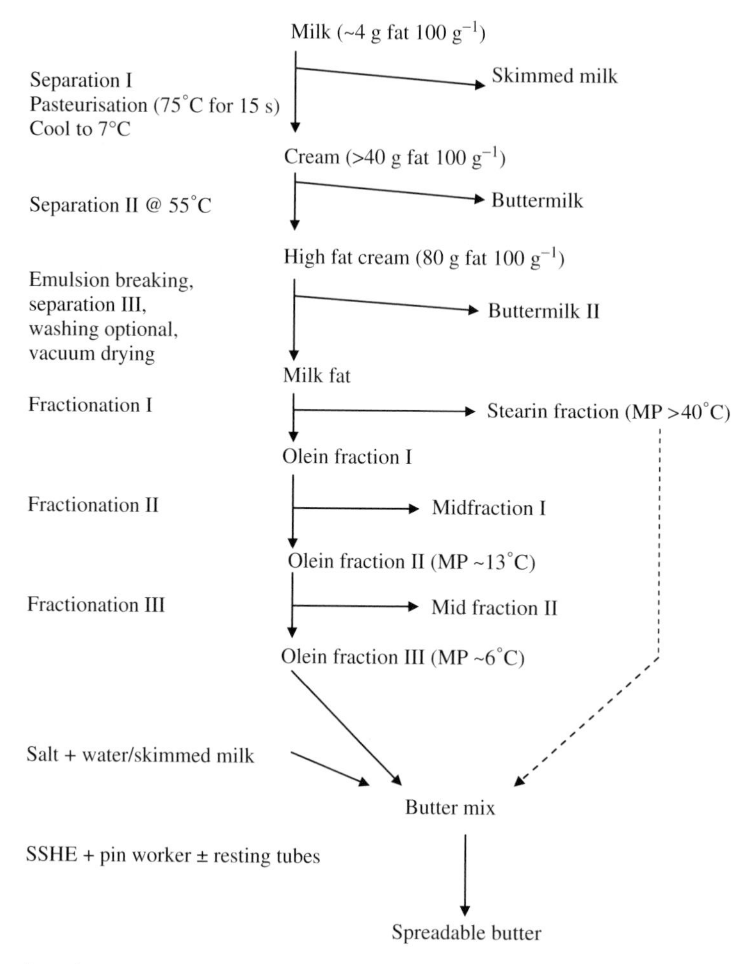

Fig. 5.5 Outline of process for making spreadable butter based on fractionated milk fat. *Note*: The use of olein fraction III will produce a product with a better range of spreadability than that using fraction II.

5.10 Packaging

Butter packing may be in bulk or retail packs, 250-g packs being the most common for the latter. Since butter is relatively stable and the profitability less than for many other dairy products, butter has been used as a sink product for surplus milk fat. As such, the production has commonly been out of balance with market needs, this being particularly so in those countries where the dairy industry is geared more for export than for supply to the domestic market. This butter would be filled into bulk packs.

In older and smaller butter plants, the butter, possibly batch-produced, might be extruded into 25-kg cartons, each carton being weighed manually as it is filled. The original system of using a loose parchment lining has been replaced by blue-pigmented polyethylene bags, as the latter give better protection and are more readily detected if part of the bag were to become detached and incorporated into the butter.

In modern plants, the freshly churned butter is collected first in a butter silo, essentially a steeply sloped tank with an auger to help feed the butter to the pump, providing a break in the product flow so that any interruption in the packing does not cause the buttermaker to be shut down. The bulk butter packing uses two-stage filling to ensure accuracy and minimum give-away; smaller bulk packs may be produced to comply with manual handling restrictions.

The shelf-life of the bulk butter may be extended considerably by storing it frozen, preferably below −18°C. When needed, the butter must then be thawed and brought up to a chill temperature, for instance, 2 to 4°C, before it can be handled. This tempering process could take several days and consequently take up a considerable plant area as well as reduce flexibility in meeting customer demands. One method to minimise the tempering time is to pass the blocks of butter through a microwave heater.

Stored bulk butter has a relatively coarse crystal structure and will not flow readily. Thus the structure must be broken down in order to restore plasticity. Two steps are used to achieve this: first, the blocks are shived, reducing the butter to thumb-sized pieces by a process analogous to grating, and second, the pieces of butter are then re-blended.

The re-blending operation may be carried out on a small-scale basis using a sigma or Z-blender, essentially a batch blender with a rounded W-shaped cross section, in the troughs of which are two Z-shaped contra-rotating rotors that have a small clearance with the walls to create a shearing as well as mixing action. This method is relatively labour intensive.

Large-scale blending operations use continuous re-workers, virtually identical to the workers on Fritz-type buttermakers. As with the buttermakers, there is the option to inject water, brine or other solutions into the re-worker to correct moisture, salt or pH to the desired levels to ensure a consistent product and maximise yield. The re-worker would discharge into a butter silo feeding one or more butter packing machines. These machines work by volumetric displacement but may be linked in with check weighers to minimise product loss.

Most retail butter is packed in either parchment or a parchment-aluminium foil laminate. Parchment is cheaper, but is permeable to moisture vapour and ultra-violet so that the surface of the butter can suffer from both surface drying and oxidative rancidity, the latter being reduced by the application of pigments such as titanium dioxide to the outer surface of the parchment. Foil laminate protects from ultra-violet and only permits moisture vapour and gas interchange at the seams, thus aiding a longer shelf-life.

Some specialist butters, for instance, where the presentation takes the form of a curl, may use transparent films. Though the film gives good protection against moisture loss, the risk of oxidative rancidity on the surface is higher than for parchment. Pre-formed plastic containers, often polypropylene, are more expensive and tend to be used for soft butters and hybrid products that would be too easily damaged in film wraps.

Butter portions, typically less than 20 g in weight, for catering and institutional use, are filled either into foil laminates, where the consistency on filling can be critical, or into plastic trays with a foil or aluminised film cover, in a form-fill seal operation.

5.11 Flavoured butters

Flavoured butters are produced by blending herbs and/or spices with butter. Batch blenders are normally used, with addition of the herbs/spices to the shived butter. It is preferable that such operations be kept separate from the main buttermaking and packing operation to avoid contamination of the normal butter by atypical odours (in the case of garlic butter) or foreign matter (parsley or black pepper). Any additives must be of good microbiological quality. The packaging system should also be designed to minimise odour transfer during distribution through the chill chain; a sealed impermeable container is preferable to a foil wrap and parchment would be unsatisfactory.

5.12 Quality issues

Raw material quality can be an issue where butter is being prepared as a recovery operation from 'downgraded' milk deemed unsatisfactory for, say, the liquid milk market. Sometimes milk may be affected by weed taints and in this case, vacuum treatment by a cream treatment unit or spinning cone evaporator may be needed, as already discussed. Where cream has been recovered from milk rinsings or cream processing for the retail market, the presence of homogenised fat globules can lead to higher fat losses in the buttermilk.

Determination of fat content in milk and cream is most readily achieved using infra-red (IR) absorption methods in most laboratories, with arguably less accurate methods based on ultrasonics or light scattering being adopted for smaller factory laboratories. All of these methods avoid the health and safety issues implicit in the Gerber method and are faster. The IR methods, particularly those based on Fourier transform IR (FTIR) technology, are more accurate than the Gerber method, assuming correct calibration.

Moisture has long been the most important parameter measured in the butter. Rapid estimation of moisture by loss in weight on boiling the butter over a Bunsen burner was commonly carried out as a routine check. For salted butters, salt was routinely estimated by titration. With the change from batch to large-scale continuous churning, the time delay on testing and the potential losses stimulated the development of on-line analytical techniques. Some success was achieved with the use of impedance for measuring moisture in unsalted butters but variations attributable to moisture were overshadowed by the effects of the salt in the aqueous phase. This salt problem was initially overcome by incorporating a second sensing system measuring the backscatter of γ-radiation produced by an americium-241 source. The subsequent development of near-IR (NIR) methodology has now overtaken this, with in-line absorbance measurement employing a flow cell with polytetrafluoroethylene (PTFE) windows in the output line from the buttermaker, as illustrated in Table 5.2 and Figure 5.6. Rapid moisture content of butter can now be determined in the laboratory using IR absorbance in machines designed for solid and semi-solid products, such as the Foss FoodScan™ available as a laboratory or process-floor model as shown in Figure 5.7.

Sensory properties of butter are variable, depending on the source, season and type of butter produced. Product to be sampled is best first brought to 10 to 12°C, and bulk packages are sampled using a slightly tapered semi-circular corer known as a 'trier'. The sensory properties of the butter may be described in terms of appearance, texture and taste:

- The appearance should be smooth but not oily, with no free moisture evident on the cut surface. Colour should be even and characteristic of that type of butter. The actual colour will vary considerably, from almost white for some lactic butters produced in winter through to a pronounced creamy yellow for sweet cream butters produced from summer pasture milk, especially from channel island cows.

Table 5.2 Examples of the performance of on-line and off-line analyses of butter, taken from manufacturers' data.

Method of analysis	Component	Range (g 100 g^{-1})	Accuracy (1 SD[a])	Repeatability (%)
On-line	Moisture	15–19	0.1	<0.05
	Fat	78–83	0.2	<0.05
	Salt	0–2	0.06	<0.05
	Component	*Range (g 100 g^{-1})*	*Accuracy*[b]	*RSQ*[c]
Off-line	Moisture	13.2–18.1	0.11	0.99
	Fat	79.4–85.2	0.16	0.99
	Salt	0.84–1.88	0.05	0.96
	Solids-not-fat (SNF)	0.87–3.27	0.13	0.96

[a]Standard deviation (SD).
[b]Standard Error of Prediction, corrected for slope and bias (1 SD absolute).
[c]Linear correlations between results and values obtained by reference method. Reproduced by permission of Q-Interline UK and Foss UK.

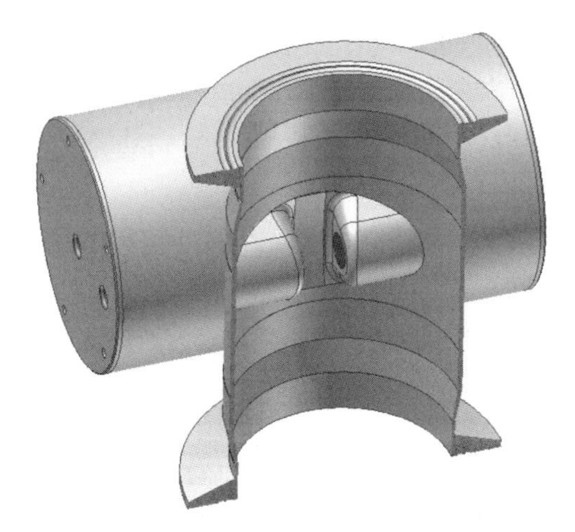

Fig. 5.6 Section through TEFWIN cell for in-line Fourier Transform-Near Infrared (FT-NIR) analysis of butter. With permission of Q-Interline.

- The texture of the butter should be smooth, with a steady melt on the palate. The consistency should be neither crumbly, indicative of under-working (as found with some batch-produced butters), nor gummy as a result of over-working. Graininess may be the result of slow cooling or temperature cycling while grittiness is usually attributable to poorly dispersed salt crystals.
- Taste will vary from a very subtle flavour with unsalted sweet cream butter to the cultured flavours of lactic and whey butters. The flavour will be modified further by any added salt. In general, the flavour should be clean, that is free of any bitter, feed-related, musty, scorched or dirty off-notes.

Butter grading is regarded as being largely an internal matter in the United Kingdom, but may be given more emphasis in other countries/states, for instance, in Wisconsin

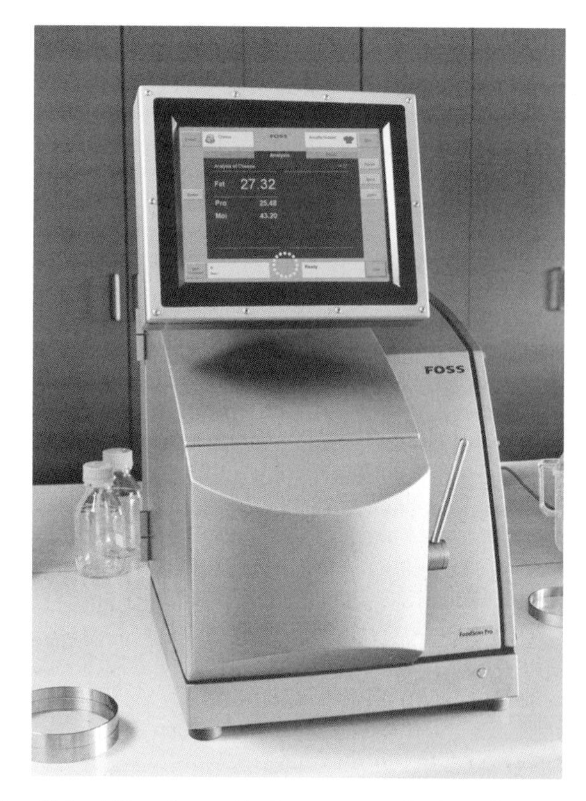

Fig. 5.7 Foss FoodScan™ Pro NIR analyser with IP65 waterproof cabinet for analysis of butter and other solid and semi-solid food products. With permission of Foss UK.

(Anonymous, 2008b). See Bodyfelt *et al.* (1988) and Clark *et al.* (2008) for further discussion of sensory analysis and grading of butter.

5.13 Concluding comments

Modern buttermaking has the potential to produce consistent products to meet consumer demands for high-quality yellow fats. Although texture remains a problem, there is the potential to produce dairy-based spreads with better spreadability when taken from the refrigerator.

References

Anonymous (2008a) *Cream Processes for Continuous Butter Production*, Simon SAS, Cherbourg.

Anonymous (2008b) Chapter ATCP 85 Butter grading and labelling, *Register*, **628**, 659–661, http://www.legis.state.wi.us/rsb/code/atcp/atcp085.pdf.

Augustin, M. A. & Versteeg, C. (2006) Milk fat: physical, chemical and enzymatic modification. *Advanced Dairy Chemistry – Lipids*, (eds. P. F. Fox & P. L. H. McSweeney), Volume 2, pp. 293–332, Springer, New York.

Bodyfelt, F. W., Tobias, J. & Trout, G. M. (1988) *The Sensory Evaluation of Dairy Products*, Van Nostrad Reinhold, New York.

Boutonnier J. L. & Dunant C. (1985) *Laits et Produits Laitiers*, (eds. F. M. Fouquet & Y. Bonjean-Linczowski), Volume 2, pp. 443–504, Lavoisier, Paris.

British Standards (1969) *Specification for Vacuum Salt for Butter and Cheese Making and Other Food Uses*, BS 998, British Standards Institution, London.

Clark, S., Costello, M., Drake, M. A. & Bodyfelt, F. W. (2008) *The Sensory Evaluation of Dairy Products*, 2nd edn., Springer-Verlag, New York.

Danmark, H. & Bagger, L. H. (1989a) Effects of temperature treatment of sweet cream on physical properties of butter. I – Factors affecting hardness and consistency. *Milchwissenschaft*, **44**, 156–160.

Danmark, H. & Bagger, L. H. (1989b) Effects of temperature treatment of sweet cream on physical properties of butter. II – Factors affecting initial moisture and fat loss. *Milchwissenschaft*, **44**, 281–283.

Dixon, B. D. (1970) Spreadability of butter: developments with the modified Alnarp cream cooling method. *Australian Journal of Dairy Technology*, **25**, 82–84.

EU (2007) Council Regulation (EC) No. 1234/2007 of 22 October 2007 establishing a common organisation of agricultural markets and on specific provisions for certain agricultural products (Single CMO Regulation). *Official Journal of the European Union,* **L 299**, 1–149.

Frede, E., Precht, D., Pabst, K. & Philipzck, D. (1992) Effects of feeding rapeseed products to dairy cows on the hardness of milk fat. *Milchwissenschaft*, **47**, 505–511.

Frede, E., Precht, D. & Peters, K. H. (1983) Consistency of butter. IV – Improvement of the physical quality of summer butter with special reference to the fat content of buttermilk on the basis of crystallisation curves. *Milchwissenschaft*, **38**, 711–714.

Herrmann, M., Godow, A. & Hasse, T. (1995) Alternative butter production with scraped surface heat exchanger. *Deutche Milchwirtschaft*, **46**, 62–67.

Hill, J. (2003) The Fonterra Research Centre. *International Journal of Dairy Technology* **56**, 127–132.

Kawanari, M. (1992) Study on the continuous manufacturing of butter from high fat cream. *Reports of Research Laboratory, Technical Research Institute, Snow Brand Milk Products Milk Co.*, **98**, 35–110.

McDowall, F. H. (1953) *The Buttermakers Manual*, New Zealand University Press, Wellington.

Munro, D. S. (1986) *Alternative Processes. Continuous Buttermaking*, Document No. 204, pp. 137–144, International Dairy Federation, Brussels.

NIZO (1976) Netherlands dairy research institute. **UK Patent Application**, 1 478 707.

Pedersen, A. (1997) Inversion of creams for butter products. *Dairy Industries International*, **62(7)**, 39–41.

Samuelsson, E. & Petersson, K. I. (1937) Årsskrift för Alnarps lantbruksmejeriog trädgårdsinstitut, cited by Mortensen, B. K. (1983) *Developments in Dairy Chemistry* (ed. P. F. Fox), Volume 2, pp. 159–194, Applied Science Publishers, London.

Walstra, P. & Jenness, R. (1984) *Dairy Chemistry and Physics*, Wiley, New York.

Wiechers S. C. & DeGoede B. (1950) *Continuous Buttermaking*. North-Holland, Amsterdam.

Wilbey R. A. (1994) Production of butter and dairy based spreads. *Modern Dairy Technology* (ed. R. K. Robinson), Volume 1, 2nd edn., pp. 107–158, Chapman & Hall, London.

Wilbey R. A. (2002) Microbiology of cream and butter. *Dairy Microbiology Handbook* (ed. R. K. Robinson), 3rd edn., pp. 123–174, Wiley-Interscience, New York.

6 Anhydrous Milk Fat Manufacture and Fractionation

D. Illingworth, G.R. Patil and A.Y. Tamime

6.1 Introduction

Anhydrous milk fat (AMF) is defined and the manufacturing processes from possible feedstocks such as direct-from-cream (DFC) and butter are outlined with reference to previous works on this subject. Each stage of the processes is listed by reference to the equipment used, with the main factors that influence the overall efficiency of the process. Quality considerations are covered by reference to the various analyses that are used to monitor AMF production such as free fatty acid (FFA) levels, measurement of oxidative deterioration such as peroxide value (PV) or anisidine value (AV), and levels of trace metals such as copper and iron that catalyse oxidation. The control of levels of dissolved and headspace oxygen during manufacture and subsequent fractionation is also discussed.

Various options for fractionation of AMF are described, and the chemical composition and physical properties of milk fat (MF) fractions are discussed in this chapter in relation to those of the parent AMF. However, ghee is the most widely used traditional milk product in the Indian subcontinent. The traditional as well as the modern methods of manufacture and packaging of ghee are extensively reviewed in this chapter. The characteristics of ghee in terms of chemical composition, physical properties, flavour, texture as influenced by various technological and storage parameters including the shelf-life of the product, its nutritional aspects and use as medicine are also detailed in this chapter.

6.2 Definitions and properties

The Codex Alimentarius standard for milk fat products (FAO/WHO, 2006) states that AMF, MF, anhydrous butteroil, butteroil and ghee are fatty products derived exclusively from milk and/or products obtained from milk by means of processes that result in almost total removal of water and non-fat solids. Table 6.1 is derived from the above standard and shows the main criteria by which these products are classified with respect to their fat/water content, and also the main quality parameters, such as FFA and PV, that are used to classify them into one of the aforementioned products. Ghee is included in Table 6.1, and will be discussed in detail later in this chapter. AMF is composed of 99% triacylglycerols (TAGs), that is, glycerol esterified with fatty acids. Unlike other fats, AMF has significant levels of very short-chain fatty acids such as butyric, from which the name 'butter' is derived. More details will emerge about the composition of AMF during the discussion of fractionation. The Codex Alimentarius standard also recommends limits for the heavy metals iron and copper of 0.2 and 0.05 $mg\,kg^{-1}$, respectively, as well as allowing the addition of a range of antioxidants and antioxidant synergists at proscribed levels of addition to all MF product categories other than AMF.

Table 6.1 Composition of milk fat products.

	AMF or anhydrous butteroil	MF	Butteroil	Ghee
Minimum MF (g 100 g^{-1})	99.8	99.6	99.6	99.6
Maximum water (g 100 g^{-1})	0.1	NR	NR	NR
Maximum FFA as oleic acid (g 100 g^{-1})	0.3	0.4	0.4	0.4
Maximum (PV) (milliequivalants of O_2 kg^{-1} fat)	0.3	0.6	0.6	0.6
Taste and odour	Acceptable for market requirements after heating a sample to 45°C.			
Texture	Smooth and fine granules to liquid depending on the temperature.			

AMF, anhydrous milk fat; FFA, free fatty acid; PV, peroxide value; NR, not reported.

6.3 Production statistics

The manufacture of AMF is widely practised in virtually all countries that have dairying as a major industry, particularly in the countries of the European Union (EU), notably Belgium and the Netherlands, and in both Australia and New Zealand. However, other countries such as France, the United States and Switzerland also produce AMF, but in relatively small volumes when compared to its production in the countries listed earlier.

In general, the production figures are obscured as no distinctions are made between the various MF categories or butter production. In 2003, the United States exported almost 6000 tonnes of an unknown volume of production; New Zealand was reported to have exported up to 80 000 tonnes per annum and Australia was reported to have exported some 60 000 tonnes per annum of AMF/butteroil (J. Thomas, personal communication). According to government statistics in the 2005–2006 season, Australia exported around 10 000 tonnes of butteroil, up from about 6300 tonnes in the previous season. Annual overall production of butter and AMF combined is over 100 000 tonnes. In the EU, export figures are given of the destination of butteroil and butter concentrates (this term is used to describe both MF and MF fractions), as shown in Table 6.2 (Anonymous, 2003a).

Figures issued by the European Commission, that is, Directorate General for Agriculture, reveal that from 1999 to 2002 the world production of butter, including AMF, made from cream rose by almost 1 million tonnes from just under 7 million tonnes. The largest single producers are India and the EU, followed by the United States, Russia and Pakistan. By world standards, New Zealand and Australia are behind these countries in production, but export more than 90% of their production with New Zealand having the largest export volume of all the countries listed (Figure 6.1). Standard 68a of the International Dairy Federation (IDF, 1977) also defines the standards of identity for milk fat in its various forms.

6.4 Anhydrous milk fat/butteroil manufacture processes

6.4.1 *Principles*

The principle behind the manufacture of MF products that include AMF and all its variants is the removal of the water and water-soluble components of milk to leave only pure fat or oil (Figure 6.2) (Anonymous, 2003b). In milk, the fat is enclosed within a membrane that

Table 6.2 European Union (EU) destination of exports of butteroil and butter concentrates (×1000 tonnes).

	Year					
	1997	1998	1999	2000	2001	2002
Exports to EU member countries						
Total	122.5	135.7	123.6	127.7	127.4	112.6
Exports to different countries						
Russia	0.6	1.1	1.0	2.2	2.5	4.0
North Africa	4.3	5.6	7.1	7.0	10.7	12.3
Central and South America	9.7	5.0	9.4	7.8	6.8	8.8
Middle East	13.7	10.5	14.6	13.0	11.4	14.4
Asia	4.7	1.9	3.6	3.1	4.6	5.3
Miscellaneous	5.7	2.9	6.7	10.9	7.6	6.6
Total	38.7	27.0	42.4	44.0	43.6	51.4

Compiled from Anonymous (2003a).

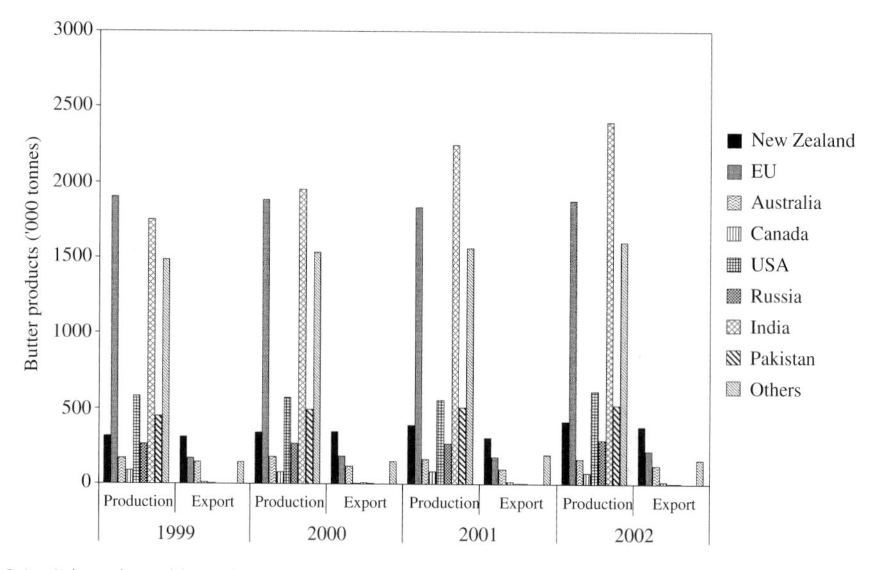

Fig. 6.1 Selected world production and export figures for butter products between 1999 and 2002. Source: European commission, Directorate General for Agriculture (http://ec.europa.eu/agriculture/agrista/2004/table_en/42021.pdf).

consists of proteins, phospholipids and cholesterol. This membrane protects the fat against attack by bacteria that can lead to both lipolysis and oxidation, and maintains the fat in a stable oil-in-water emulsion. To recover the fat, it is necessary to rupture the membrane and invert the emulsion.

The process involves a series of concentration and separation steps that concentrate the fat so that the inversion can be achieved (Figure 6.3) (Anonymous, 2003b). The various pieces

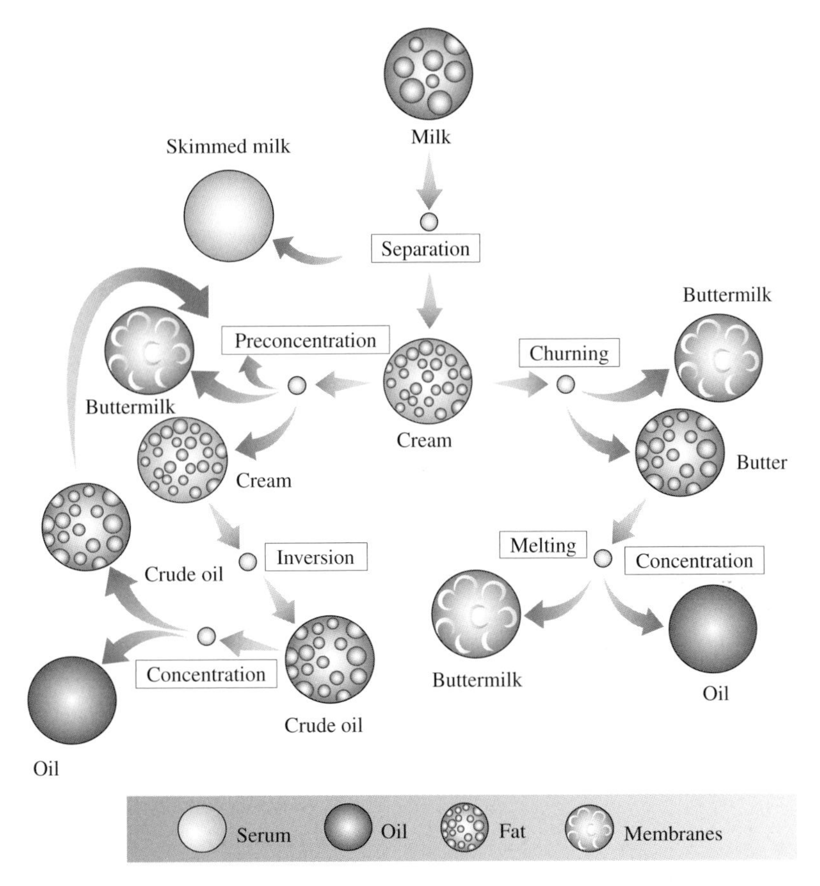

Fig. 6.2 Basics of anhydrous milk fat (AMF) production – concentration of milk fat, phase inversion and concentration of oil. Reproduced with permission of Tetra Pak A/B, Lund, Sweden.

of equipment including homogenisation devices, separators and dehydration vessels that can be used in a process plant are described in detail by Illingworth & Bissell (1994). They also describe the various parameters that are important in ensuring efficient operation of the plant, for example, the temperatures and the pressures required for phase inversion. These parameters were also the subject of an extensive study by Watt (1982).

6.4.2 *Manufacturing options*

If milk with 3–4 g fat 100 g^{-1} is concentrated to 40 g fat 100 g^{-1} by removal of the skimmed milk, and then further concentrated to about 75 g fat 100 g^{-1} by separation and removal of the serum phase, the high-fat emulsion can be directly inverted to release the fat. Further concentration, to remove the remaining milk solids-not-fat (SNF) and polishing of the fat using separators, produces 99.5 g fat 100 g^{-1} butteroil that can then be dehydrated, resulting in AMF (Figure 6.4) (Anonymous, 2003b).

Alternatively, 40 g fat 100 g^{-1} cream can be churned to butter (80–82 g fat 100 g^{-1}), a process that also involves inverting the cream emulsion, removing buttermilk in the process. The butter can then be melted, and the MF separated from the serum, through a

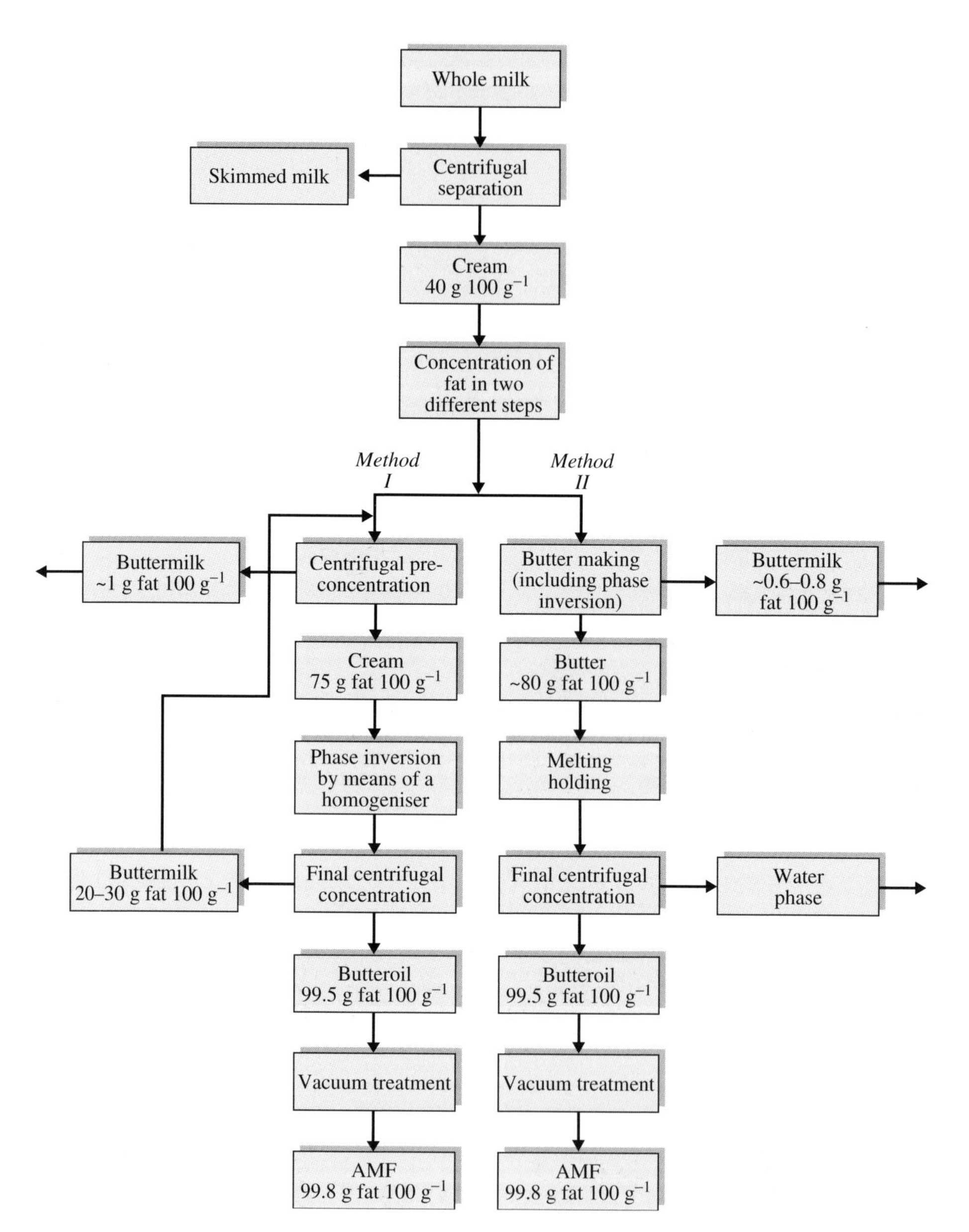

Fig. 6.3 Block diagram showing principles of anhydrous milk fat (AMF) production. Reproduced with permission of Tetra Pak A/B, Lund, Sweden.

process similar to the DFC process (Figure 6.5) (Anonymous, 2003b). Using this route to produce AMF, both salted and unsalted sweet-cream butters together with lactic butter can be used as starting materials, either from fresh production or, more commonly, using butters that have been stored frozen for some time. Each butter type requires different handling and the emulsion inversion is slightly different (Illingworth & Bissell, 1994). Both process variants result in heavy-phase streams that can be variously described as buttermilk. However, the

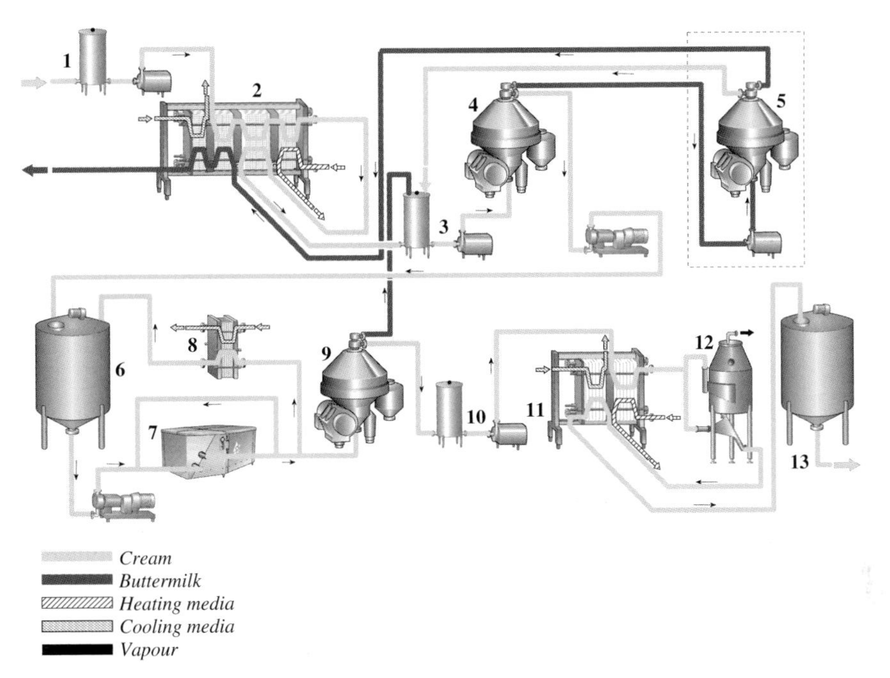

Fig. 6.4 Production line for anhydrous milk fat (AMF) direct from cream. 1, balance tank; 2, plate heat exchanger for heating or pasteurisation; 3, balance tank; 4, pre-concentrator; 5, separator (optional) for buttermilk from the pre-concentrator (4); 6, buffer tank; 7, homogeniser for phase inversion; 8, plate heat exchanger for cooling; 9, final concentrator; 10, balance tank; 11, plate heat exchanger for heating/cooling; 12, vacuum chamber; 13, storage tank. Reproduced with permission of Tetra Pak A/B, Lund, Sweden.

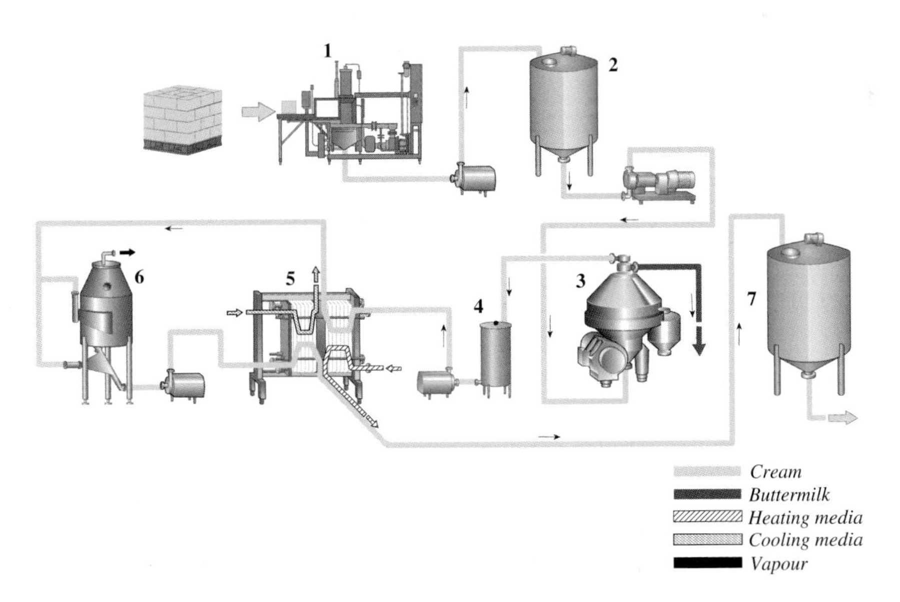

Fig. 6.5 Production line for anhydrous milk fat (AMF) from butter. 1, melter and heater for butter; 2, holding tank; 3, concentrator; 4, balance tank; 5, plate heat exchanger for heating/cooling; 6, vacuum chamber; 7, storage tank. Reproduced with permission of Tetra Pak A/B, Lund, Sweden.

by-products (sera) from the various butter feedstocks are all different with both salted and lactic butters resulting in contaminated buttermilk (salty or acidic) with different downstream handling required (i.e. these processes are not reviewed in this chapter).

6.4.3 *Quality of milk fat during and post manufacture*

Prevention of oxidative deterioration

As was noted in Table 6.1, all forms of MF have a PV specification applied to them with AMF having the tightest specification of all. Oxidation processes begin virtually as soon as the membranes around the MF globules in milk are ruptured, allowing air to come into contact with the fat. During manufacture and, most importantly, during the packing and storage of MF products, it is essential that the fat is protected as far as possible from the ravages of oxidation. Nitrogen blanketing of any tanks where AMF is held is one method of achieving this, but it is also important that pipe unions and pump seals that may allow air to be sucked into the process stream are suitable for the purpose. The edible oil industry uses nitrogen sparging – small bubbles of nitrogen are streamed into the liquid fat – to scavenge any dissolved oxygen that may be present. Measurement of dissolved oxygen at the time of packing is important, and to ensure a long shelf-life a maximum level of 3% is recommended (3 Pa oxygen partial pressure at atmospheric pressure of 100 Pa), where normal air is 20.9% O_2 partial pressure. This corresponds to about 5.5 mg O_2 kg^{-1} AMF, given that AMF (40°C) saturated with air (20.9% O_2) contains about 38.5 mg O_2 kg^{-1}. During packing, any air, for example, in drums, should be replaced with nitrogen before filling is commenced, and packaging should ideally be filled from the bottom to prevent any incorporation of air during filling.

Neutralisation

Enzymatic hydrolysis of MF in milk or cream produces FFA. The solubility of these acids in water depends on their chain lengths with short-chain acids being water soluble and long-chain ones more soluble in oil. Alkali refining was originally used by the vegetable oil industry and carried out batchwise in open kettles until a continuous process was developed in 1932 (Anderson, 1962).

The manufacturers of DFC AMF plants offer an option that allows neutralisation to be carried out in-line after the oil concentrating separator in order to avoid contamination of MF serum (Figure 6.6) (Anonymous, 2003b). Sodium hydroxide solution (6–10 g 100 g^{-1}) is injected and mixed into the fat stream using a static mixer or other device to ensure good mixing. The butter oil and caustic mixture is held in the holding tube for 10–20 s to allow the reaction to take place. Hot water (60–70°C) is then added with a further 5–10 s holding time before the stream passes on to the polishing separator.

Polishing is usually sufficient to reduce any residual soaps to 35–90 mg kg^{-1}, but there is little margin for overcoming process fluctuations. A second washing and polishing step would reduce alkalinity to perhaps 10–50 mg kg^{-1} consistently, but at the cost of another separator and polishing losses. If the neutralisation option is used, the resultant fat then falls outside the definition of AMF and into one of the butteroil categories.

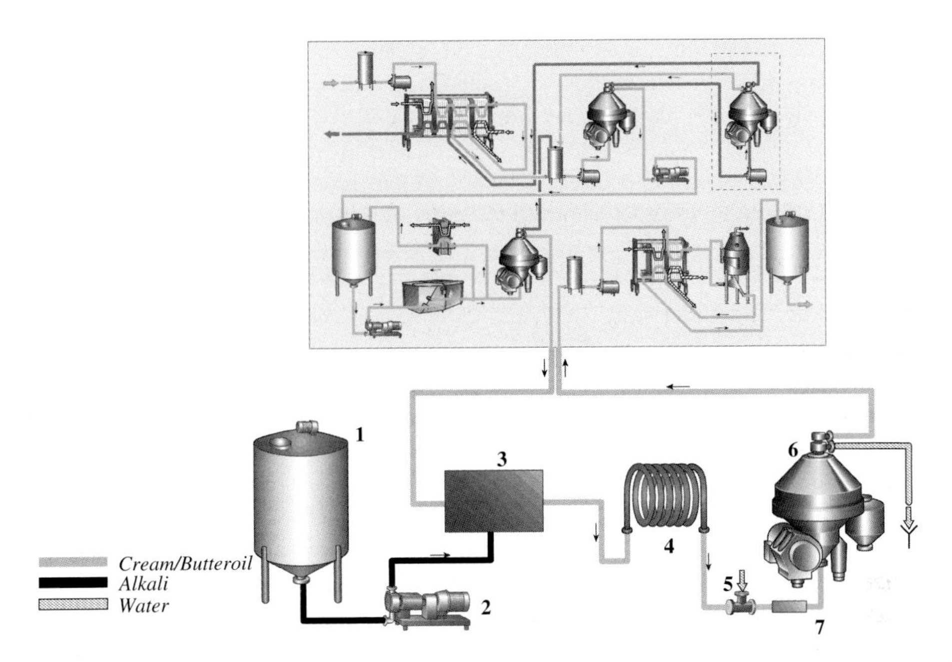

Fig. 6.6 Neutralisation of free fatty acids (FFA) can be one of the refining processes in the production of anhydrous milk fat (AMF). 1, tank for alkali; 2, dosing pump; 3, mixing equipment; 4, holding cell; 5, water injection; 6, separation of saponified free fatty acids; 7, oil/water mixer. Reproduced with permission of Tetra Pak A/B, Lund, Sweden.

Cholesterol removal

Removal of cholesterol is not a routine operation in the manufacture of AMF. Two processes have been cited. One, which does not seem to have any effect on the flavour of AMF, is to mix the fat with β-cyclodextrin, a modified starch. This molecule surrounds the cholesterol molecule, allowing it to be filtered from the fat. The second process was developed by S.A. Fractionnement Tirtiaux of Belgium; it involves steam distillation in what is effectively a deodoriser as used in the physical refining of edible oils. The high temperatures required for this process also destroy the natural carotene colour of the AMF and remove any natural antioxidants; what is more critical is that the natural flavour for which AMF is prized is also distilled off.

6.5 Milk fat fraction

Fat fractionation is widely used in the edible oil industry, and has been applied very successfully to MF. It involves crystallisation of fat under controlled conditions of temperature and agitation. MF is used to denote milk fat used as the feedstock for fractionation, but freshly made using the DFC process before proceeding directly with the fractionation process.

To speak of AMF implies a final, packaged product manufactured either DFC or from stored butter. AMF may be used as a fractionation feedstock but, because it also may be stored for some time post manufacture, the fractions are not as fresh as those from MF.

MF is valued mainly for its natural flavour; its complex TAG composition makes it inadequate for many uses because it results in a wide melting range, usually considered as – 40°C to +40°C, although there are individual TAGs with melting points higher than this. High-melting fractions are generally considered to have melting points in excess of 38°C, and are used as speciality bakery fats in products such as puffs, pastries and croissants (Rodenburg, 1973; Munro & Illingworth, 1986). Low-melting fractions usually have melting points generally in the range 21–28°C. The uses for MF fractions are detailed in Chapter 7.

Both AMF and butter suffer many disadvantages in their physical properties or melting characteristics compared with those of margarines that can be formulated precisely for a given application. Since the middle of the twentieth century there has been a rapid growth in MF fractionation that has resulted in its becoming more competitive in both specialist bakery use and in cold spreading applications. Now, in the twenty-first century, MF in its fractionated forms is widely used commercially for a variety of end uses in many countries (Kaylegian *et al.*, 1993; Kaylegian & Lindsay, 1994; Munro *et al.*, 1998; Hartel & Kaylegian, 2001). In addition, fractionation is the only technology that allows a pure spreadable butter to be achieved without the addition of vegetable oils (Deffense, 1987).

6.5.1 *Process options*

Several other process options that are possible for fat fractionation have been applied to MF with varying degrees of success and are described in detail by Illingworth (2002).

Melt fractionation without additives

This is probably the most simple and the most widely used fractionation process for MF; this type of fractionation has been in use for over 40 years in Europe and for at least 20 years in New Zealand. Both single-stage and multi-stage processes are possible using the same equipment. Although Australia and the United States produce significant quantities of MF products and have experimented with fractionation, to date only the United States has a commercial MF fractionation plant. Melt fractionation of MF has been the subject of several reviews in recent years (Deffense, 1993, 2000; Kaylegian & Lindsay, 1994).

In order to ensure that the fractionation process proceeds without problems, the quality of the fat to be fractionated is of critical importance (Illingworth & Hartel, 1999). Care must be taken to ensure that water and traces of protein are absent since they interfere with the nucleation and crystallisation processes. In addition, the fat must be deaerated since dissolved air can also interfere with the nucleation process. Thus, the polishing and dehydration stages of the MF manufacturing process must be controlled closely.

The most widely used process is that developed by S.A. Fractionnement Tirtiaux – the first people to recognise both the importance of controlling the temperature of the crystallising fat, and the value of using this to dictate the overall efficiency of the process (Tirtiaux, 1976). Controlled agitation of the crystallising mass assists in producing crystals of optimum size and number to facilitate good filtration. Because of the slow nature of TAG crystallisation, long crystallising times (24–36 h) are necessary. A schematic diagram of the arrangement

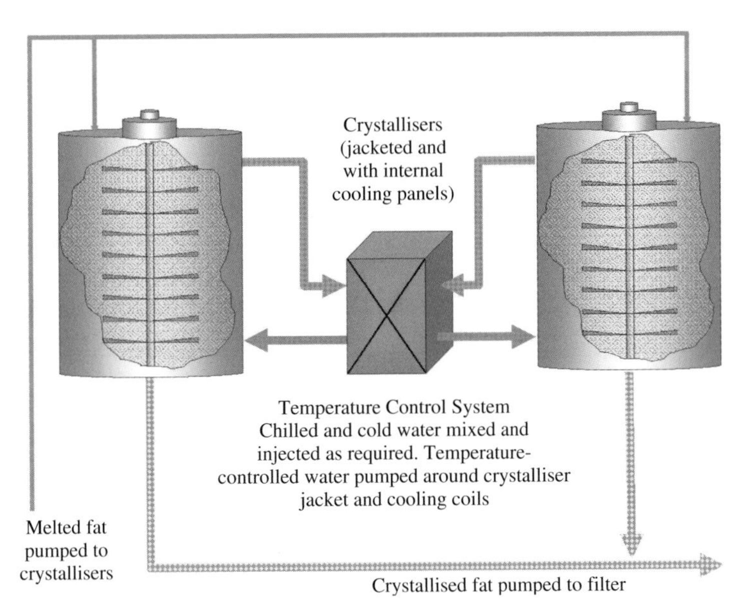

Fig. 6.7 Schematic diagram of the Tirtiaux crystallisation process showing the temperature control regime for ensuring the crystallisation proceeds in a controlled manner. Reproduced with permission of S. A. Fractionnement Tirtiaux, Fleurus, Belgium.

of crystallisers and temperature control systems is shown in Figure 6.7, and it also shows how water at two temperatures is mixed to allow very precise control of the temperature as the crystallising MF is cooled to the final temperature where filtration is carried out. This arrangement is very similar to the Tirtiaux crystallisation process (Gibon & Tirtiaux, 2000) referred to earlier.

The de Smet process uses similar control principles, but applies them in a different way (van Dam *et al.*, 1998), achieving optimum crystallisation in much less time (8 h) than the Tirtiaux process. The temperature of the crystallising fat is reduced in a series of steps leading to exponential cooling. The number and size of the steps are dictated by the fat being crystallised and the final fractionation temperature desired.

Because fats tend to have an inherent stability in the liquid state (i.e. ability to supercool), it is necessary to use temperature reduction and agitation to overcome this and allow the crystallisation process to begin via the essential steps of nucleation and growth. Removal of sensible heat prior to any nuclei being formed may be done rapidly, allowing fast agitator speeds. While nuclei are being formed, it is essential that the cooling rate and the agitator speed are controlled so that the number of nuclei is optimised, and the small crystals that form initially are not damaged by collision to form more crystals. During the final growth stage, while controlling the temperature and bringing the crystallising mass to the final fractionation temperature, the agitator speed must be sufficient to keep crystals in suspension, yet slow enough to prevent crystal damage and formation of a second crop of smaller crystals that may impede efficient separation.

To assist the nucleation stage, seed material may be added to the crystallising mass at a certain point to overcome the inherent stability of the liquid phase as mentioned earlier. If high-melting TAGs are present in the crystallising mass, this step is not always necessary, but once these have been removed, as in a multi-stage process where much softer olein fractions are sought, then the addition of seed material may be advisable (Gibon, 2005).

For the separation of stearin crystals from the liquid olein, various filter systems are available commercially, but probably the most commonly used today in melt fractionation is some form of plate-and-frame membrane filter. Unlike the simple plate-and-frame filters used, for example, for filtering edible oil after the bleaching process, the plates are made with flexible membranes (Figure 6.8) that allow the crystalline stearin to be squeezed to improve the efficiency of the separation and increase the yield of the olein fraction – an important consideration when a multi-stage fractionation is being used. Two types of filter press are available: (a) low-pressure presses that use compressed air and (b) high-pressure presses that rely on hydraulic pressurisation. The operation of the press is relatively simple; the plates are held together under pressure to create a series of chambers that are then filled through either corner or central channels. The chambers gradually become filled with solid crystals that form into a stearin cake as the olein passes through a filter cloth supported by the plates. The dimples in the surface of the flexible membranes allow the olein to drain towards the drain holes at the bottom. In turn, these connect to the main olein drain channel. When all chambers are filled, the feed and olein channels are blown clear with nitrogen, then pressure is applied (600–3000 kPa) to inflate the membranes and squeeze more olein from the cake. The pressure is then relieved and the press is opened allowing the cakes to drop from the press (Figure 6.9).

Filters, such as the Statofrac filter from Krupp (Willner *et al.*, 1990), routinely use pressures of up to 3000 kPa to achieve efficient filtration. As a result, the quality of stearins that can be obtained from dry or melt fractionation, coupled with membrane filtration, is comparable with those obtained from solvent fractionation, and as might be expected, the yield of olein is markedly increased over vacuum or detergent separation.

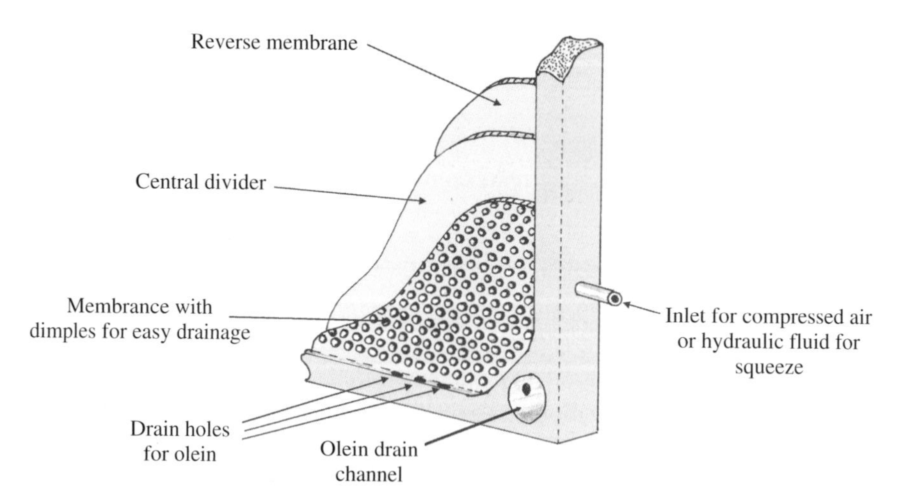

Fig. 6.8 A section of a membrane filter plate showing part of the flexible membrane that supports a filter cloth, together with one of the olein channels and the port that allows the membrane to be expanded under pressure. Reproduced with permission of S. A. Fractionnement Tirtiaux, Fleurus, Belgium.

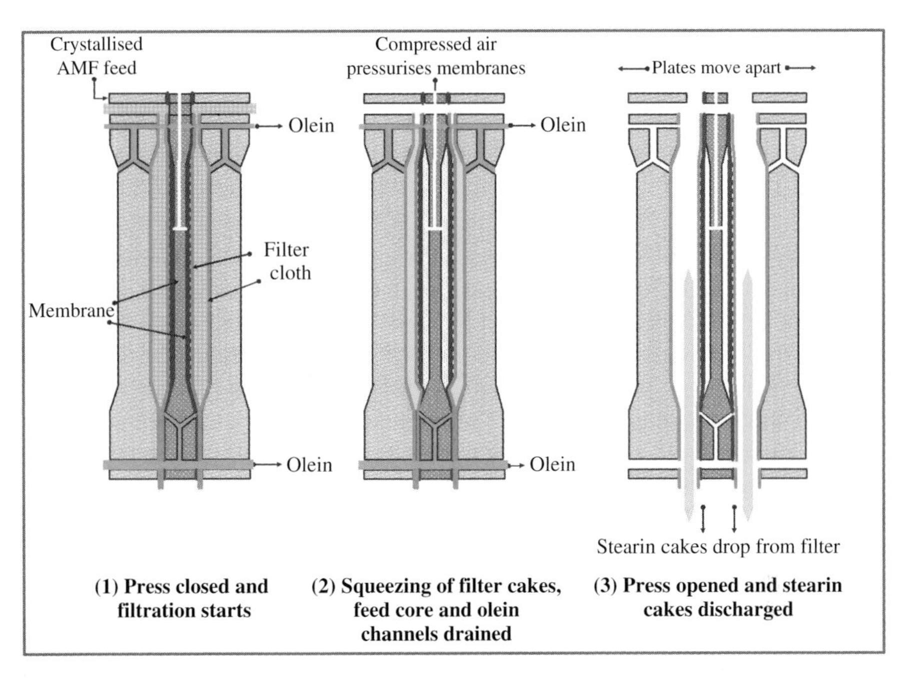

Fig. 6.9 Diagram showing the operation sequence for a membrane filter press used for filtering partially crystallised fat during fractionation. 1, filling the filter with filtration of the olein from the crystals; 2, squeeze applied to remove more olein; 3, filter is opened to allow stearin filter cakes to drop into hopper below. Redrawn from Gibon (2005).

Melt fractionation with additives (detergent separation)

One of the problems associated with melt fractionation is the entrainment of olein in the stearin crystals and the filter cake. By mixing the crystallised fat with an aqueous detergent solution, the problem is largely overcome as the crystals are wetted by the detergent and pass into the aqueous phase where they can be separated. This was first demonstrated by Lanza in 1905. Tetra Pak (previously known as Alfa Laval) has marketed this process under the name Lipofrac (Fjaervoll, 1970a, 1970b) for many years. It was used both experimentally and commercially for MF in New Zealand in the 1970s and early 1980s (Jebson, 1970). The process fell out of favour when the International Dairy Federation discouraged the use of detergent-separated fractions, with a number of countries legislating against the use of MF fractions manufactured by this process in spreads and other fat products.

Solvent fractionation

Fractionation using a solvent, such as acetone, is the most efficient of all fractionation options. However, all the electrical equipment (motors, wiring and lighting) for the plant require flame-proof installation and the equipment itself (crystallisers and filters) must be sealed to prevent solvent vapours entering the environment. Ideally, the building housing the plant should be operated at reduced pressure. As a result, solvent fractionation is a very expensive option. Its major advantage is washing the crystals during filtration, which allows entrained olein TAGs in the stearins to be removed and improves the separation. Its other major advantage over the batch operation that is currently the mode of melt fractionation is that it lends itself to continuous processing, which is not currently feasible for melt fractionation because of high viscosities in the liquid phases, particularly at low temperatures. However, a process for continuous fractionation of MF from the melt has been proposed (Breitschuh, 1998; Breitschuh *et al.*, 1999).

The most selective solvent for fat fractionation is acetone. Acetone fractionation of MF was the basis of a New Zealand patent for a spreadable butter in the 1970s (Norris, 1976). To improve the economics of acetone fractionation the solvent must be recoverable for reuse. Undesirable flavours, such as skatole and indole, tend to remain with the acetone and, thus, eventually would contaminate the fractions. It is also necessary to completely eliminate the traces of solvent from the fractions. In the edible oil industry, where fractions undergo a full refining, the processing results in bland oils or fats. To apply this to MF fractions would remove all the desirable flavour compounds that are essential in preserving their dairy identity.

Super-critical fractionation

Fractionation from super-critical carbon dioxide was considered by some researchers as an ideal process for MF fractionation (Bhaskar, 1997; Bhaskar *et al.*, 1993, 1998) with the main attraction being the removal of cholesterol at a time when cholesterol was linked specifically to the cause of heart disease and arterial deterioration.

Current commercial fractionation technologies using melt crystallisation without additives are far cheaper and easier to operate, and yield a range of fractions with properties that can be applied to most applications.

Other fractionation technologies

Illingworth (2002) describes other fractionation technologies that have been mooted for edible oils including MF. However, to date, none of these has emerged as commercially viable other than the use of nozzle centrifuges for the separation stage of the process (Eyer, 2000; Gibon, 2005). Separation by basket centrifuge has also been claimed to be superior to other separation means (Dijkstra & Maes, 1983; Maes & Dijkstra, 1985).

6.5.2 *Fraction properties*

A three-stage fractionation of MF from the melt at 22°C, 14°C and 7°C without any additives (Figure 6.10) can be used to demonstrate how the properties of the fractions differ in their relationship to each other, and to those of the parent MF (Illingworth *et al.*, 1990). MF fractionated at 22°C yields a stearin fraction (H) and an olein fraction (S). The latter is then refractionated at 14°C to produce a second stearin (SH) and second olein (SS). The second olein is then fractionated at 7°C to produce a third stearin (SSH) and a third olein (SSS).

Chemical composition of MF and its fractions

The first thing to consider is how the various fatty acids are distributed during fractionation. Gas chromatography of the fatty acid methyl esters allows the H, SH and SSH fractions

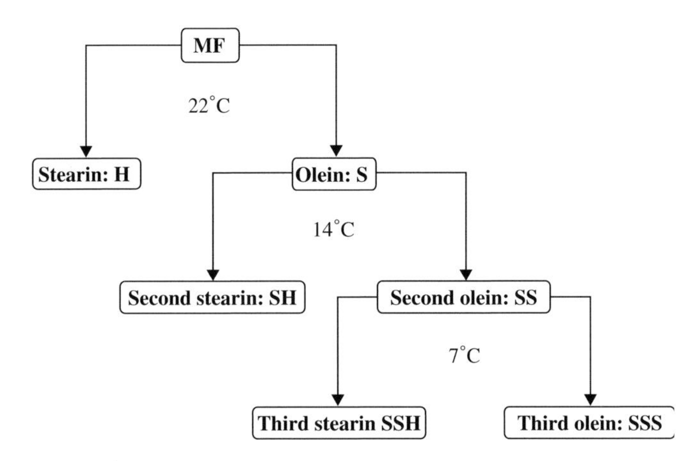

Fig. 6.10 A typical three-stage fractionation scheme for milk fat (MF) at 22°C, 14°C and 7°C to produce a range of fractions with different properties.

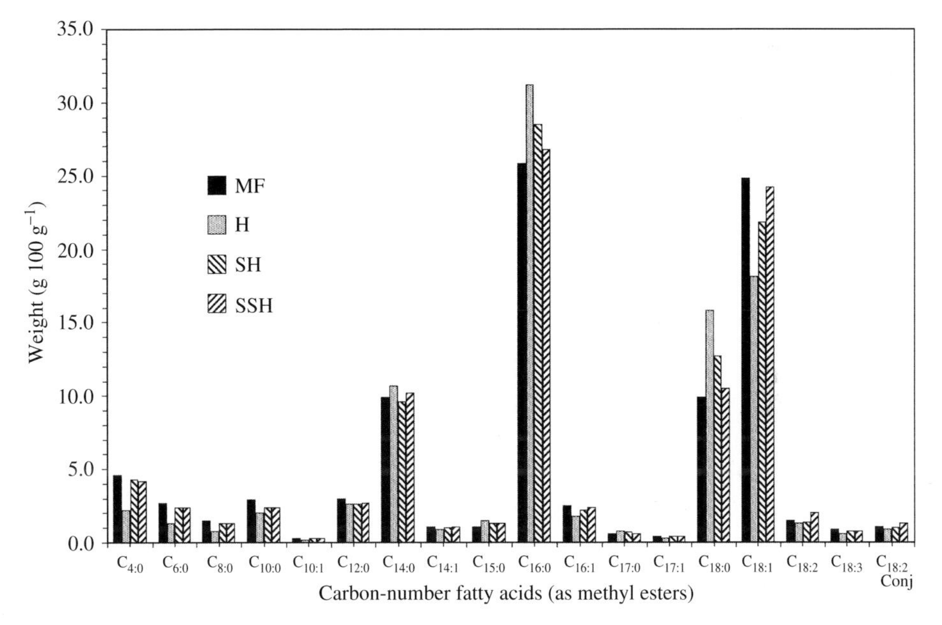

Fig. 6.11 Fatty acid methyl ester composition of typical stearin fractions obtained from three-stage fractionation of MF at 22°C, 14°C and 7°C.

obtained from the three-stage process with the parent MF to be compared (Figure 6.11). The H stearin, while being depleted in the short- and medium-chain fatty acids and long-chain unsaturated fatty acids (e.g. oleic, $C_{18:1}$), is enriched in long-chain fatty acids, such as palmitic ($C_{16:0}$) and stearic ($C_{18:0}$). The distinctions among MF, SH and SSH are less pronounced.

Similarly, the olein fractions from each stage, S, SS and SSS show a gradual depletion in $C_{16:0}$ and $C_{18:0}$ with enrichment in $C_{18:1}$ (Figure 6.12). It is also possible to distinguish how different classes of fatty acids are distributed during fractionation (Figures 6.13 and 6.14 and Table 6.3).

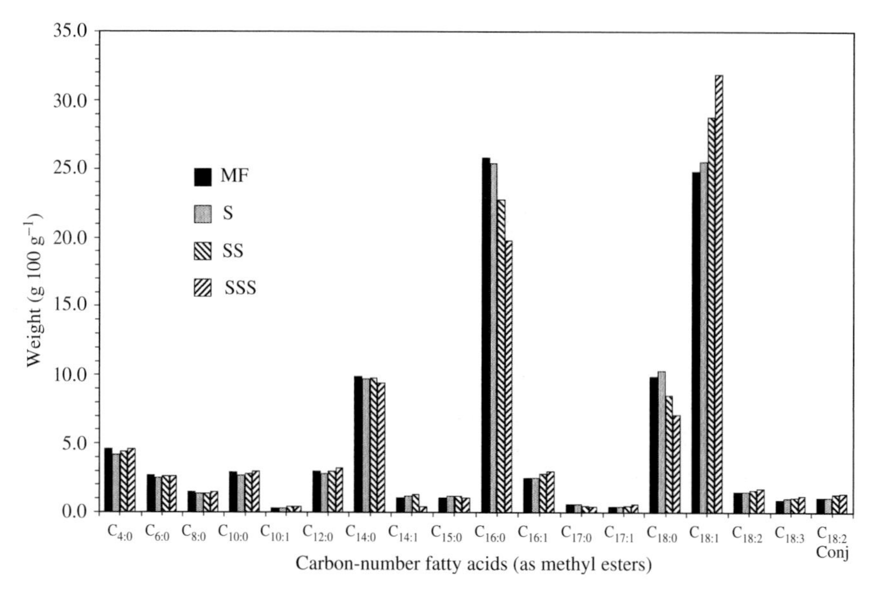

Fig. 6.12 Fatty acid methyl ester composition of typical olein fractions obtained from three-stage fractionation of MF at 22°C, 14°C and 7°C.

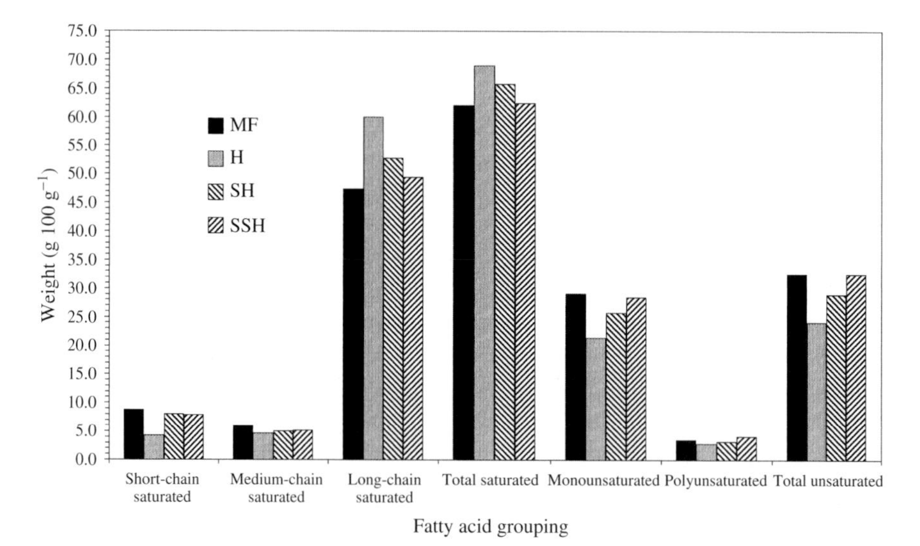

Fig. 6.13 Distribution of fatty acid types between stearins after three-stage fractionation of MF at 22°C, 14°C and 7°C.

As might be expected, there is a tendency for the stearin fractions to be richer in the saturated fatty acids, particularly the long-chain ones that tend to be in TAGs with higher melting points. Short-chain saturated fatty acids tend to be evenly distributed and have levels similar to that of the parent MF, apart from the H fraction, the only fraction that is significantly depleted. Medium-chain saturated acids are more or less evenly distributed between all the fractions. On the whole, unsaturated fatty acids tend to favour the olein

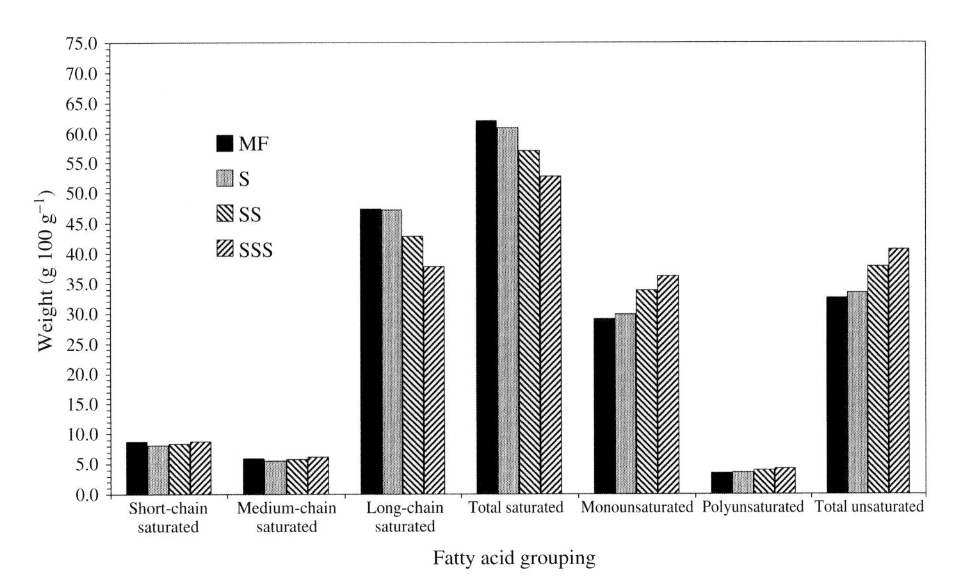

Fig. 6.14 Distribution of fatty acid types between oleins after fractionation of MF at 22°C, 14°C and 7°C.

Table 6.3 Fatty acid (FA) distribution in anhydrous milk fat (AMF) and fractions.

	Saturated FA				Unsaturated FA		
Fraction	Short-chain	Medium-chain	Long-chain	Total	Mono-	Poly-	Total
AMF	8.8	5.9	47.3	62.0	29.1	3.5	32.6
Stearin (H)	4.3	4.6	60.0	68.9	21.3	2.8	24.1
Olein (S)	8.1	5.5	47.2	60.8	29.9	3.6	33.5
2nd refractioned stearin (SH)	8.0	5.0	52.8	65.8	25.7	3.2	28.9
2nd refractioned olein (SS)	8.4	5.8	42.8	57.0	33.8	4.0	37.8
3rd refractioned stearin (SSH)	7.9	5.1	49.4	62.4	28.4	4.1	32.5
3rd refractioned olein (SSS)	8.7	6.2	37.8	52.7	36.3	4.3	40.6

fractions, although because of the nature of the TAGs in MF, there are unsaturated fatty acids in all the fractions. This is also a reflection of the entrainment or contamination of the stearin fraction at each stage with TAGs that are really part of the olein fraction, but which cannot be fully eliminated because of the constraints of the filtration system.

Gas chromatography also demonstrates how the many TAGs present in MF are distributed during fractionation. The separation by melting point especially for the H and S fractions is very clear (Figures 6.15 and 6.16), with again less differentiation between the fractions obtained at the lower fractionation temperatures.

Thus, in terms of their chemical compositions, fractions obtained from MF appear to be very similar to the parent fat. For the most part, the compositions are within the accepted variation of MF composition that occurs because of seasonal and other environmental changes. In fact, many producers use this in marketing MF fractions, claiming that baked goods made from them may be described as 'made from butter'.

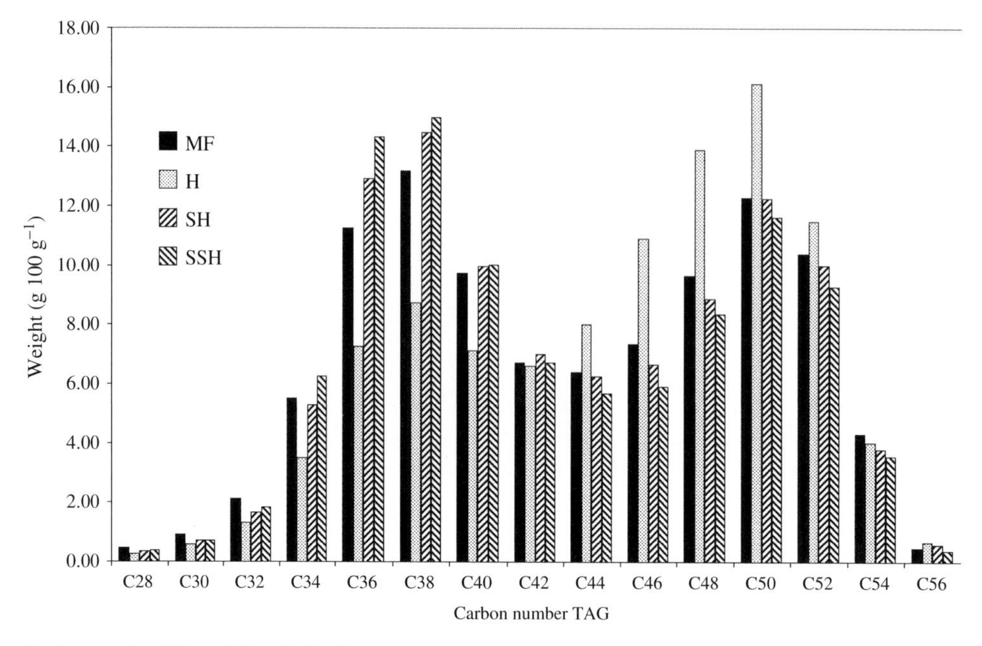

Fig. 6.15 Triacylglycerol (TAG) distribution in stearins from three-stage fractionation of milk fat at 22°C, 14°C and 7°C.

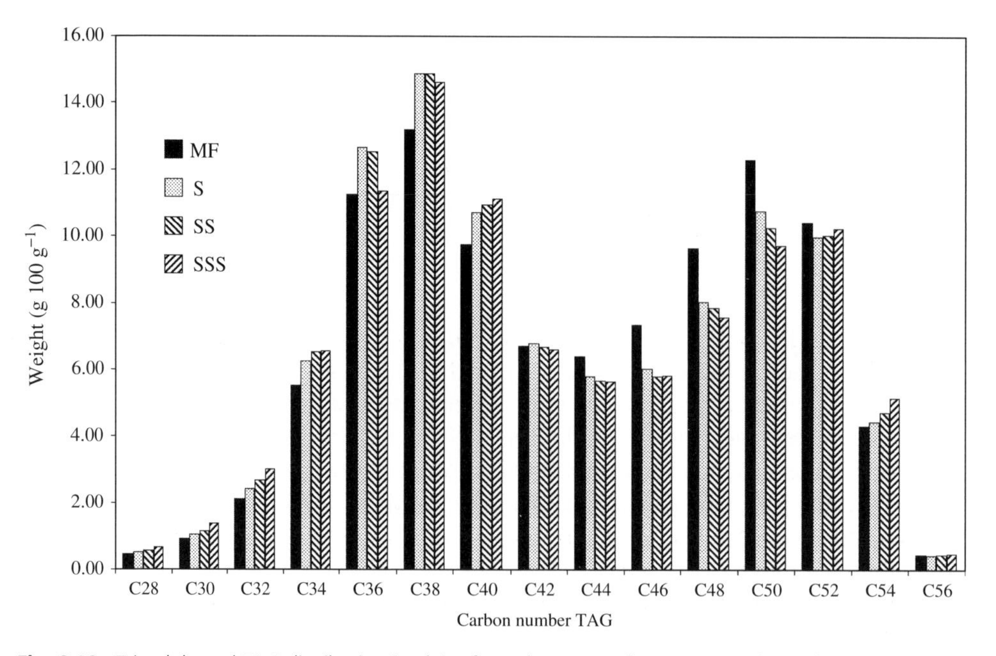

Fig. 6.16 Triacylglycerol TAG distribution in oleins from three-stage fractionation of milk fat at 22°C, 14°C and 7°C.

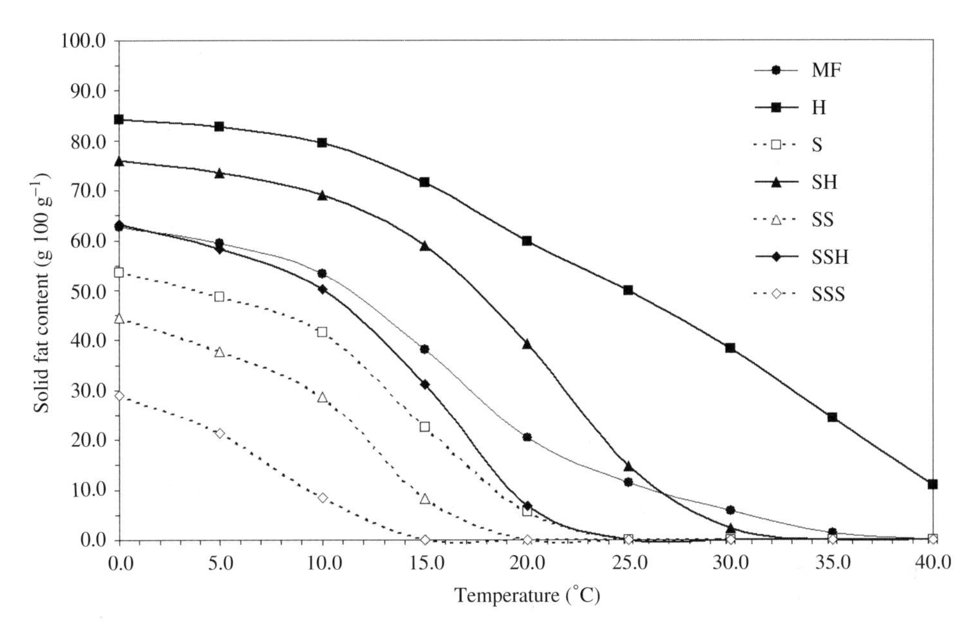

Fig. 6.17 Solid fat content (SFC) profiles of milk fat and fractions obtained from three-stage fractionation at 22°C, 14°C and 7°C.

Physical properties of MF and its fractions

It is the physical properties of the fractions as shown by their melting behaviour that truly distinguish them from each other. The clear distinction in the melting properties is self-evident from solid fat content (SFC) profiles (Figure 6.17) as determined by low-resolution-pulsed nuclear magnetic resonance (NMR) (MacGibbon & McLennan, 1987), and shows how a range of fats, all with a true dairy identity, can be manufactured from a single MF source. Compared with the parent MF, the H fraction has a higher SFC at all temperatures with SH and SSH showing steeper melting behaviour than either MF or H. In fact, SSH is softer than MF at all temperatures other than 0°C. All the olein fractions have SFC melting profiles that reflect the fractionation temperature used.

Differential scanning calorimetry (DSC) as described by MacGibbon (1988) is another technique than can be used to differentiate between MF and its fractions. For example MF, (Figure 6.18) is shown to contain three distinct TAG types: high-, intermediate- and low melting. Fractionation of the MF demonstrates that high-melting TAGs concentrate into the H fraction, shown by the large peak. The smaller peak is due to lower-melting TAGs that are entrained in the crystals of the H fraction. The S fraction has two melting peaks corresponding to the intermediate- and low-melting TAGs that do not crystallise under the conditions used.

These conditions for multi-stage fractionation of MF are only one of many combinations of processing conditions that can be used. The importance of the fractionation temperature and temperature control during any of the stages cannot be underestimated. For any stage, the fractionation temperature can be changed to enable fractions with specific melting behaviours to be made and, therefore, to suit the intended application(s) for the fractions as will be discussed in Chapter 7. In addition, more stages can be added to refine the separation, but this will of course affect the cost of the process.

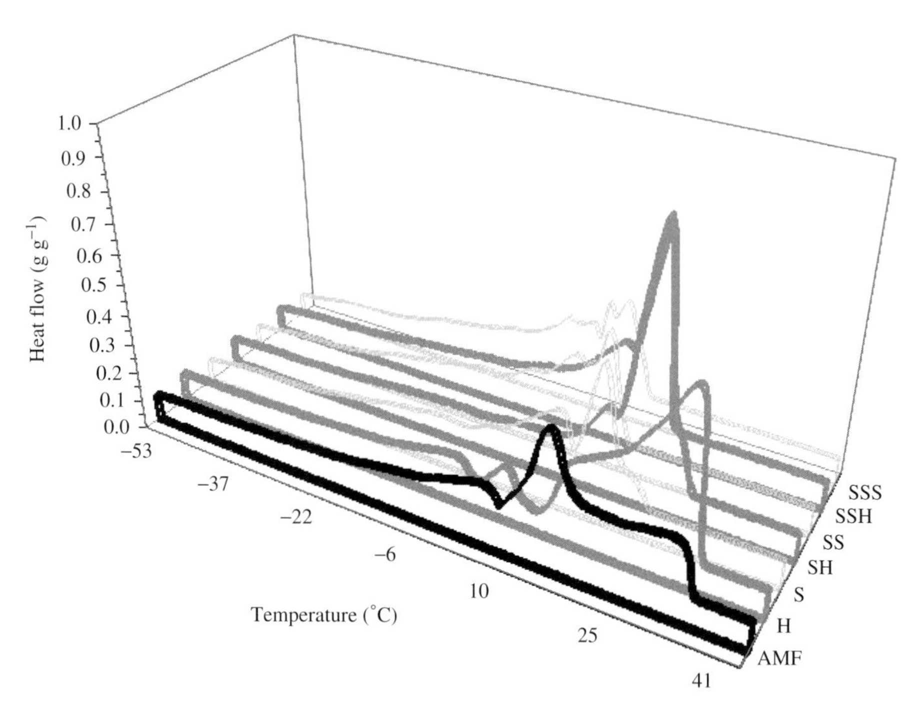

Fig. 6.18 Differential scanning calorimetry (DSC) profiles of MF and fractions obtained from fractionation at 22°C.

Costs for fractionation from the melt are significant (Kellens, 2000), but not as expensive as other modification processes, such as hydrogenation and interesterification, used in the edible oil industry but not applicable to MF because they affect the flavour.

6.6 Ghee

6.6.1 *Introduction*

Ghee, also called *desi* (indigenous) ghee, is the most widely used milk product in the Indian subcontinent. Beginning from almost Vedic times (3000 to 2000 BC), there is ample recorded evidence to show that *makkhan* (indigenous butter) and ghee were extensively used by the early inhabitants of India both in their dietary and religious practices. The *Rigveda*, which is the oldest collection of Hindu hymns, contains numerous references on ghee (Achaya, 1997). Ghee occupies a very significant place in the Indian diet. Ancient Sanskrit literature describes ghee (synonyms are *ghrit, ghritam, nai, sarpi* or *sarpish, havi* or *havish* and *ajya*) as the food fit for the gods and a commodity of enormous value. Ghee, therefore, is considered an ideal offering to deities and is used for performing *yagna* in all the religious ceremonies of the pre-Vedic and Indo-Aryan traditions. Two highly auspicious beverages (*madhuparka* and *panchagavya*) of ritual significance have ghee as one of the five essential components. From *ayurved* to food craft to philosophy, ghee constitutes an important part of Indian life. Nevertheless, India utilises 28% of the milk produced for the manufacture of ghee and, in

2002, it was estimated that the annual production figure was 900 000 tonnes, which was valued at Rs. 135 000 million (Kurup, 2002).

The products similar to ghee have been available in other parts of the world probably since equally ancient times and known as *samna* in Egypt (Abou-Dhonia & El-Agamy, 1993), *meshho* in ancient Assyrian empire (2400 BC to 612 BC) (Abdalla, 1994), *samin* in Sudan (Hamid, 1993), *maslee* or *samn* in Middle East, *rogan* in Iran (Urbach & Gordon, 1994), and *samuli* in Uganda (Sserunjogi *et al.*, 1998). Ghee is also gaining popularity in Australia, Arabian countries, the United States, the United Kingdom (UK), Belgium, New Zealand, Netherlands and many other African and Asian countries.

Several definitions of ghee have been proposed (Ganguli & Jain, 1972; IDF, 1977; Singh & Ram, 1978; Chand *et al.*, 1986; Munro *et al.*, 1992; Kumar & Singhal, 1992; Bajwa & Kaur, 1995; Singh *et al.*, 1996), most of which are ambiguous and fail to adequately distinguish ghee from anhydrous butterfat or butteroil. Codex Alimentarius (FAO/WHO, 1997, 2006) defined ghee as a product exclusively obtained from milk, cream or butter by means of processes, which result in almost total removal of water and non-fat solids, with an especially developed flavour and physical structure. Since ghee is primarily distinguished from butteroil by its flavour characteristics, the definition proposed by Codex Alimentarius (FAO/WHO, 1997, 2006) describes the product more closely. However, considering ghee as a product in its own right, it can more appropriately be defined as a pure clarified fat exclusively obtained from milk, cream or butter, by means of processes involving application of heat at atmospheric pressure, which result in the almost total removal of moisture and SNF and which gives the product a characteristic flavour and physical structure and texture.

A major portion of ghee is utilised for culinary purposes, for example, as a dressing for various foods and for cooking and frying of different foods. It is considered as the supreme cooking or frying medium. In the Indian context, ghee is a product of the sacred cow, 'born of fire', and hence pure and confers purity to other foods when used in making them. Almost the entire range of Indian sweets, prepared with admixtures of milk, cereals, fruits, vegetables and nuts, are preferably cooked using ghee as the medium. Ghee is one of the four basic elements in the Indian cooking. Sweets and meals cooked in ghee enjoy special status and are recognised for their distinguished flavour attributes derived from ghee. Meals for special occasions are cooked in ghee.

In many rural families, ghee mixed with cane sugar (*boora, khand* or *shakkar*) is relished almost on all occasions as a befitting dessert to a good meal or even as a snack food. In its table use, ghee is served in hot melted form and used for garnishing rice or spreading lightly on *chapatis*. In India, ghee is considered as a sacred item and used also in religious rites (Rajorhia, 1993). Ghee is also used in *Ayurvedic* system of medicine. There exists a separate therapy called *Govaidak*, which uses several ghee preparations for the treatment of various diseases (Adhvaryu, 1994).

Uses of various products related to ghee have been documented from different parts of the world. *Meshho*, a traditionally Assyrian product, is added to dishes mainly as a garnish (Abdalla, 1994). In Sudan, *samin* is mainly used as a topping for mullah, a type of sauce normally made from a variety of ingredients. *Samin* is also drunk as is, usually in small quantities such as a coffee cupful every morning. Other uses of *samin* are that it is fed to children in a pure form or mixed with food, as a relish and as a topping for coffee or tea, or for therapeutic purposes. A mixture of honey and *samin* is believed to be very nutritious and an effective aphrodisiac (Hamid, 1993). In Uganda, *samuli* is basically used for cooking and frying various foods.

A considerable amount of ghee is consumed in many parts of the world. In India, the bulk of ghee is produced by indigenous methods. The consumption of ghee (*meshho*) by an average Assyrian family is estimated to be about 60 kg every year (Abdalla, 1994). In Sudan, in the mid-1980s, the total annual consumption of *samin* was estimated at 4500 tonnes in the Khartoum province alone (Hamid, 1993). The consumption figures for most other regions of the world, where ghee and related indigenous products are popular, are not readily available. Rajorhia (1980) reported that consumer preferences for ghee in India vary from region to region (Table 6.4).

6.6.2 *Methods of manufacture*

Ghee may be produced through heat clarification of cream or via conversion into butter, followed by heat desiccation. The first step of ghee manufacture involves the preparation of the raw material, that is, whole milk, *malai* (clotted cream), cream or *makkhan* or butter. If milk is the starting material, it is normally allowed to ferment to produce *dahi* (fermented milk) before it is churned to produce *makkhan*. It is also preferable to use soured cream; otherwise the resultant ghee is regarded as flat and tasteless (Urbach & Gordon, 1994). To convert butter or cream to ghee, heat is applied at controlled temperatures at the various stages of processing. Warner (1976) described the stages of heat clarification of butter or cream into ghee. Initially, the temperature is gradually raised to about the boiling point of water while stirring to control frothing. In the second stage, most of the free water evaporates, which requires a considerable amount of heat. As most of the water evaporates, the rate of heating is controlled and the temperature is maintained at about 103°C to prevent the charring of the SNF, which results in the development of bitter flavours and/or a brown colour.

Overheating could drive off desirable volatile flavour materials and also impair the formation of suitable grains upon cooling. The impairment of crystal formation appears to be associated with the possible volatilisation of some short-chain FFA, which changes the

Table 6.4 Regional preference for ghee flavour and texture in India.

Region	Physical character	Comments
Northern India	Flavour	Slightly acidic, mildly curdy
	Texture	Fine- to medium-size grains (half to three quarters solid portion)
Western India	Flavour	Mildly curdy (very curdy in Saurashtra)
	Texture	Coarse grains, i.e. size of 0.3 to 0.6 mm
Southern India	Flavour	Mild to highly cooked and aromatic, higher level of free fatty (butyric) acid, preference for special herb flavours in Tamil Nadu and Karnataka
	Texture	Medium sized grains in Tamil Nadu, coarse grains in Andhra Pradesh and Karnataka
Eastern India	Flavour	Slightly to definitely cooked flavour
	Texture	Medium grains (one quarter liquid and three quarters solid)

Data compiled from Rajorhia (1980).

normal composition of the fat in ghee (Warner, 1976). Finally, the temperature is raised to between 105 and 118°C with constant agitation in order to remove the water bound to the SNF and to develop the characteristic flavour. In general, a temperature range of 110–120°C is preferred. However, the final temperature to which ghee is heated during manufacture depends upon the region of the country; normally the temperature is around 110°C (or below) in north India and 120°C (or even higher) in south India. A lower heating temperature improves the colour, but decreases the keeping quality of the ghee obtained due to its greater residual moisture content. A higher temperature, on the other hand, tends to reduce the vitamin A content (with acid butter) and darken the colour, but increases the keeping quality of the finished product (Sethna & Bhatt, 1950).

Methods of ghee manufacture vary with respect to the material used (milk, cream and butter), the intermediate treatment of raw materials and the handling of the semi-finished or fully formed ghee. There are four methods for the production of ghee: (a) the indigenous milk butter (MB) method, (b) the direct cream (DC) method, (c) the cream butter (CB) method and (d) the pre-stratification (PS) method as illustrated in Figure 6.19. In India, where commercial ghee manufacture is well developed, the most common method of producing ghee is by a batch process, whereby butter or cream is heated in steam-jacketed stainless steel vessels of 500–1000 kg capacity (Achaya, 1997). Because of commercial importance of ghee for the Indian dairy industry, considerable refinement and mechanisation have taken place in its manufacturing processes. Of the various indigenous milk products, ghee production has received maximum research and development (R&D) inputs. Attempts have been made to modify, scale up and adopt the traditional batch process for commercial production. Increased awareness about energy management in the past motivated the research works to develop energy-efficient and continuous methods for ghee manufacture (Punjrath, 1974), which use either an oil separator (Bhatia, 1978) to separate serum and fat phase or scrapped surface heat exchangers (Abichandani *et al.*, 1995). Both the processes save energy and yield a comparable product. Comparison of various methods of manufacture of ghee is given in Table 6.5. Some such processes, developed for ghee production on cottage and commercial scales are described in the following text.

Indigenous (desi) milk-butter (MB) method

The indigenous methods (Figure 6.19) of ghee making usually involve the following different processes: (a) direct churning of raw milk, (b) lactic acid fermentation of heat-treated milk for converting milk into *dahi* (i.e. Indian fermented milk) followed by churning, or (c) skimming off the thick clotted cream layers (*malai*) formed at the air–liquid interface of milk, which is heated above 90°C followed by grinding of clotted cream, its dispersal in water and finally, churning.

The direct churning of raw milk method, as applied in the homes in India, involves the souring of raw milk in earthenware vessels that have been used previously as a milk container, and which contain bacteria within the pores of the wall of the container. After addition of more milk over successive days, the *dahi* is churned to obtain *makkhan* (Munro *et al.*, 1992; Podmore, 1994). Hand-driven wooden beaters are usually employed for separating the *makkhan*. After accumulating a sufficient quantity over a period of few days, the *makkhan* is melted in a metal pan or an earthenware vessel on an open fire until almost all the moisture has been removed. During the initial stage of heating the *makkhan*, extensive frothing takes place, which must be controlled to avoid losses associated with boil-over. As the moisture evaporates and frothing

Table 6.5 Comparison of manufacture of ghee by different methods.

Particular	Milk butter (*desi*)	Cream butter (CB)	Direct cream	Prestratification	Continuous
Fat recovery (%)	88–90	88–92	92	>93	>93
Aroma	Strong nutty	Pleasantly rich	Pleasantly rich	Pleasantly rich	Mild
Flavour	Acid	Normal	Normal	Normal	Flat
Texture	Packed coarse grains	Slushy fine grains (cow) or packed fine grains (buffalo)	Fine grains	Fine grains	Greasy
Clarification using heat	Easy, economic and prestratification possible	Easy, economic and prestratification possible	Easy and economic	Easy and economic	Easy and economic
Essential equipment	Butter churn	Cream separator and butter churn	Cream separator and butter churn	Cream separator and butter churn	Scraped surface heat exchanger (SSHE)
By-product(s)	Buttermilk and ghee residues	Skimmed milk, buttermilk and ghee residues	Skimmed milk, buttermilk and ghee residues	Skimmed milk, buttermilk and ghee residues	Skimmed milk, buttermilk and ghee residues
Adaptability	Small scale	Large scale	Large scale	Large scale	Very large scale
Keeping quality	Poor	Good	Good	Good	Good

Data compiled from Rajorhia (1993) and Aneja *et al.* (2002).

subsides, the temperature begins to rise above 100°C. When it reaches about 100°C, slight caramelisation of curd particles is noticeable. At this stage, frothing completely subsides and the emission of moisture bubbles also ceases. Extent of frothing may be used as an index to judge when to terminate the heating. Heating is discontinued as soon as the curd particles attain the desired golden yellow or brown colour. After this, the contents of the kettle are left undisturbed till the residue settles down. Clarified fat is decanted off by tilting the kettle. Ghee is then pooled for final packing. The ghee produced by this method has a highly desirable flavour, body and texture. However, this method leaves behind a large quantity of ghee residue and also leads to low fat recoveries (88–90%). Nevertheless, village ghee constitutes the major share of base material used for the blending operations at ghee grading and packing centres functioning under the Agricultural Marketing Grading (AGMARK) scheme in India.

Creamery butter (CB) method

This is a three-step process of ghee manufacture whereby cream is separated from the milk, churned into butter or butter granules and then processed into ghee (Figure 6.19). This is the

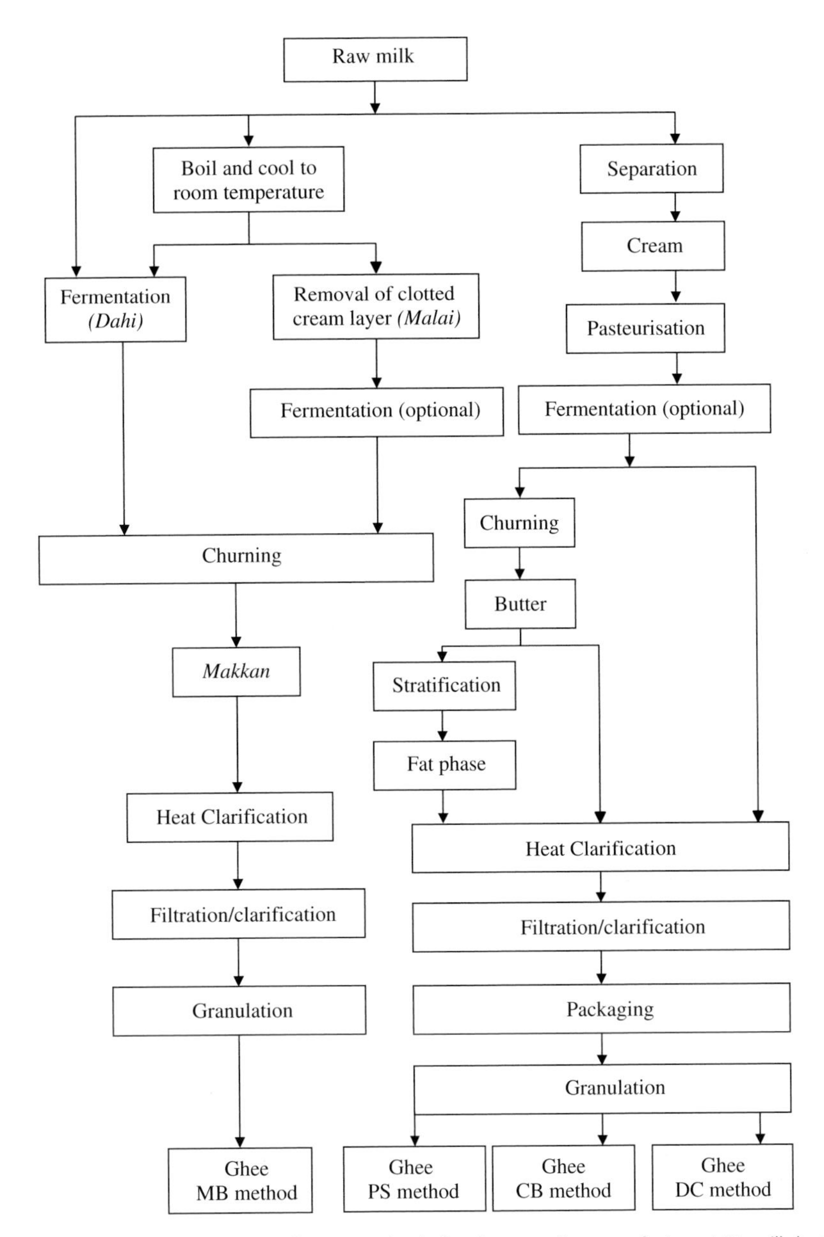

Fig. 6.19 Flow diagram illustrating different methods for the manufacture of ghee. MB, milk butter; CB, cream butter; DC, direct cream; PS, prestratification.

standard method adopted in almost all large dairies in India. A typical plant assembly for the CB method comprises the following:

- A cream separator
- Butter churn
- Butter melting equipment

- Steam-jacketed, stainless steel ghee kettle with agitator and process controls
- Ghee filtration devices, such as disc filters or oil clarifier
- Storage tanks for cream, butter and ghee
- Pumps and pipelines interconnecting these facilities
- Crystallisation tanks
- Product filling and packaging lines

First, the butter mass is melted at 60°C, pumped into the ghee heating kettle, and the steam pressure increased to raise the temperature to 90°C. This temperature remains constant as long as the moisture is being driven off. The contents are constantly agitated throughout the process of conversion of butter into ghee, to prevent scorching. The 'scum', which collects on the top surface of the product, is removed from time to time with the help of a perforated ladle. Usually, there is profuse effervescence accompanied by a crackling sound in the early stages of boiling, but both gradually decrease as the moisture content is reduced. When practically all the moisture has been driven out, the temperature of the liquid mass suddenly shoots up and the heating at this stage has to be carefully controlled. The end point shows the disappearance of effervescence, appearance of finer air bubbles on the surface of the fat and browning of the curd particles. At this stage a characteristic ghee flavour also emanates, and this is an indication that it has been heated sufficiently. The final temperature of heating/clarification usually ranges from 110 to 120°C. The ghee is then pumped, via an oil filter or clarifier, into settling tanks, which is cooled by circulating water at 60°C inside the jacket of the tank. Afterwards, the product is transferred for packaging and granulation (Rajorhia, 1993). A fat recovery of 88–92% has been reported when ghee is manufactured by the CB method.

DC method

This process omits the need for production of butter because cream is directly converted into ghee (Figure 6.19). The fresh cream obtained by centrifugal separation of whole milk, cultured cream or washed cream is heated to 115°C in a stainless steel, jacketed ghee kettle fitted with an agitator, steam control valve, pressure and temperature gauges and a movable, hollow, stainless steel tube centrally bored for emptying out the contents. Alternatively, provision can be made for a tilting device on the ghee kettle to decant off the product. Heating is discontinued as soon as the colour of the ghee residue turns to golden yellow or light brown. The ghee residue is used for animal feeding. One of the limitations of the DC method is that it requires a long heating time to remove the moisture, and the ghee produced has a slightly greasy texture. A high content of serum solids in the cream may also produce a highly caramelised flavour in the ghee and lead to about 4–6 g 100 g^{-1} loss of butter fat in the ghee residue or during handling operations, depending upon the fat percentage in the cream. A negative correlation ($r = -0.44$) exists between the SNF content of cream and recovery of ghee, whereas a positive correlation ($r = 0.64$) between the SNF content of cream and fat loss has been well established (Pal & Rajorhia, 1975). The use of plastic cream or washed cream with about 75–80 g fat 100 g^{-1} is recommended for minimising both fat loss and steam consumption. De *et al.* (1978) observed that, as the fat level in the cream increased from 40 to 80 g 100 g^{-1}, the ghee residue gradually decreased from 10.4

to 3.0 kg 100 kg^{-1} cream and the fat recovered in the ghee increased from 85 to 96.6%. However, when low-SNF cream is used, the final product will have a flat flavour. Ripening of the washed cream or plastic cream with lactic starter culture to an acidity level of 0.20 g 100 mL^{-1}, or acidification of the cream with citric acid to the same level prior to clarification as above, improves the flavour of the finished ghee (De, 1980).

Pre-stratification (PS) method

This method, also known as the *clarified butter method, induced-stratification method* or the *stratification method*, takes advantage of gravity settling for removing most of the moisture from butter and saves the large thermal energy required for removing moisture from butter. This method is particularly suitable if large quantities of butter are available (van den Berg, 1988). The method involves melting the butter mass at 80–85°C, pumping the mass into a vertical storage tank and then leaving it undisturbed for 30 min (Figure 6.19). During this period, butter stratifies into three distinct layers. The top layer consists of a curdy mass, the middle layer is of fat and the bottom layer consists of buttermilk serum (Ray & Srinivasan, 1976). The buttermilk serum, which consists of about 80 g 100 g^{-1} of the moisture and 70 g 100 g^{-1} of the SNF contained in the butter, is drained off from the bottom of the vat and the fat layer is drawn into a double-jacketed steam kettle where it is heated to 105–110°C to evaporate any traces of moisture and for developing the typical ghee flavour (Rajorhia, 1993). Removal of the buttermilk eliminates the need for prolonged heating for evaporation of the moisture and results in the formation of a significantly low quantity of ghee residue, thereby reducing the fat losses. The PS method has been reported to offer the advantages of economy in fuel consumption up to 35–50%, and saving in time and labour up to 45%. The PS method gives a slightly higher ghee yield than direct clarification; the product is lower in moisture content, FFA and acidity and less susceptible to peroxide development during storage (Ray & Srinivasan, 1976). This method is further reported to produce ghee with a mild flavour (Rajorhia, 1993).

Continuous ghee making

To overcome the problems of batch methods of ghee making, such as limitations of scale of operation and excessive exposure of the plant operators to the stress of heat and humidity, dairy plants have evolved continuous ghee making systems to suit their requirements. Punjrath (1974) described the development of two prototype continuous ghee making plants. The first process uses three successive stages of scraped surface heat exchangers (SSHE), followed by flashing into vertical vapour separators. Molten butter is pumped through the successive stages. Ghee leaving the final stage is passed through a centrifugal clarifier to remove residue, and the clarified ghee is stored in a tank for final packing. Another continuous process for manufacture of ghee from cream involves concentration of the fat and breaking of the fat-in-water emulsion mechanically with the help of centrifugal force in the clarifixator and concentrator. The use of heat is limited to the development of flavour and removal of traces of moisture. The plant requires low energy input and can be adapted for much larger volumes. The process also results in lower SNF losses and better fat recovery; the separated SNF by concentrating the cream is used for milk standardisation or for the manufacture of skimmed milk powder.

Abichandani *et al.* (1995) developed a continuous 'ghee' making system comprising (a) a continuous butter melting unit and (b) a straight sided horizontal thin film SSHE for converting the molten butter into ghee. The butter melting unit, which can be hooked up with a continuous ghee making machine, works on the principle of tank-type heat exchanger. It is a jacketed shell where the solid material is fed at one end, and is mechanically conveyed along the length of the heat exchanger having condensing steam in its jacket. The heat transfer is very rapid due to the turbulence created by the rotor, and the molten butter is then passed onto a thin film SSHE. The heat exchanger utilises steam to heat the inner shell of the jacket, which is in direct contact with the product. The centrifugal action of the rotor blades causes the product to spread uniformly as a thin film on the inner shell of the heating surface. It results in rapid evaporation of water from the product. The vapour is removed through an outlet, and may be reused for pre-heating the butter. Ghee is drawn from a 'well' located at the bottom of the downstream end of the SSHE and led to a balance tank. The ghee residue is clarified by pumping the product through a centrifugal clarifier. The clarified ghee is then stored in a tank for final packing. Freshly manufactured ghee produced by the continuous process was reported to be of quality comparable to that made by the batch process (Bector *et al.*, 1996). To produce 100 kg of ghee, the plant utilises 32 kg of steam and 0.45 kWh of electrical energy.

However, various approaches to continuous ghee making were aimed at achieving the following features in the process:

- high heat transfer coefficient and hence, compact design;
- improved flow characteristics of the product inside the processing equipment;
- only small hold-ups of raw material in the plant at any given time; hence, no chances of a whole batch getting spoiled;
- the system adaptable to automation and cleaning-in-place (CIP);
- simple, robust and hygienic design;
- minimum strain on the operator;
- total absence of fouling; hence, heat transfer coefficient can be maintained throughout the run of the production stages;
- easy capacity control with no foaming problem, that is, cream can be handled conveniently;
- short residence time and, as a consequence, less heat damage to the product;
- ghee retains 10–15% more vitamin A as compared to that made through conventional process;
- uniform product quality as the process is continuous;
- no spillage loss;
- sanitary operation, as the process takes place in a completely closed environment;
- better economic operation (Abichandani *et al.*, 1995; Patel *et al.*, 2006).

In spite of several advantages claimed for the continuous ghee making systems, these systems are not widely adopted by the industry in India.

Improved industrial method

Patel *et al.* (2006) reported an improved method for the production of ghee with a special focus on reducing fat and SNF losses, which is being commercially used at Panchmahal

Dairy, Godhra, India. The process involves the use of a serum separator, spiro-heater and a continuous butter making machine.

In brief, this method of ghee production consists of heating cream (30–40 g fat 100 g^{-1}) at 90–92°C for 15 s, chilling to 10–12°C, and storing in insulated and jacketed tanks. The cream is then pumped to a continuous butter making machine to obtain 'white' butter. The resultant buttermilk along with serum from the serum separator is chilled in a plate cooler and diverted for use in standardisation of fresh milk.

The 'white' butter is pumped via a screw conveyor to the spiro-heater where it is melted by circulating hot water; the melted butter is transferred to tanks fitted with an agitator and hot water circulation in the jacket where it is subjected to serum separation. The serum is separated, chilled in a plate cooler and pooled with sweet buttermilk for use in milk standardisation. Melted butter (i.e. low in moisture and high in fat) and the serum solids are collected in a butter melting tank from where it is pumped to a primary settling tank, and then continuously pumped to different ghee kettles (boilers) for ghee manufacturing in a normal way during which the residual moisture is evaporated at 113°C. After some holding time, the ghee clarification is carried out at 95°C using a centrifugal ghee clarifier to remove fine particles of residue(s). Clarified ghee is pumped into ghee settling tanks where the product is cooled with water circulation in the jacket of the tank to 50°C and packed.

The quality of ghee made by this new method was reported to be at par with AGMARK standards. The benefits obtained by adopting the new method commercially are as listed below:

- steam saving (e.g. 250–300%);
- increased ghee production per unit time;
- saving in fat and SNF losses because the serum separated from fresh 'white' butter having 0.8 g fat 100 g^{-1} and 8.0 g SNF 100 g^{-1} was used in milk processing for standardisation;
- reduced load on effluent treatment plant because of serum separation;
- improved hygienic conditions as the circuit for heating through plate heat exchanger and conveying the melted butter is totally closed and a continuous one;
- better working conditions because less scraping of the ghee boilers is required due to lower ghee residues;
- lower cost of production due to saving in fat/SNF, steam, electricity, manpower and water.

Conversion of butter oil or AMF into ghee

In many countries, any surplus of MF is conserved in the form of butter oil, rather than butter, because of savings in cold storage capacity and the anticipated long shelf-life at ambient temperature. AMF lacks the characteristic flavour, aroma and texture that are otherwise required in ghee, although the chemical composition is almost identical to that of ghee. Wadhwa & Jain (1991) have suggested a scheme for producing ghee with an almost *desi*-like ghee flavour from butter oil by the addition of skimmed milk *'dahi'* (20 g 100 g^{-1}) or freeze dried *'dahi'* powder (5 g 100 g^{-1}) to butter oil, and heating the mixture to 120°C for 3 min. It is also possible to incorporate synthetic flavour concentrate in various proportions to butter oil. Wadhwa & Jain (1985a) simulated ghee flavour in butteroil using a synthetic mixture containing 3 mg kg^{-1} of δ-C_{10} lactone, 15 mg kg^{-1} of δ-C_{12} lactone, 5 mg kg^{-1} of decanoic acid and 10 mg kg^{-1} of nonanone-2. The simulated flavour in butter oil was similar to that of CB ghee. Another simple method has been reported by Wadhwa & Bindal (1995), which makes use of ghee residue for flavouring the butter oil, and also enhances its keeping quality. This method is recommended as the simplest and most economical method for flavour simulation. Ghee flavour has also been simulated in butter oil using adjunct microorganisms and synthetic compounds (Yadav & Srinivasan, 1992).

Microwave process

Mehta & Wadhwa (1999) applied microwave technology to the preparation of ghee from high-fat cream or butter with enhanced shelf-life and good sensory properties. The fatty acid and flavour profiles of microwave-processed ghee samples were similar to those of conventionally processed samples. Microwave-processed ghee conformed to legal specifications with respect to moisture, FFAs, Reichert–Meissl value, Polenske value and butyro-refractometer reading. There was no loss of vitamin A, E, phospholipids, conjugated linoleic acid (CLA) and polyunsaturated fatty acids (PUFA) in microwave-processed ghee. Microwave radiation did not accelerate lipid oxidation through free radical generation during the preparation of ghee.

6.6.3 *Packaging*

Ghee is susceptible to deterioration from exposure to light, air and metal ions. Therefore, it is generally packed in metal-coated cans (non-toxic and non-tainting) of various capacities ranging from 250 g to 15 kg. Due to large institutional consumption of ghee, about 60% of the ghee produced in India is packed in 15 kg metal containers. The remainder is packed in consumer-size metal containers (500 mL, and 1, 2 or 5 L size), paper/plastic stand-up packs (200 mL, 500 mL and 1 L size) and pillow pouches of metallised polyester and plastic polylaminated film (500 mL and 1 L size) (Aneja *et al.*, 2002). Though metal cans are very expensive, they protect the product against tampering and allow transport to far places without any wastage, and they can be printed with attractive and colourful designs. Food-grade plastic containers and polyethylene pouches are also used. Ghee packed in flexible pouches is placed in paper cartons that contain some cushioning matter to absorb vibrations during transportation or rough handling. Various packaging materials, such as polymer-coated cellophane, polyester, nylon-6, food-grade polyvinyl chloride (PVC) or various laminates, have been used for packaging of ghee (Rao, 1981). Ideally, ghee should be stored in a cool and dry place for a long shelf-life.

6.6.4 *Chemical composition*

Chemically, ghee is a complex lipid consisting of mixed glycerides, FFAs, phospholipids, sterols, sterol esters, fat-soluble vitamins, carbonyls, hydrocarbons, carotenoids (only in cow's ghee), small amounts of charred casein and traces of calcium, phosphorus, iron and zinc. The composition varies depending upon several factors such as species of mammals, breed of cows or buffalos, feeding pattern and season or stage of lactation. The chemical composition of buffalo's and cow's MF and/or ghee are summarised in Table 6.6, and the main components are discussed briefly in the following text.

Table 6.6 Chemical composition of buffalo and cow milk fat and/or ghee.

	Mammalian species	
Constituent	Buffalo	Cow
Saponifiable constituents		
Fat (g 100 g^{-1})	99.0–99.5	99.0–99.5
Moisture (g 100 g^{-1})	<0.5	<0.5
Charred casein, salts of copper, iron and other minor components	Trace	Trace
Saturated fat (g 100 g^{-1})	46	NA
cis-Monoene (g 100 g^{-1})	29	NA
trans-Monoene (g 100 g^{-1})	7	NA
Diene (g 100 g^{-1})	13	NA
Polyene (g 100 g^{-1})	5	NA
Triglycerides[a]		
Short chain[b] (g 100 g^{-1})	43–49	36–40
Long chain[b] (g 100 g^{-1})	54.7	62.4
Trisaturated[b] (g 100 g^{-1})	32–47	32–42
High melting[b] (g 100 g^{-1})	6–12	3–7
Unsaturated[b] (g 100 g^{-1})	56	54.5
Partial glycerides[a]		
Diglycerides (g 100 g^{-1})	4.5	4.3
Monoglycerides (g 100 g^{-1})	0.6	0.7
Phospholipids (mg 100 g^{-1})	42.5	38.0
Unsaponifiable constituents		
Total cholesterol (mg 100 g^{-1})	275.0	330.0
Lanosterol (mg 100 g^{-1})	8.27	9.32
Lutein (μg g^{-1})	3.1	4.2
Squalene (μg g^{-1})	62.4	59.2
Carotene (μg g^{-1})	0.0	7.2
Vitamin A (μg g^{-1})	9.5	9.2
Vitamin E (μg g^{-1})	26.4	30.5
Ubiquinone (μg g^{-1})	6.5	5.0
Flavour compounds		
Total free fatty acids (mg g^{-1})	5.83–7.58	5.99 – 12.29
Headspace carbonyls (μM g^{-1})	0.027	0.035

(continued)

Table 6.6 *(continued).*

Constituent	Mammalian species	
	Buffalo	Cow
Volatile carbonyls ($\mu M\ g^{-1}$)	0.26	0.33
Total carbonyls ($\mu M\ g^{-1}$)	8.64	7.20
Total lactones ($\mu g\ g^{-1}$)	35.4	30.3

NA, not available.
[a] Based on the percentage of total glycerides.
[b] Values pertaining to MF obtained by clarifying at 100°C and filtering with filter paper.
Data compiled from Achaya (1997), Bector & Narayanan (1975a), Bindal & Jain (1973c), Ramamurthy & Narayanan (1966, 1975) and Gaba & Jain (1974, 1975, 1976).

Triglycerides

The bulk of ghee is made up of triglycerides. The average content of long-chain triglycerides is higher in cow's ghee (62.4 g 100 g^{-1}) than in buffalo ghee (54.7 g 100 g^{-1}). On the other hand, the proportion of short-chain triglycerides is higher in buffalo ghee (45.3 g 100 g^{-1}) than in cow ghee (37.6 g 100 g^{-1}); this is because buffalo ghee contains higher amounts of butyric acid than cow ghee (Ramamurthy & Narayanan, 1975). The $C_{4:0}$ and $C_{6:0}$ acids occur exclusively in the short-chain triglyceride fraction, whereas $C_{16:0}$ is largely concentrated in the long-chain triglycerides. Buffalo ghee contains higher (40.7 g 100 g^{-1}) trisaturated glycerides than cow ghee (39.0 g 100 g^{-1}). The fatty acids in trisaturated glycerides and whole MF are similar, indicating an apparent random distribution of fatty acids in MF. The average proportion of high-melting triglyceride is much higher in buffalo ghee (8.7 g 100 g^{-1}) than in cow ghee (4.9 g 100 g^{-1}). This is because buffalo ghee has larger proportions of long-chain saturated fatty acids – palmitic and stearic acids (Table 6.7). For the same reason, buffalo ghee is distinctly harder than cow ghee. Levels of unsaturated glycerides are high in buffalo ghee (56.0 g 100 g^{-1}) compared with cow ghee (54.5 g 100 g^{-1}), which makes the former more prone to autoxidation than the latter. Buffalo ghee made during the winter season contains higher amounts of $C_{12:0}$, $C_{14:0}$ and total saturated acids, and lower amounts of $C_{14:1}$, $C_{16:1}$ and $C_{18:1}$ and total unsaturated acids than the ghee made during the summer and/or the monsoon season (Joshi & Vyas, 1976a). Ghee from mastitic and late lactation milk contains a lower concentration of triglycerides and high concentrations of polyenoic acids (Agarwal & Narayanan, 1976).

Feed significantly affects the composition of ghee. Feeding cottonseed or groundnut cake increases the unsaturation of MF, and decreases the lower chain fatty acids. Thyroprotein supplementation increases the concentration of unsaturated fatty acids (Patel, 1979). Low roughage diet to the animals have been reported to cause a decrease in short-chain fatty acids ($C_{4:0}$–$C_{14:0}$), and an increase in long-chain fatty acids, C_{18} and above. Pasture grass, which is rich in linolenic acid ($C_{18:3}$), causes an increase in the levels of $C_{18:3}$ acids in MF by about 0.5 to 1.0 g 100 g^{-1}. Hay or silage, which is rich in palmitic acid and low in $C_{18:3}$ acid causes an increase in palmitic acid content of MF. Feeding oils or fats has been shown to cause a general decrease in short-chain fatty acids accompanied by an increase in long-chain fatty acids. Certain specific fatty acids present in the feed lipids may also appear in MF. For instance, when rape seed oil, which typically contains erucic acid ($C_{20:1}$), is fed to the animal, this fatty acid appears in MF. Similarly, coconut oil, which is rich in lauric and

Table 6.7 Fatty acid composition of ghee (g 100 g^{-1}) made from milks of different species of mammals.

Fatty acid	Cow	Buffalo
$C_{4:0}$	3.6	6.7
$C_{6:0}$	3.1	3.0
$C_{8:0}$	2.0	1.1
$C_{10:0}$	3.2	3.4
$C_{10:1}$	0.6	0.4
$C_{12:0}$	3.9	3.6
$C_{12:1}$	0.1	–
$C_{14:0}$ (br)	0.3	0.2
$C_{14:0}$	12.9	13.9
$C_{14:1}$	1.0	1.0
NI	1.2	1.3
$C_{15:0}$	2.2	1.0
$C_{16:0}$ (br)	0.4	0.1
$C_{16:0}$	23.8	28.4
$C_{16:1}$	1.0	0.2
$C_{17:0}$	0.8	0.9
$C_{18:0}$	11.8	12.1
$C_{18:1}$	25.5	20.5
$C_{18:2}$	1.2	0.9
$C_{18:3}$	0.8	0.8
$C_{20:0}$	0.2	0.1

(br), branched chain; NI, not identified.
Data compiled from Ramesh & Bindal (1987).

myristic acids, causes an increase in the levels of these fatty acids in MF. Feeding cottonseed, soya and groundnut oils increases unsaturation of MF in addition to the general changes mentioned earlier (Ramamurthy, 1980).

Addition of urea to replace 20 g 100 g^{-1} of digestible proteins has been found to decrease butyric ($C_{4:0}$) and caproic ($C_{6:0}$) fatty acids (10–30 g 100 g^{-1}) and increase $C_{8:0}$ to $C_{16:0}$ fatty acids. The physical nature of the feed can affect the fatty acid composition of MF through the changes in rumen bacterial flora. For example, it has been demonstrated that feeding cattle with ground and pelletted roughages decreased the level of short-chain fatty acids in MF (Ramamurthy, 1980).

Partial glycerides

The buffalo and cow ghee contained 4.5 and 4.3 g 100 g^{-1} and 0.6 and 0.7 g 100 g^{-1} of diglycerides and monoglycerides, respectively. In both species, these partial glycerides were poorer in $C_{4:0}$, $C_{6:0}$, $C_{8:0}$ and $C_{8:1}$ fatty acids, and richer in $C_{14:0}$ and $C_{16:0}$ fatty acids than the respective whole fats. Monoglycerides contained much lower quantities of $C_{4:0}$, $C_{6:0}$, $C_{18:0}$ and $C_{18:1}$ fatty acids, but higher quantities of $C_{14:0}$ and $C_{16:0}$ fatty acids than diglycerides (Ramamurthy & Narayanan, 1974). Presence of β-ketoglycerides has been demonstrated in buffalo MF (Gaba & Jain, 1977).

Phospholipids

The phospholipids content of ghee has been reported to vary considerably. Buffalo ghee, on average, contains slightly higher amounts (42.5 mg 100 g^{-1}) of phospholipids than cow ghee (38.0 mg 100 g^{-1}) (Ramamurthy & Narayanan, 1966). In addition, the phospholipids content of ghee is also dependent on its method of preparation. Narayanan *et al.* (1966) reported that the ghee made from butter had more phospholipids than that prepared from cream. Ghee prepared by direct heating of butter at 120°C contained less phospholipids than the ghee prepared by pre-stratification and subsequent heating of the fat and serum at 120°C (Pruthi *et al.*, 1972a). The temperature at which the ghee is filtered also affects the phospholipids content of ghee. Ghee filtered at 110°C contained more phospholipids than when filtered at 60°C. The phospholipids content of liquid (31.0 mg 100 g^{-1}) and solid fraction (489.8 mg 100 g^{-1}) also vary significantly (Pruthi, 1984). Increasing the momentary heating temperatures from 120°C to 150°C results in a 12-fold increase in the phospholipids content, probably due to the combined effect of liberation of phospholipids from the phospholipid–protein complex, and the more efficient removal of moisture (Pruthi, 1980). Heating time also influences the phospholipids content of ghee, which increases by about 13 times when heating time at 120°C is increased from 0 to 90 min. On further heating, however, there is a progressive decrease in the phospholipid content of the ghee due to thermal oxidation resulting in dephosphorylation. Heating ghee for 40 min and more at 120°C results in the development of a deep dark brown colour and a fishy odour, probably due to interaction of amino groups of phosphatidyl ethanolamine and phosphatidyl serine with aldehydes (Kuchroo & Narayanan, 1977).

Ghee prepared from ripened cream or stored butter contained more phospholipids than when prepared from fresh material. The acidity developed during ripening or storage appeared to facilitate removal of moisture from the ghee residue, forcing more phospholipids into the oil phase (Rajput & Narayanan, 1968). Ghee prepared from fresh cream (cow) at 120°C contained 34.3 mg phospholipids 100 g^{-1} fat, whereas that prepared from the corresponding ripened cream contained 45.3 mg 100 g^{-1} (Singh & Ram, 1978). Winter ghee contained a higher amount of phospholipids (28.94 mg 100 g^{-1}) than summer ghee (16.91 mg 100 g^{-1}) with monsoon ghee at an intermediate level (22.82 mg 100 g^{-1}) showing the highly significant effect of season (Joshi & Vyas, 1977). Phospholipid content was correlated positively with free and saturated fatty acid contents and melting point and negatively with liquid portion and unsaturated fatty acid content. Fat obtained from mastitic milk and colostrum also contained higher phospholipids (Agarwal & Narayanan, 1976).

The percentage proportion of the major phospholipids present in ghee prepared at 120°C are phosphatidyl choline (27.9), phosphatidyl ethanolamine (24.6), sphingomyelin (32.4), phosphatidyl inositol (4), phosphatidyl serine (3), lysophosphatidyl ethanolamine (5) and lysophosphatidyl choline (3) (Kuchroo & Narayanan, 1977).

Unsaponifiable constituents

The unsaponifiable matter in buffalo ghee ranges from 391 in autumn to 410 mg 100 g^{-1} in spring (average ~398 mg 100 g^{-1}), and in cow ghee from 428.4 in summer to 465.0 mg 100 g^{-1} in spring (average ~449.7 mg 100 g) (Bindal & Jain, 1973a).

The cholesterol content has been reported to be lower in buffalo ghee than in cow ghee. The mean cholesterol content of ghee prepared from cow's milk from Hariana, Sahiwal, and Sahiwal X Friesian, and from Murrah buffaloes was 303, 310, 328 and 240 mg 100 g^{-1}, respectively. Seasons of the year also affect the cholesterol content of ghee, with the highest (301 mg 100 g^{-1}) content in winter and lowest (291 mg 100 g^{-1}) in summer (Prasad & Pandita, 1987). Cow ghee contains higher free cholesterol than buffalo ghee (283 mg 100 g^{-1} against 212 mg 100 g^{-1}), whereas esterified cholesterol is higher in buffalo ghee (Bindal & Jain, 1973b). The method of preparation of ghee also affects its cholesterol content. Cow ghee made by the MB method contains significantly lower (292 mg 100 g^{-1}) total cholesterol than ghee prepared by the DC method (306 mg 100 g^{-1}). Mastitic MF contains much more cholesterol (cow 729 and buffalo 625 mg 100 g^{-1}) than normal milk, whereas the clarification temperature has no effect on the cholesterol content of ghee (Bindal & Jain, 1973c).

The minor unsaponifiable constituents ($\mu g\ g^{-1}$ fat) of cow ghee and buffalo ghee are lutein (4.2 and 3.1), vitamin A (28.3 and 39.83 IU 100 g^{-1} fat), squalene (59.2 and 62.4) and ubiquinone (5.03 and 6.51), respectively. The buffalo ghee contains higher vitamin A, squalene and ubiquinone than cow ghee, whereas the concentration of lutein is significantly higher in cow ghee. Lanosterol content (mg 100 g^{-1}) ranges between 5.8 and 13.0 in cow ghee and between 4.8 and 12.1 in buffalo ghee (Bindal & Jain, 1973d). Higher concentrations of carotene, vitamins A and E and cholesterol in mastitic and other abnormal MFs have been reported (Agarwal & Narayanan, 1977). Thyroprotein supplementation increases the cholesterol content in both species (Patel, 1979).

6.6.5 *Flavour*

The chemistry of the highly prized, characteristic ghee flavour has been extensively studied. The characteristic flavour of ghee is due to a complex mixture of compounds produced during the various stages of processing (Abraham & Srinivasan, 1980; Yadav & Srinivasan, 1984, 1992; Achaya, 1997). The flavour of ghee analysed through gas liquid chromatography (GLC) has revealed a wide spectrum consisting of more than 100 flavour compounds (Wadhwa & Jain, 1990). Free fatty acids, carbonyls and lactones are the major groups of compounds contributing to ghee flavour. A maximum number of compounds (36) are found in ghee prepared by *desi* (MB) method followed by ghee prepared by industrial methods (22–29), and the lowest number of compounds is found in butteroil (16). Of the identified volatile compounds, maltol, 5-hydroxymethyl furfuraldehyde, dihydrodihydroxypyranone, 1,3-butanediol and 1-octanol have been identified only in *desi* ghee. The concentration of acetic acid was found to be remarkably higher in *desi* ghee than in industrial ghee. Also, the levels of identified fatty acids, methyl ketones, aldehydes, lactones and alcohols were high in *desi* ghee compared with industrial ghee and butter oil (Wadodkar *et al.*, 2002). The compounds identified to be associated with the ghee flavour are shown in Table 6.8.

Table 6.8 Compounds associated with the flavour of ghee.

Compound	Number of different types
Carbonyls	
Monocarbonyls	
Alkan-2-ones	39 (38)[a]
Alkanals	9 (7)
Alk-2-enals	8 (11)
Alk-2,4-dienals	9 (9)
Unidentified	8 (11)
Dicarbonyls (polar)	5
Diacetyl[b]	10
Methylglyoxal[b]	
α-Keto glutaric acid[b]	
Furfural[c]	
Hydroxy methyl furfural[c]	
Unidentified	5
Methyl ketones	
2-Pentanone	
2-Heptanone	
2-Nonanone	
2-Undecanone	
2-Tridecanone	
2-Pentadecanone	
Lactones	
δ-Lactone	6
γ-Lactone	1
n-Saturated δ and γ-lactone	24
Others (Partially characterised)[d]	20
Free fatty acids	
Of even carbon $C_{4:00-18:00}$	12
Esters	
Ethyl esters of $C_{3:00}$, $C_{5:00}$,$C_{6:00}$,$C_{8:00}$, $C_{14:00}$, $C_{16:00}$, $C_{18:00}$ and $C_{18:1}$ fatty acids, propyl esters of $C_{4:00}$ fatty acids	
Alcohols	
1-Hexanol, ethyl, nonanol, ethanol, n-butanol	
Diols	
1,3-Butanediol	
Propanediol	

Table 6.8 *(continued).*

Compound	Number of different types
Hydrocarbons	
Saturated and unsaturated ($C_{9:00}$–$C_{20:00}$)	12
Cycloalkane (cyclopentane)	1
2-Keto hydrocarbons ($C_{3:00}$–$C_{9:00}$)	

[a] Figures in parentheses for monocarbonyls are for ghee from goat milk.
[b] Detected in ghee clarified at 100°C.
[c] Detected in ghee clarified at 120°C.
[d] Including mostly unsaturated lactones.
Adapted from Sserunjogi *et al.* (1998).

Carbonyl compounds

Carbonyl compounds contribute substantially to the flavour of ghee (Singh *et al.*, 1979; Wadhwa & Jain, 1990). Both monocarbonyls and dicarbonyls (polar carbonyls) have been identified in ghee. The carbonyl content of fresh cow ghee (0.035 $\mu M\ g^{-1}$) has been found to be higher than in buffalo ghee (0.027 $\mu M\ g^{-1}$) (Gaba & Jain, 1976), whereas the total carbonyls in fresh *desi* buffalo ghee were higher (8.64 $\mu M\ g^{-1}$) than that in cow ghee (7.2 $\mu M\ g^{-1}$). The percentage proportion of different classes of headspace monocarbonyls in cow ghee consisted of alkan-2-ones (~85), alkanals (11), alk-2-enals (2) and alk-2,4-dienals (2). The percentage proportions of the above monocarbonyl classes in buffalo ghee were about 79, 19, 1 and 1, respectively. On storage, although no new carbonyls were detected, there was a marked increase in the alkanals, alk-2-enals and alka-2,4-dienals, which may be the reason for the deterioration in ghee flavour during storage (Gaba & Jain, 1974). In about 100 days of storage at 37°C, an off-flavour developed with a significant 4- to 5-fold rise in gas-stripped carbonyls. After 200 days storage, all ghee samples developed pronounced off-flavours, and levels of such carbonyls increased 9- to 10-fold (Gaba & Jain, 1976). The level of alkan-2-ones declined significantly during storage with concomitant increase in the levels of other monocarbonyl classes. Gaba & Jain (1975) have characterised dicarbonyls of ghee and detected six components in fresh and stored ghee; two were tentatively identified as diacetyl and methylglyoxal.

The *desi* ghee made from buffalo's milk contains significantly higher concentrations of carbonyl compounds, and has a better flavour than ghee made under controlled conditions, probably due to flavour contributed by the adventitious microorganisms getting entry through contamination. The role of dairy starter culture microorganisms in flavour development has been well established (Yadav & Srinivasan, 1985a, 1985b, 1985c). Besides, uncontrolled milk fermentation by a variety of microorganisms including yeasts and moulds may also influence the flavour of the ghee.

Lactones

Lactones have a coconut-like aroma, which is associated with the characteristic flavour of ghee (Urbach *et al.*, 1972; Yadav & Srinivasan, 1992; Wadodakar *et al.*, 1996). Delta-lactones have been reported to constitute the major components of ghee volatiles, whereas the γ-lactones only constituted 5–10% of the total lactones. The ratio of δ-lactone to γ-lactone has been found to be 6:1–20:1 depending on the type of ghee. Of the total lactone content, the saturated lactones have been reported to constitute 65%, while the remaining 35%, mainly unsaturated lactones, have not been identified (Yadav & Srinivasan, 1992). The δ-lactones (δ-octalactone (C_8), δ-decalactone (C_{10}) and δ-dodecalctone (C_{12})) seem to be the most important compounds influencing the flavour of ghee. By far, δ-octalactone has the lowest flavour threshold of all the lactones identified in MF.

Wadhwa & Jain (1985b) reported that the levels of δ-lactones plus γ-lactones and total lactones were higher in buffalo ghee than cow ghee. The levels of δ-lactones plus γ-lactones were higher in ghee prepared by the DC method than in that prepared by the creamery butter or *desi* ghee methods, with mean levels ($\mu g\ g^{-1}$) being 27.21 *versus* 19.66 and 17.43, respectively, in cow ghee and 29.75 *versus* 25.11 and 23.45, respectively, in buffalo ghee. Total lactone levels were also higher in ghee (both cow and buffalo) made by the DC method than in that made by the creamery butter or *desi* ghee methods. This trend is opposite to the increasing flavour trend from DC to creamery butter to *desi* method. Apparently, lactones are only one component of ghee flavour and the others, namely, FFAs and carbonyls, are probably more dominant. The lactone level in butter (12 $\mu g\ g^{-1}$) increased 1.9-, 2.4-, 2.8- and 3.0-fold on clarifying at 110, 120, 140 and 180°C, which showed that heat plays a major role in the formation of lactones (Wadhwa & Jain, 1984). The near doubling of the lactone level on clarification of butter at 110–120°C contributes to the pleasing flavour of the product. On storage, the γ-lactones (12 $\mu g\ g^{-1}$) showed a decrease of 20–28 after 100 days and 48–71 after 200 days. On the other hand, the δ-lactones showed marginal increases (2–12.5 $\mu g\ g^{-1}$) on the storage of ghee. Variations in the ratios of different lactones may contribute towards off-flavour (oxidative rancidity) development in ghee (Wadhwa *et al.*, 1980).

Free fatty acids

FFAs are undesirable components in MF products as they are primarily responsible for the rancid flavour development; they also contribute to the normal flavour of ghee, and their level has been closely related to the flavour quality. The lower fatty acids $C_{6:0}$–$C_{10:0}$, though present in low concentration in ghee (0.4–1 mg g^{-1}) accounting for only 5–10% of total FFA, contribute significantly to ghee flavour. The concentrations of both medium-chain ($C_{10:0}$ to $C_{14:0}$) and long-chain ($C_{15:0}$ and above) FFA were higher in cow ghee than in buffalo ghee. Also, the average total FFA level of cow ghee was higher (5.0–12.3 mg g^{-1}) than that of buffalo ghee (5.8–7.6 mg g^{-1}). The average total FFA level of *desi* ghee was higher (7.6–12.3 mg g^{-1}) than that of the ghee prepared by the creamery butter method (6.0–7.3 mg g^{-1}) or the DC method (5.8–7.3 mg g^{-1},which contains the lowest) (Wadhwa & Jain, 1990). This trend is in tune with the flavour trend of the three types of ghee. Ripening of cream also affects the concentration of FFA in ghee, being higher (Singh *et al.*, 1979) in the product prepared from ripened (0.34–0.40 g oleic acid 100 g^{-1}) than from unripened (0.23–0.28 g oleic acid 100 g^{-1}) cream or butter.

Esters

Esters may originate from the esterification of short-chain alcohols and FFA by the action of bacterial esterases (Hosono *et al.*, 1974), and they are the powerful odour compounds contributing to the flavour of ghee (Wadodakar *et al.*, 1996).

Miscellaneous compounds

Miscellaneous compounds (e.g. acrolein, dimethyl sulphide, alcohols, diols and denatured protein), which are produced from the decomposition of milk constituents like fat, proteins, amino acids, lactose and glucose during the clarification process, have been reported to play a possible role in the flavour of ghee (Yadav & Srinivasan, 1992). The flavouring compounds of butter, indole and skatole (3-methylindole), could also pass over into ghee in their original or modified form, thereby influencing its flavour (Stark *et al.*, 1976; Sarkar *et al.*, 1993).

Origin of flavours

Milk lipids are the source of a majority of the flavour compounds occurring in dairy products. The composition of milk lipids is so complex that they can produce a wide spectrum of flavour compounds. The heating process generates flavour compounds through the interaction of lipids, protein degradation products, lactose, minerals and metabolites of microbial fermentation (Achaya, 1997), and possibly through the degradation of FFA and lactose (Fennema, 1985). According to Wadhwa & Jain (1990), the various mechanisms responsible for generation of ghee flavour are as follows:

hydrolysis
Fatty acid glycerides – – – – – – – – – –> FFA

hydrolysis
β-Keto acid glycerides – – – – – – – – –>alkan-2-ones
decarboxylation

hydrolysis
δ-Hydroxy acid glycerides – – – – – – – – –> lactones
decarboxylation

autoxidation
Unsaturated fatty acid glycerides – – – – – – – –>aldehydes, ketones, alcohols

Proteins and lactose contribute to the flavour of milk products in the following manner:

transamination
Protein – – – – – – – – – – – –>α-keto acids
(amino acids)

fermentation
Lactose, citrate – – – – – – – – – – – –>dicarbonyls

browning
Lactose – – – – – – – – – – – – – –>glyoxal, furfurals
caramelisation

The source of carbonyl compounds in ghee is quite diverse. They are formed as a result of microbial metabolism of various milk constituents, oxidation of milk lipids, heat decomposition of milk carbohydrates and fat and/or from chemical reactions. When ghee is made from fermented milk or cream, acid carbonyl compounds that also contribute to the flavour of cheese are expected to form through complex reactions between some carbonyls, such as glyoxal and methylglyoxal with amino acids (Griffith & Hammond, 1989). Products like ketoglycerides and carbonylic compounds are formed from oxidation of lipids (Joshi & Thakar, 1994), whereas polar (dicarbonyl) compounds, such as diacetyl, furfural and hydroxymethylfurfural, are formed by dehydration and thermal degradation of carbohydrates (Fennema, 1985).

The production of the characteristic ghee flavour also depends upon the extent to which the flavouring metabolites, such as FFA and carbonyls, are transferred from the aqueous to the fat phase during heat clarification as a result of moisture removal. It has been hypothesised that such a transfer is facilitated by the development of acidity of the ripened material through the metabolic activity of the starter culture (Yadav & Srinivasan, 1985a). The mechanism by which this occurs is not clear. One possible explanation appears to be that, when the water is evaporated, the flavour compounds that are water soluble but have higher boiling points than water remain in the fat phase. In addition, the steam volatile monocarbonyls are more intimately associated with the fat phase than with the water phase (Wadodakar *et al.*, 1996). More flavour compounds separate with the fat phase upon churning of fermented cream and subsequently pass to ghee upon heat clarification.

6.6.6 *Physicochemical properties*

The physicochemical properties of ghee have been summarised in Table 6.9. There is wide variation in physicochemical properties of ghee depending upon the species of mammals, type of feed, season and stage of lactation. In general, cow ghee has a lower melting point, softening point, Reichert–Meissl value, saponification value, Kirschner value and unsaponifiable matter, but a higher Polenske value, iodine value and butyro-refractometer reading than buffalo ghee. These differences in physicochemical properties are due to differences in fatty acid composition of cow ghee and buffalo ghee. Nevertheless, feeding of cottonseed supplements to cows lowers the Reichert–Meissl and Polenske values and vitamin A and carotene contents of ghee, but increases iodine values, butyro-refractometer readings and

Table 6.9 Physicochemical characteristics of ghee made from different milks.

Parameter	Cow	Buffalo
Reichert–Meissl value	16.9–29.7	14.5–39.9
Polenske value	0.9–3.2	0.4–5.3
Iodine value	31.0–45.6	21.4–39.9
Butyro-refractometer reading (40°C)	42.5–47.7	40.9–46.9
Melting point (°C)	28.8–35.7	26.6–36.1
Unsaponifiable matter (g 100 g^{-1})	0.42–0.49	0.45–0.54
Saponification value	212.8–232.8	198.0–239.3
Kirschner value	22.1	28.4
Softening point (°C)	33.5–33.9	33.5–34.6
Smoking point (°C)	252.0	Not reported

Data compiled from Achaya *et al.* (1946), Angelo & Jain (1982) and Ramesh & Bindal (1987).

melting ranges (Angelo & Jain, 1982). Seasonal variation in physicochemical characteristics of ghee has also been observed. For example, winter ghee shows a higher acidity, melting point and grain size, whereas in summer ghee, the iodine value is higher and, in the monsoon season, the saponification value is higher (Singhal *et al.*, 1973). No significant seasonal variation is noted in Reichert–Meissl, Polenske and butyro-refractometer values (Joshi & Vyas, 1976a).

6.6.7 *Texture*

A good grainy texture of ghee is very much appreciated by consumers. Since MF is made up of a wide variety of triglycerides with varying melting points, ghee shows a unique property of forming grains. When ghee is stored at room temperature, it crystallises into three distinct fractions or layers namely (a) liquid, (b) granular semi solid-sitting and (c) hard flaky portion floating on the top and sticking to the sides of container. The granules of buffalo ghee are irregular clusters, whereas those of cow ghee are smaller and made up of fine divergent monocrystals. These variations are attributed to differences in the glycerides and fatty acid composition (Armughan & Narayanan, 1979).

The liquid portion of ghee varies according to storage temperature. At 24°C, ghee is small grained but of compact texture; while, at 34°C, it is in a completely liquid state. Good texture and granulation is obtained by storing ghee between 28 and 29°C for 72 h (Joshi & Vyas, 1976b). Granulation of ghee at 29°C is complete in 3 days after melting at 80°C. On the 3rd day, the grain size is maximum 420 μm for buffalo ghee and 108 μm for cow ghee with liquid fractions of 59 and 80 g 100 g^{-1}, respectively (Armughan & Narayanan, 1979). Besides the temperature of incubation, the following factors affect the formation of grains in ghee:

- Rate of cooling – slow cooling (2–3 h) to incubation temperature (28–29°C) gives better grain size.
- Cow *versus* buffalo – buffalo ghee forms bigger size crystals (0.31–0.42 mm) than cow ghee (0.10–0.24 mm).
- Temperature of clarification – higher temperature of clarification gives better grain size.
- Method of preparation – creamery butter method gives better grain size than DC.
- FFA – the presence of FFA increases the grain size markedly, but the quantity of grains is increased to a small extent.
- Seeding – seeding (1–3 g 100 g^{-1}) with grains of ghee gives grains of good appearance; the grain shape is needle like as compared to those of spherical ones obtained without seeding (Ramamurthy, 1980).

6.6.8 *Thermal oxidation*

Since 80% of the ghee produced in India is used for culinary purposes (i.e. for frying), the changes taking place in the composition of ghee during heating at different temperatures are of great interest.

Effect on saponifiable matter

Prolonged heating of ghee, especially high acid ghee, at temperatures ranging from 150 to 225°C for 2 h results in gradual decrease in triglycerides and an increase in diglycerides and monoglycerides, probably due to hydrolysis (Bector & Narayanan, 1974a, 1977). Heating of ghee up to 225°C has no appreciable effect on saturated fatty acids from $C_{4:0}$ to $C_{12:0}$, but causes a decrease in $C_{10:1}$, $C_{14:1}$, $C_{16:1}$, $C_{18:1}$, $C_{18:2}$ and $C_{18:3}$ fatty acids. Generally, the higher the temperature, the greater is the effect (Bector & Narayanan, 1974b).

Effect on unsaponifiable constituents

Prolonged heating decreases the cholesterol, carotene, vitamin A and E contents of ghee. Decrease in cholesterol content by about 12 and 27% (i.e. 33.8 and 76.1 mg 100 g^{-1}, respectively) has been observed upon heating at 150 and 225°C for 2 h (Bector & Narayanan, 1975a). At 200°C and 225°C of heating, carotene and vitamin A were destroyed completely within 15 min, but the losses in vitamin E were 65.5 and 71.0% (i.e. 17.3 and 18.7 $\mu g\ g^{-1}$, respectively). Vitamin E was completely destroyed at the end of a 30-min heating period.

Effect on carbonyl compounds and flavour

Upon heating fresh ghee at 150°C, peroxides and epoxides increase with increasing heating period. Thermally oxidised ghee contains more peroxides of lower polarity than of higher polarity. Bector & Narayanan (1975b) noted a great increase in the ratio of monocarbonyls to total carbonyls with saturated aldehydes as main monocarbonyls formed as a result of thermal oxidation.

Effect on physicochemical properties

Prolonged heating does not appreciably change the Reichert–Meissl, Polenske and saponification values, but slightly increases the FFA content and butyro-refractometer reading and decreases the iodine value of ghee when heated at 150–225°C for 2 h. The higher the temperature and greater the period of heating, the greater is the effect (Bector & Narayanan, 1974c).

6.6.9 *Shelf-life of the product*

Ghee has a long shelf-life of 6–8 months, even at ambient temperatures. Longer storage periods of up to 2 years have also been reported (Bekele & Kassaye, 1987). The storage stability of ghee is attributed to the low moisture content (ca. 0.2 g 100 g^{-1}) and the high content of phospholipids (ca. 400 mg kg^{-1}), and perhaps the free amino acids, which are liberated from the phospholipid–protein complex into the fat phase (Achaya, 1997). The low acidity of the ghee and the presence of natural antioxidants are also believed to contribute to the extension of its shelf-life (van den Berg, 1988).

In spite of its intrinsic shelf-life stabilising properties, ghee does eventually spoil mainly due to development of oxidised flavour as a result of the oxidation of the fatty acids during storage at ambient temperature and/or rancid flavours due to lipolysis (van den Berg, 1988). Ghee supports little microbial activity. The high heat treatment during manufacture of ghee destroys most of the bacteria, and the moisture content in the product is too low to allow normal growth of most of the microorganisms. However, certain *Bacillus* species, such as

Bacillus subtilis and *Bacillus megatherium*, have been found in ghee (van den Berg, 1988). These isolates were detected mostly as spores; they were most probably not growing. It was, therefore, suggested that rancidity may develop as a result of the activity of residual microbial lipases provided the ghee contains sufficient moisture.

The keeping quality of ghee is governed by several factors, such as the ripening of cream, method of manufacture, clarification temperature and the permeability of the packaging material to air and moisture (Singh & Ram, 1978). These factors are summarised below.

Effect of species of mammals

Cow ghee is apparently more shelf-stable than buffalo ghee due to the higher content of natural antioxidants in the former product (van den Berg, 1988). Although buffalo ghee has been reported to be more resistant to lipolysis than cow ghee (van den Berg, 1988), it is more prone to oxidative deterioration (Gaba & Jain, 1973).

Effects of feeding and season

Ghee prepared from cottonseed-fed animals showed that the fat had better keeping quality, presumably because of the antioxidant properties of gossypol, a phenolic substance in cotton seed (Tandon, 1977). Chatterjee (1977) observed that ghee produced and packed in winter had a longer shelf-life (10–11 months) than that packed in summer (6 months) and the rainy season (3 months).

Effects of method of preparation

The keeping quality of ghee is affected by method of manufacture. It is ~9 months by *desi* method, ~12 months for DC method and ~4 months for creamery butter method; in general, ripening of the cream reduces the keeping quality of ghee prepared from it (Singh *et al.*, 1979). Ghee prepared by the *desi* method initially has a slightly better flavour (Gaba & Jain, 1974), but deteriorates during subsequent storage (Rajput & Narayanan, 1968; Singh & Ram, 1978; Rao *et al.*, 1987).

Higher temperatures, or longer periods of heating at a particular temperature, have been shown to impart better oxidative stability (Ramamurthy *et al.*, 1968) because of greater liberation of phospholipids from phospholipid–protein complexes (Narayanan *et al.*, 1966). Greater liberation of phospholipids was observed in the ripened cream ghee, but did not lead to an improvement in storage quality (Rajput & Narayanan, 1968; Singh & Ram, 1978). Various attempts have been made to elucidate the influence of high temperature of clarification on the shelf-life of ghee. It has been suggested that during heating, especially after most of the moisture has been evaporated, antioxidants are produced from phospholipids. The heat-modified phospholipids are believed to be absorbed by the fat, and hence contribute to the keeping quality of the ghee. The total reducing capacity of the system, which involves, in particular, certain sulphur compounds responsible for a cooked flavour, appears to play a dominant role in the flavour and keeping quality of ghee (Singh *et al.*, 1979). Consequently, the ghee produced by the DC method is said to be more stable than that produced by the CB method because of the longer heat treatment and the higher phospholipid content (Unnikrishnan & Rao, 1977; van den Berg, 1988). Flavour-induced butter oil could be stored at room temperature for 70 days as against just 30 days for butter oil (Wadhwa *et al.*, 1979) suggesting the role of antioxidants produced during heating of fat in the presence of SNF.

Effect of phospholipids

The antioxidative properties of phospholipids in ghee have been well established, and it has been shown that the presence of 0.1 mg 100 g^{-1} phospholipids improves the keeping quality of ghee (Lal *et al.*, 1984). Phospholipids may exhibit antioxidant activity by binding metals, regenerating other antioxidants and providing a synergism with phenolic antioxidants (Richardson & Korycka-Dahl, 1988). The main fraction of phospholipids, which exerted antioxidant property, was found to be cephalin. This fraction also showed maximum browning, which presumably was correlated with antioxidant properties (Pruthi *et al.*, 1972b). It was demonstrated that phospholipids acts synergistically with α-tocopherol, and it has also a metal-inactivating action with copper (Bector & Narayanan, 1972).

The residue from ghee manufacture, large quantities of which are wasted, contains 3.6–13.2 g 100 g^{-1} phospholipids (Lal *et al.*, 1984). Ramamurthy *et al.* (1969) observed that the addition of ghee residue to ghee enhanced the shelf-life considerably. The antioxidant effect was found to depend on the temperature and method of clarification of ghee residue; the potency decreased in the following sequence: CB ghee residue > *desi* butter ghee residue > cream ghee residue. This was in accordance with their phospholipids content (Santha & Narayanan, 1978), which may be the factor responsible. This was established by addition of isolated phospholipids to ghee. Addition of 15–20 g 100 g^{-1} ghee residue imparted a fairly good oxidative stability to flavoured butter oil comparable with that of ghee (Wadhwa *et al.*, 1991a).

Among the lipid constituents of ghee residue, phospholipids had the maximum antioxidant property followed by α-tocopherol and vitamin A. Among non-lipid constituents, addition of amino acids, lactose, glucose, galactose and their interaction products to ghee along with protein and phospholipids increased the oxidative stability. It was concluded that the antioxidant property of ghee residue was due not only to phospholipids, but to other constituents also (Santha & Narayanan, 1979). Sripad *et al.* (1994, 1996) associated the antioxidant properties of the ghee residue with the presence of tocopherols, phospholipids and products of browning reactions. They showed that the fat-insoluble compounds of the residue had better antioxidant properties than the fat-soluble ones. Addition of browning compounds, prepared by heating amino acids and dicarbonyls, and addition of alcohol extracts containing compounds from the Maillard reaction greatly delayed the development of oxidative rancidity in ghee, indicating that these compounds may be useful in increasing the shelf-life of the product (Nath & Ramamurthy, 1988a).

Instant mixing of ghee with an equal amount of ghee residue at 120°C yielded free fat carrying about 2 mg 100 g^{-1} phospholipids. This fat, when added to ghee at a 5 g 100 g^{-1} level brought the phospholipids concentration to about 0.1 g 100 g^{-1} (Pruthi *et al.*, 1973). This concentration of phospholipids in ghee more than doubled the induction period and was suggested as a commercial method of increasing the shelf-life of ghee. Pruthi & Yadav (1998) gave another method of preparation of phospholipid-rich ghee, which involves addition of 1.0 g 100 g^{-1} dipotassium hydrogen phosphate to butter at 115°C during ghee manufacture.

Addition of phospholipids from other sources has also been tried. Gram seed lecithin was found to be less effective than cephalin (Gupta *et al.*, 1979). The antioxidative effectiveness of whole phospholipids from seeds was in the following order: sunflower > groundnut > soya bean > cottonseed. This was also the order of decreasing phosphatidylethanolamine content, and of various phospholipids classes studied, phosphatidylethanolamine had the best effect on oxidative rancidity in ghee (Bhatia *et al.*, 1978). Sunflower seed oil phosphatides were found to be more effective than many synthetic antioxidants in controlling oxidative and lipolytic deterioration of ghee during storage (Kaur *et al.*, 1982)

Effect of moisture, trace elements and amino acids

Water was found to act as an antioxidant when present at levels of 2.5–5.0 g 100 g^{-1}. This effect was further enhanced in the presence of phenolic compounds, but sodium thioglycolate had a negative effect (Gupta *et al.*, 1978). Addition of 0.1–1.0 mg kg^{-1} of copper showed considerable adverse effect on the shelf-life of ghee (Chaudhary *et al.*, 1980a, 1980b). Migration of added copper and iron into cream was found to depend on the method of preparation; the use of the DC process resulted in ghee with higher levels of copper and iron, both of which enhanced oxidative rancidity (Unnikrishnan & Rao, 1977). Enrichment of the ghee with β-carotene, in the presence of contaminating copper, slightly increased the rate of oxidation (Kempanna & Unnikrishnan, 1985). Glycine and alanine were found to be potent antioxidants. Among aromatic amino acids, tryptophan was most potent and, among the sulphur-containing ones, cysteine was best. Acidic amino acids were slightly better antioxidants than basic ones. In general, the antioxidant properties of amino acids were attributed to their ability to chelate with pro-oxidative metals, and therefore the structure decided the antioxidant properties (Gupta *et al.*, 1977).

Effect of synthetic antioxidants

Addition of synthetic antioxidants individually (0.005–0.02 g 100 g^{-1}) and also in combinations of two (mixture not exceeding 0.02 g 100 g^{-1}) with or without phospholipids to ghee revealed that the efficiency of the antioxidants for protecting the buffalo ghee against oxidative deterioration at 80°C was in the following order: propyl gallate (PG) > octyl gallate (OG) > dodecyl gallate (DG) > butylated hydroxytoluene (BHT) > butylated hydroxyanisole (BHA). On simultaneous addition of 0.01 g 100 g^{-1} of each of two antioxidants, the 'protection factor' values were as follows: BHA + PG > BHT + PG > BHA + BHT > BHT + OG > BHA + OG > BHT + DG > BHA + DG. Milk phospholipids increased the protective effects of the individual antioxidants and the mixtures, but only had a synergistic effect on the BHA + PG and the BHA + BHT mixtures (Kuchroo & Narayanan, 1972). In another study, BHT was also found to be more effective than BHA and the two showed a synergistic effect (Kuchroo & Narayanan, 1973; Rao *et al.*, 1985). Under commercial conditions, BHA could improve the aroma, flavour and shelf-life of ghee (Chatterjee, 1977). The antioxidant potentialities of certain other compounds tested were in the following order: hydroquinone > catechol > resorcinol, and again: palmitoyl ascorbate > PG > OG > BHA. A phenolic group was found necessary; when a second group was present the compound proved a better antioxidant, the order being para > ortho > meta. tert-Butylhydroquinone (TBHQ) showed the greatest antioxidant potency and, at a very low level (0.005 g 100 g^{-1}), for imparting a longer shelf-life to flavoured butter oils (Wadhwa *et al.*, 1991b). The strong effect of hydroquinone could be due to the formation of a stable quinine, which can terminate the chain reaction (Gupta *et al.*, 1979).

Effect of naturally occurring plant materials

A large variety of naturally occurring plant materials, such as soya bean, safflower, *amla* (*Phylanthus amblica*) fruits, curry (*Murraya koenigi*) leaves and betel (*Piper betel*) leaves have been shown to improve the shelf-life of ghee, and it is possible to replace synthetic antioxidants with natural ones to improve the keeping quality of the product (Rao & Singh, 1990).

Betel or curry leaves at 1 g 100 g^{-1} concentration may be used instead of BHA and BHT for extending the shelf-life of ghee, since their addition during clarification yielded products with a higher resistance to oxidation and higher sensory score than those carrying synthetic antioxidants (Patel & Rajorhia, 1979). A sorghum (*Sorghum bicolor*) grain powder extract (SGPE) prepared by heating it with ghee at a ratio of 1:1 (wet basis) to 120°C has been shown to enhance shelf-life of ghee by giving protection against autoxidation. The SGPE contained 320 mg 100 g^{-1} phospholipids and 220 mg 100 g^{-1} water-extractable phenolic compounds on a dry-matter basis. The presence of at least six phospholipids in a chloroform–methanol extract of SGPE and 12 water-soluble phenolic compounds in SGPE was confirmed. Guleria *et al.* (1983) also reported that the powder of tomato seeds when added at 5 g 100 g^{-1} to fats inhibited rancidity, and was as effective an antioxidant as the standard 0.01 g 100 g^{-1} butylated hydroxytoluene or butylated hydroxyanisole.

The use of mango (*Mangifera indica L*) seed kernels enhanced the oxidative stability of ghee suggesting that the phospholipids and phenolic compounds of mango seed kernel were transferred to ghee (Parmar & Sharma, 1990; Dinesh *et al.*, 2000; Puravankara *et al.*, 2000). A pre-extract prepared by heating mango seed kernel powder and ghee at a ratio of 1:1 (wet basis) at 120°C contained 430 mg 100 g^{-1} phospholipids and 224 mg 100 g^{-1} water-extractable phenolic compounds. The presence of at least 11 phospholipids in a chloroform–methanol extract of mango seed kernel powder and 8 water-soluble phenolic compounds in pre-extract was confirmed. Addition of pre-extract to ghee at 4, 6, 8 and 10 mL 100 mL^{-1} levels increased the phospholipids content of ghee over the control to 11.2, 20.9, 26.4 and 34.4 mg 100 g^{-1} of ghee, and those of water-extractable phenolic compounds to 7.4, 11.1, 15.5 and 20.7 mg 100 g^{-1} of ghee, respectively (Parmar & Sharma, 1990).

The use of extracts from herbs and spices as natural antioxidants in ghee has also been reported. Merai *et al.* (2003) studied the effect of antioxygenic compounds extracted from two varieties of *Tulsi* (*Ocimum sanctum* and *Ocimum tenuiflorum*) on oxidative stability of ghee. The addition of pre-extracted powder of *Krishna Tulsi* (*Ocimum tenuiflorum*) leaves at a level of 0.6 g 100 mL^{-1} into creamery butter ghee was found almost equally effective as that of BHA at a level of 0.02 g 100 g^{-1} in preventing autoxidation. The phenolics present in the *tulsi* leaves appeared to be the main contributory factors in extending the oxidative stability of ghee. Amr (1990) studied the role of some aromatic herbs such as rosemary (*Rosmarinus officinalis* L.), sage (*Artemisia herba-alba* Asso.), fennel (*Foeniculum vulgare* Mill.) or rue (*Ruta graveolens* L.) in extending the stability of sheep ghee. Ghee prepared from ripened unsalted butter from Awassi sheep's milk containing 7.5 g 100 g^{-1} rosemary (DM basis) showed an antioxidant effect equivalent to that of BHA/BHT, whereas the other herbs were less effective. The effectiveness of traditional methods of preparing ghee with spices for enhancing the keeping quality was examined by Tanzia & Prakash (2000). Ghee prepared with 1 g 100 g^{-1} of cardamom, clove, fenugreek, pepper or turmeric had higher keeping quality than the control ghee, with turmeric being more effective than the other spices. Semwal *et al.* (1997) studied the antioxidant activity of turmeric and its fractions extracted using various solvents, and observed that the ground spice and its water-soluble fraction had antioxidant activity.

Effect of abnormal MF

Clarified butter fat from mastitic and other abnormal milks was reported to be more prone to oxidative spoilage than normal MF because of a high concentration of PUFA and FFA (Agarwal & Narayanan, 1979).

Effects of packaging material

The type of container did not affect the ghee quality during storage. Keeping quality and texture of ghee packed in polyethylene plastic containers, brown glass bottles (Singh & Ram, 1978), lacquered metal containers or low-density polyethylene (LDPE) (Chauhan & Wadhwa, 1987) and metallised plastic pouches (Pantulu & Ramamurthy, 1988) were found to be almost similar.

Effect of dissolved oxygen

Shekar & Bhat (1983) observed that the rate of development of peroxides in cow ghee and buffalo ghee increased with increasing dissolved oxygen in the milk. Reducing the dissolved oxygen content of the initial milk enhanced the keeping quality of ghee prepared from it.

6.6.10 *Nutritional aspects*

Ghee is a source of fat-soluble vitamins (A, D, E and K) and essential fatty acids (Chand *et al.*, 1986). Ghee is a rapid source of energy as compared with other vegetable oils as the lower chain fatty acids of ghee are quickly absorbed and metabolised (Nhavi & Patwardhan, 1946; Basu & Nath, 1946). Ghee improves the digestibility of proteins and improves the absorption of minerals resulting in improved growth rate. Improvement of digestibility of protein by 36% and biological availability by 62% was observed when cow ghee was added to a diet that was suboptimal for vitamin A (Kehar *et al.*, 1956). Mineral absorption from the diet also increases with ghee consumption. An increase in retention of calcium up to 45% and phosphorus up to 57% has been observed upon consumption of cow ghee (Steggerda & Mitchell, 1951; Kehar *et al.*, 1956). Ghee contains anticarcinogens, such as CLA, butyric acid and/or sphingomyelin.

CLA is a mixture of positional and geometric isomers of linoleic acid with conjugated unsaturation. It has been suggested that CLA can be formed by free radical oxidation of linoleic acid followed by reprotonation of the radicals by proteins, such as the whey proteins (Ha *et al.*, 1989; Fritsche & Steinhart, 1998). It has been postulated that during the manufacture of ghee, milk protein provides hydrogen, which reacts with the double bonds of linoleic acid during heating under anaerobic conditions, and catalyses the formation of CLA (Aneja & Murthi, 1990). An increase in the CLA content of MF from base level of 0.5–0.6 up to 1.0 g 100 g^{-1} was observed in the *desi* ghee due to fermentation. Also, the CLA content of ghee can be increased up to 5-fold from the base level, by increasing the temperature of clarification of *desi* ghee from 110 to 120°C (Aneja & Murthi, 1990). *cis*-9, *trans*-11-Octadecadienoic acid was found to account for about 90% of total CLA in various dairy products (Chin *et al.*, 1992). Butter and ghee are the richest sources of *cis*-9, *trans*-11-octadecadienoic acid, and their consumption has been suggested to influence the levels of CLA in blood serum and human milk (Parodi, 1994; see also Chapter 2). The anticarcinogenic effect of CLA has been demonstrated in several animal models (Chin *et al.*, 1992). It has been shown in the animal studies that CLA suppresses the tumor when provided as 1.0 to 1.5 g 100 g^{-1} dietary supplement (Belury *et al.*, 1996). CLA has been reported to be cytotoxic to human malignant melanoma, colorectal and breast cancer cells *in vitro*, and is more inhibitory to cancer cell proliferation than β-carotene (Parodi, 1994). Furthermore, ghee consumption also improves the CLA content of human milk. It has been observed that breast milk from women of the Hare Krishna religious sect contained twice as much CLA as milk from conventional Australian mothers (11.2 mg g^{-1} *versus* 5.8 mg g^{-1}). This difference was attributed to the large amount of butter and ghee consumed by Hare Krishna women (Fogerty *et al.*, 1988).

Hypocholesterolemic effects

Arteriosclerosis, a process in which cholesterol containing fatty deposits accumulates on the inner walls of arteries, is believed to be due to consumption of higher levels of saturated fat and dietary cholesterol. Due to higher level of saturated fats in ghee, its consumption is suspected to be one of the reasons for high incidence of cardiovascular diseases among immigrant populations of South East Asians settled in Western and African countries (Jacobson, 1987). On the contrary, ghee was found to have hypocholesterolemic effect upon consumption and is even used in *Ayurvedic* treatments of heart diseases (Pandya, 1996). It is observed that there is lower prevalence of cardiovascular heart disease (CHD) in Indian men with a higher ghee intake. Significantly lower prevalence of CHD with consumption of ghee >1 kg $month^{-1}$ has been observed (Gupta & Prakash, 1997). Hypocholesterolemic effect upon consumption of higher levels of ghee has also been shown in rat feeding studies (Kumar *et al.*, 1999a). The serum lipid profiles of these animals showed a dose-dependent decrease in total cholesterol, low-density lipoprotein and very low-density lipoprotein and triglycerides when ghee was present at levels higher than 2.5 g 100 g^{-1} in the diet. Although the exact mechanism of hypocholesterolemic effect of ghee is not known, it is believed that it is mediated by increasing the secretion of biliary lipids (Kumar *et al.*, 2000). Other possible mechanism is the antiatherogenic effect of CLA. Ghee is one of the richest natural sources of CLA and as low as 0.05 g 100 g^{-1} level of CLA has been shown to considerably reduce total cholesterol, low-density lipoprotein and triglycerides (Nicolosi *et al.*, 1993).

Cholesterol oxidation products (COP)

Ghee contains 0.3–0.4 g 100 g^{-1} cholesterol, which has a tendency to autoxidise spontaneously in air and to peroxidise *in vivo*. Cholesterol oxides are reported to be cytotoxic, mutagenic and may be atherogenic (Nath *et al.*, 1996). Cholestantriol and hydroxycholesterol are particularly important (Prasad & Subramanian, 1992). COPs may be the cause of atherosclerotic lesions, whereas deposits of cholesterol merely represent a secondary process (IDF, 1996). In rat feeding studies, it has been observed that feeding high amounts of oxidised ghee affects the erythrocyte ghost membrane structure, which results in changes in membrane fluidity and increased platelet aggregation, and decreased proliferation of lymphocytes through alterations in the structure of the lymphocyte membranes (Niranjan & Krishnakantha, 2000a, 2000b, 2001). Therefore, the presence of COPs in foods has recently evoked much interest; however, current recommendations restrict their levels in foods (Nielsen *et al.*, 1996).

Ghee manufactured and stored under normal conditions does not contain COPs. They are formed only when the ghee gets autooxidised to such an extent (PV increased to >10) that it becomes unacceptable for consumption (Nath & Ramamurthy, 1988b). Using ghee for frying for short periods does not cause the production of cholesterol oxides; however, they are formed after frying for 15 min. The oxidation of cholesterol may be inhibited by using antioxidants. In the presence of effective antioxidants, cholesterol oxides are produced only when the ghee is used for frying for a long time (e.g. 60 min) (Nath *et al.*, 1996). Intermittent frying increases the concentration of COP. All major COPs are increased with frying time. Most atherogenic COP, such as 25-hydroxycholesterol and cholestantriol, are formed more in intermittently heated ghee (8.1–9.2 g 100 g^{-1} of the total COP) than in fried ghee samples (7.1 g 100 g^{-1}) (Kumar & Singhal, 1992).

Jacobson (1987) suggested cholesterol oxides in Indian ghee to be the possible cause of unexplained high risk of atherosclerosis in Indian immigrant populations. However, consumption of diets containing up to 10 g 100 g^{-1} ghee as the sole fat source does not significantly increase serum lipid profile risk factors of cardiovascular diseases in rats (Kumar *et al.*, 1999a).

6.6.11 *Ghee as a medicine*

Ghee has been recognised as Indian medicine in Ayurveda and is being used for treatment of various disorders from time immemorial. There are about 55–60 medicated ghee types reported in *Ayurvedic* literature. Medicated ghee is always prepared with selective fortification with herbs so as to acquire all the required fat-soluble therapeutic components of the herbs (Saxena & Daswani, 1996). Different medicated ghees with their main application are listed in Table 6.10 and, in most of the treatments, so far, it is not well established whether ghee or the components of ghee or the herb extracts have the disease-curing ability. One such study has revealed that the effect of herbs and herb extracts were high when used along with ghee as compared to its usage in powder or tablet form (Joshi, 1998). The claims on the efficacy of some of the medicated ghee, such as *Bramhi Ghrita* for treatment of learning and memory disorders (Achliya *et al.*, 2004a), hepatoprotective effect of *Amalkadi Ghrita* against hepatic damage (Achliya *et al.*, 2004b) and the immunomodulatory activity of *Haridradi Ghrita* (Fulzele *et al.*, 2003) have been validated in experimental animals. Significant reduction in scaling, erythema, pruritus and itching in psoriasis patients with marked improvement in the overall appearance of skin has also been reported using medicated ghee (Kumar *et al.*, 1999b).

Table 6.10 Different types of medicated ghee used in *Ayurvedic* treatments.

Medicated ghee	Treatment	Medicated ghee	Treatment
Arjuna ghrit	Heat diseases	*Dhanyak ghrit*	Urinary tract infection
Anantadhya ghrit	Syphilis	*Dhanvantar ghrit*	Diabetes
Amruta ghrit	Leprosy	*Narach ghrit*	Ascites
Amrutadi ghrit	Leprosy	*Patoladhya ghrit*	Eye diseases
Amrutprash ghrit	Anti-ageing	*Palanbhedi ghrit*	Piles
Asta mangal ghrit	Child diseases	*Panchcoal ghrit*	GI disorders
Ashok ghrit	Leucorrhoea	*Panchgavya ghrit*	Hysteria
Ashwagandha ghrit	Gastro intestinal (GI) disorders	*Panchtikta ghrit*	Psoriasis
Kalyan ghrit	Madness	*Phal ghrit*	Female disorders
Indukant ghrit	GI disorders	*Bindu ghrit*	Digestive disorders
Kamdev ghrit	Leprosy	*Brahmi ghrit*	Memory disorders and hysteria
Kumar kalyan ghrit	Child diseases	*Shatavari ghrit*	Female disorders
Kulthadya ghrit	Stone	*Manha shiladi ghrit*	Asthma
Kushadhya ghrit	Stone	*Mahakalyanak ghrit*	Madness
Kantakari ghrit	Cough	*Maha badrick ghrit*	Leucoderma
Chavyadi ghrit	Piles	*Maha chaitas ghrit*	Hysteria
Changeri ghrit	Immunopotentiation	*Maha trifla ghrit*	Eye diseases
Chitrak ghrit	Spleen and liver disorder	*Rohtik ghrit*	Spleen and liver disorders
Jirkadhya ghrit	Improves digestion	*Varunadi ghrit*	Piles
Tikta ghrit	Leucoderma	*Som ghrit*	Infertility
Trifla ghrit	Eye diseases	*Chagladhya ghrit*	Tuberculosis
Dashmulshatpal ghrit	Cough	*Yastimadhu ghrit*	Ulcers
Dadimadi ghrit	Anemia	*Vasa ghrit*	Asthma
Durtradya ghrit	Leprosy		

Adapted from Pandya & Kanawjia (2002).

6.7 Conclusion

MF is unique among all edible fats and oils. It is extracted from cow's milk and, unlike other oils and fats, is prized in its native state. Whereas a vegetable oil may need to undergo refining to render it suitable for use as a food item, MF requires none of this, and indeed, any application of the standard refining techniques to MF renders it far less valuable because these techniques remove natural flavours, colour and naturally occurring antioxidants that are characteristic of milk fat. Thus, the process of extracting milk fat from milk is designed to preserve all of these properties.

With its very specific fatty acid profile with significant amounts of short-chain fatty acids, milk fat fatty acid composition varies only with seasonal or lactational changes, rendering the physical properties of milk fat inadequate in applications such as baking and confectionery. Butter, the form in which milk fat is seen by most consumers, does not spread directly from the refrigerator. Where the edible oil producer can turn to a selection of modification processes to overcome these failings, only a physical fractionation process without additives or any other step that might compromise flavour is suitable to modify milk fat properties so that the range of applications for use is increased. Thus, fractionation from the melt with some form of filtration process has become the ideal method for achieving this. The milk fat and subsequent fractions can be protected against deterioration and the natural nature of the parent milk fat is retained so that any food product may be described as 'being made from pure milk fat or butter'.

Ghee is the most widely consumed milk product in the Indian subcontinent. Of the four methods used for the production of ghee (e.g. (a) the indigenous MB process, (b) the DC method, (c) the CB method and (d) the prestratification (PS) method) the CB method, either batch or continuous process, is the most widely used system for the commercial manufacture of ghee. The chemical composition and physical properties of ghee are influenced by the milk of different species, type of feed, season and method of manufacture of ghee. Ghee has a characteristic flavour, which is due to a complex mixture of compounds produced during the various stages of processing. More than 100 flavour compounds, FFAs, carbonyls and lactones are the major groups of compounds contributing to ghee flavour. Ghee also has a characteristic grainy texture, which is very much appreciated by consumers. The characteristic flavour and texture of ghee are major features that distinguish such products from AMF, and also ghee has a longer shelf-life than AMF. In the Indian subcontinent, ghee in combination with various herbs has been used as medicine against several ailments.

6.8 Acknowledgements

Author DI would like to thank Dr A. K. H. MacGibbon and Mr W. F. van de Ven (both of Fonterra Research Centre, Palmerston North, New Zealand) for their assistance in writing this chapter.

References

Abdalla, M. (1994) Milk in the rural culture of contemporary Assyrians in the Middle-East. *Milk and Milk Products from Medieval to Modern Times*, (ed. P. Lysaght), pp. 27–39, Canongare Press, Edinburgh.

Abichandani, H., Sarma, S.C. & Bector, B.S. (1995) Continuous ghee making system–design, operation and performance. *Indian Journal of Dairy Science*, **48**, 646–650.

Abou-Donia, S.A. & El-Agamy, S.I. (1993) Ghee. *Encyclopedia of Food Science, Food Technology and Nutrition*, (eds. R. Macrae, R.K. Robinson & M.J. Sadler), Vol. 6, pp. 3992–3994, Academic Press, London.

Abraham, M.J. & Srinivasan, R.A. (1980) Effect of ripening cream with selected lactic acid bacteria on the quality of ghee. *Journal of Dairy Research*, **47**, 411–415.

Achaya, K. T. (1997) Ghee, vanaspati and special fats in India. *Lipids Technologies and Applications*, (eds. F. D. Gunstone & F. B. Padley), pp. 369–390, Marcel Dekker Inc., New York.

Achaya, K. T., Katrak, B. N. & Banerjee, B. N. (1946) The analytical constants of ghee. *Current Science*, **15**, 107–108.

Achliya, G., Barabde, U., Wadodkar, S. & Dorle, A. (2004a) Effect of *Bramhi Ghrita*, an polyherbal formulation on learning and memory paradigms in experimental animals. *Indian Journal of Pharmacology*, **36**, 159–162.

Achliya, G. S., Wadodkar, S. G. & Dorle, A. K. (2004b) Evaluation of hepatoprotective effect of *Amalkadi Ghrita* against carbon tetrachloride-induced hepatic damage in rats. *Journal of Ethnopharmacology*, **90**, 229–232.

Adhvaryu, R. P. (1994) *Sarangdhar Samhita: Ghrit and Taila*, Sahitya Samkul Publishers, Surat, India.

Agarwal, V. K. & Narayanan, K. M. (1976) Influence of mastitis on physico-chemical status of milk lipids I. Glycerides, free fatty acids and phospholipids. *Indian Journal of Dairy Science*, **29**, 83–87.

Agarwal, V. K. & Narayanan, K. M. (1977) Influence of mastitis on physico-chemical status of milk lipids III. Unsaponifiable constituents. *Indian Journal of Dairy Science*, **30**, 343–346.

Agarwal, V. K. & Narayanan, K. M (1979) Influence of mastitis on physico-chemical status of milk lipids IV. Oxidative and hydrolytic properties. *Indian Journal of Dairy Science*, **32**, 84–88.

Amr, A. S. (1990) Role of some aromatic herbs in extending the stability of sheep ghee during accelerated storage. *Egyptian Journal of Dairy Science*, **18**, 335–344.

Anderson, A. J. C. (1962) Section 2: neutralisation or de-acidification. *Refining Oils and Fats for Edible Purposes*, (ed. P. N. Williams), 2nd edn., pp. 28–213, Pergamon Press, Oxford.

Aneja, R. P., Mathur, B. N., Chandan, R. C. & Banerjee, A. K. (2002) Fat-rich products. *Technology of Indian Milk Products*, (ed. P. R. Gupta), pp. 183–190, Dairy India, New Delhi.

Aneja, R. P. & Murthi, T. N. (1990) Conjugated linoleic acid contents of Indian curds and ghee. *Indian Journal of Dairy Science*, **43**, 231–238.

Angelo, I. A. & Jain, M. K. (1982) Physico-chemical properties of ghee prepared from the milk of cows and buffaloes fed with cottonseed. *Indian Journal of Dairy Science*, **35**, 519–525.

Anonymous (2003a) European Union: destination of exports of butteroil and butter concentrates. *ZMP Dairy Review*, 111.

Anonymous (2003b) *Dairy Processing Handbook*, 2nd edn., pp. 293–299, Tetra Pak Processing Systems A/B, Lund, Sweden.

Armughan, C. & Narayanan, K. M. (1979) Grain formation in ghee (butter fat) as related to structure of triglycerides. *Journal of Food Science and Technology*, **16**, 242–247.

Bajwa, U. & Kaur, A. (1995) Effect of ghee (butteroil) residue and additives on physical and sensory characteristics of cookies. *Chemie, Mikrobiologie, Technologie der Lebensmittel*, **17**, 151–155.

Basu, K. P. & Nath, H. P. (1946) The effects of fats on calcium utilisation in human beings. *Indian Journal Medical Research*, **34**, 27–31.

Bector, B. S., Abichandani, H. & Sarma, S. C. (1996) shelf-life of ghee manufactured in continuous ghee making system. *Indian Journal of Dairy Science*, **49**, 398–405.

Bector, B. S. & Narayanan, K. M. (1972) The role of milk phospholipids in the autoxidation of butterfat. Part III. Synergistic and metal inactivation action. *Indian Journal of Dairy Science*, **25**, 222–227.

Bector, B. S. & Narayanan, K. M. (1974a) Effect of thermal oxidation on the glycerides of ghee. *Indian Journal of Dairy Science*, **27**, 292–293.

Bector, B. S. & Narayanan, K. M. (1974b) Effect of thermal oxidation on the fatty acid composition of ghee. *Journal of Food Science and Technology*, **11**, 224–226.

Bector, B. S. & Narayanan, K. M. (1974c) Effect of thermal oxidation on the physico-chemical constants of ghee. *Indian Journal of Dairy Science*, **27**, 90–93.

Bector, B. S. & Narayanan, K. M. (1975a) Comparative stability of unsaponifiable constituents of ghee during thermal oxidation. *Indian Journal of Nutrition and Dietetics*, **12**, 178–180.

Bector, B. S. & Narayanan, K. M. (1975b) Carbonyls in thermally oxidized ghee. *Indian Journal of Dairy Science*, **28**, 211–215.

Bector, B. S. & Narayanan, K. M. (1977) Effect of free fatty acids on thermal oxidation behaviour of ghee. *Indian Journal of Animal Sciences*, **47**, 185–189.

Bekele, E. & Kassaye, T. (1987) Traditional Borana milk processing-efficient use of subtle factors needs further research work. *International Livestock Centre for Africa (ILCA) Newsletter*, **6**(4), 4–5.

Belury, M. A., Nickel, K. P., Bird, C. E. & Wu, Y. (1996) Dietary conjugated linoleic acid and modulation of phorbol ester skin tumor promotion. *Nutrition and Cancer*, **26**, 149–157.

van den Berg, J. C. T. (1988) *Dairy Technology in the Tropics and Subtropics*, Pudoc, Wageningen, Netherlands.; Cited from Sserunjogi, M. L., Abrahamsen, R. K. & Narvhus, J. (1998) Current knowledge of ghee and related products. *International Dairy Journal*, **8**, 677–688.

Bhaskar, A. R. (1997) *Supercritical Fluid Processing of Milk fat: Modelling Fractionation and Applications*, PhD Dissertation, Cornell University, Ithaca.

Bhaskar, A. R., Rizvi, S. S. H., Betoli, C., Fay, L. B. & Hug, B. (1998) A comparison of physical and chemical properties of milk fat fractions obtained by two processing technologies. *Journal of the American Oil Chemists' Society*, **75**, 1249–1264.

Bhaskar, A. R., Rizvi, S. S. H. & Sherbon, J. W. (1993) Anhydrous milkfat fractionation with continuous countercurrent supercritical carbon dioxide. *Journal of Food Science*, **58**, 748–752.

Bhatia, T. C. (1978) *Ghee Making With Centrifugally Separated Butterfat and Serum*, MSc Thesis, Kurukshetra University, Kurukshetra, India.

Bhatia, I. S., Kaur, N. & Sukhija, P. S. (1978) Role of seed phosphatides as antioxidants for ghee (butter fat). *Journal of the Science of Food and Agriculture*, **29**, 747–752.

Bindal, M. P. & Jain, M. K. (1973a) A note on the unsaponifiable matter of ghee and its quantitative relationship. *Indian Journal of Animal Sciences*, **43**, 900–902.

Bindal, M. P. & Jain, M. K. (1973b) Studies on the cholesterol content of cow and buffalo ghee. *Indian Journal of Animal Sciences*, **43**, 918–924.

Bindal, M. P. & Jain, M. K. (1973c) Minor unsaponifiable constituents of ghee prepared from cow and buffalo milk. *Indian Journal of Animal Sciences*, **43**, 1054–1067.

Bindal, M. P. & Jain, M. K. (1973d) A simple method for estimation of lanosterol in ghee. *Indian Journal of Dairy Science*, **26**, 76–78.

Breitschuh, B. (1998) *Continuous Dry Fractionation of Milk Fat–Application of High Shear Fields in Crystallization and Solid-Liquid Separation*, Doctor of Technical Science Dissertation, Swiss Federal Institute of Technology, Zürich.

Breitschuh, B., Drost, M. & Windhaab, E. J. (1999) Process development for continuous crystallisation and separation of fat crystal suspension. *Chemical Engineering and Technology*, **22**, 425–428.

Chand, R., Kumar, S., Srinivasan, R. A., Batish, V. K. & Chander, H. (1986) Influence of lactic bacteria cells on the oxidative stability of ghee. *Milchwissenschaft*, **41**, 335–336.

Chatterjee, K. L. (1977) Shelf-life of ghee produced and stored under commercial conditions. *Indian Dairyman*, **29**, 797–803.

Chaudhary, S. M., Vyas, S. H., Upadhyay, K. G. & Thakar, P. N. (1980a) Keeping quality of ghee in relation to its copper content. *Gujarat Agricultural University Research Journal*, **6**, 98–104.

Chaudhary, S. M., Vyas, S. H., Upadhyay, K. G. & Thaker, P. N. (1980b) Organoleptic study on butter and ghee prepared from cream with added copper. *Gujarat Agricultural University Research Journal*, **6**, 130–132.

Chauhan, P. & Wadhwa, B. K. (1987) Comparative evaluation of ghee in tin and polyethylene packages during storage. *Journal of Food Processing and Preservation*, **11**, 25–30.

Chin, S. F., Liu, W., Storkson, J. M., Ha, Y. L. & Pariza, M. W. (1992) Dietary sources of conjugated dienoic isomers of linoleic acid, a newly recognised class of anticarcinogens. *Journal of Food Consumption and Analysis*, **5**, 185–197.

van Dam, P. H., Eshuis, J. J., Hogervorst, W. & Noomen, S. (1998) *Fractional crystallization of triglyceride fats*. **World Patent Application**, WO 9835001.

De, S. (1980) Indian dairy products. *Outlines of Dairy Technology*, pp. 382–466, Oxford University Press, New Delhi.

De, S., Unnithan, K. P. & Guglani, J. L. (1978) Studies on outturn of ghee from butter and cream. *Indian Journal of Dairy Science*, **31**, 376–378.

Deffense, E. (1987) Multi-step fractionation and spreadable butter. *Fett Wissenschaft Technologie*, **13**, 3–8.

Deffense, E. (1993) Milk fat fractionation today: A review. *Journal of the American Oil Chemists' Society*, **70**, 1193–1201.

Deffense, E. (2000) Dry fractionation technology in 2000. *European Journal of Lipid Science and Technology*, **102**, 234–236.

Dijkstra, A. J. & Maes, P. J. (1983) Separating solid and liquid phases in edible oils – by centrifugal sieving or decantation with short residence time. **European Patent Application**, EP 88949.

Dinesh, P., Boghra, V. R. & Sharma, R. S. (2000) Effect of antioxidant principles (phenolics and phospholipids) isolated from mango (*Mangifera indica L.*) seed kernels on oxidative stability of ghee (butter fat). *Journal of Food Science and Technology*, **37**, 6–10.

Eyer, H. (2000) Milchfett fraktionieren für neue Produkte (Milkfat fractionation for new products). *Schwiezerische Milchzeitung/Le Latier Romand*, **38**, 19.

FAO/WHO (1997) *Draft Revised Standard for Milk Fat Products*, (A–2) 37–39, Food and Agriculture Organization of the United Nations, Rome.

FAO/WHO (2006) Codex Alimentarius standard for butter, A-1-1971, revised 1–1999, amended 2003 and 2006, Food and Agriculture Organization of the United Nations, Rome. http://www.codexalimentarius.net/web/standard_list.jsp.

Fennema, O. R. (ed.) (1985) *Food Chemistry*, 2nd edn., Marcel Dekker Inc., New York.

Fjaervoll, A. (1970a) Anhydrous milkfat fractionation offers new applications for milk fat. *Dairy Industries*, **35**, 239.

Fjaervoll, A. (1970b) Fractionation of milkfat. *XVIII International Dairy Congress*, **IE**, 239.

Fogerty, A. C., Ford, G. L. & Svomos, D. (1988) Octadeca-9,11-dienoic acid in foodstuffs and in the lipids of human blood and breast milk. *Nutrition Reports International*, **38**, 937–944.

Fritsche, J. & Steinhart, H. (1998) Amounts of conjugated linoleic acid (CLA) in German foods and evaluation of daily intake. *Zeitschrift fur Lebensmittel Untersuchung und -Forschung*, **206**, 77–82.

Fulzele, S. V., Satturwar, P. M., Joshi, S. B. & Dorle, A. K. (2003) Study of the immunomodulatory activity of *Haridradi ghrita* in rats. *Indian Journal of Pharmacology*, **35**, 51–54.

Gaba, K. L. & Jain, M. K. (1973) A note on the flavour changes in ghee on storage: their sensory and chemical assessment. *Indian Journal of Animal Sciences*, **43**, 67–70.

Gaba, K. L. & Jain, M. K. (1974) Comparative appraisal of the total carbonyls in fresh and stored desi ghee. *Indian Journal of Dairy Science*, **27**, 81–89.

Gaba, K. L. & Jain, M. K. (1975) A note on isolation and characterization of dicarbonyls of ghee. *Indian Journal of Animal Sciences*, **45**, 696–697.

Gaba, K. L. & Jain, M. K. (1976) Head-space carbonyls in fresh and stored desi ghee. *Indian Journal of Dairy Science*, **29**, 1–6.

Gaba, K. L. & Jain, M. K. (1977) Presence of ketoglycerides in buffalo butterfat. *Indian Journal of Dairy Science*, **30**, 84–85.

Ganguli, N. C & Jain, M. K. (1972) Ghee: its chemistry, processing and technology. *Journal of Dairy Science*, **56**, 19–25.

Gibon, V. (2005) Crystallisation: practices and future developments. Paper presented at *Fractionation–Current Status and Future prospects in a Low-trans World*, 22–23 November, SCI Oils & Fats Group, Ghent, Belgium.

Gibon, V. & Tirtiaux, A. (2000) Winterisation, dewaxing, fractionation - crystal clear. *World Conference and Exhibition on Oilseed Processing and Utilization*, November 2000, Cancun.

Griffith, R. & Hammond, E. G. (1989) Generation of Swiss cheese flavour components by the reaction of amino acids with carbonyl compounds. *Journal of Dairy Science*, **72**, 604–613.

Guleria, S. P. S., Vasudevan, P., Madhok, K. L. & Patwardhan, S. V. (1983) Use of tomato seed powder as an antioxidant in butter and ghee. *Journal of Food Science and Technology*, **20**, 79–80.

Gupta, R. & Prakash, H. (1997) Association of dietary ghee intake with coronary heart disease and risk factor prevalence in rural males. *Journal of the Indian Medical Association*, **95**(3), 67–69, 83.

Gupta, S., Sukhija, P. S. & Bhatia, I. S. (1977) Role of amino acids as antioxidants for ghee. *Indian Journal of Dairy Science*, **30**, 319–324.

Gupta, S., Sukhija, P. S. & Bhatia, I. S. (1978) Effect of moisture content on the keeping quality of ghee. *Indian Journal of Dairy Science*, **31**, 266–271.

Gupta, S., Sukhija, P. S. & Bhatia, I. S. (1979) Role of phenolics and phospholipids as antioxidants for ghee. *Milchwissenschaft*, **34**, 205–206.

Ha, Y. L., Grimm, N. K. & Pariza, P. Y. (1989) Newly recognised anticarcinogenic fatty acids: identification and quantification in natural and processed cheeses. *Journal of Agricultural and Food Chemistry*, **37**, 75–81.

Hamid, A. D. (1993) *The Indigenous Fermented Foods of the Sudan: A Study in African Food and Nutrition*, CAB International, Wallingford.

Hartel, R. W. & Kaylegian, K. E. (2001) Advances in milk fat fractionation: technology and applications. *Crystallization Processes in Fats and Lipid Systems*, (eds. N. Garti & K. Sato), pp. 329–355, Marcel Dekker, New York.

Hosono, A., Elliot, J. A. & McGugan, W. A. (1974) Production of ethylesters by some lactic acid and psychrotrophic bacteria. *Journal of Dairy Science*, **57**, 535–539.

IDF (1977) *Anhydrous Milkfat, Anhydrous Butteroil or Anhydrous Butterfat, Butteroil or Butterfat, Ghee: Standards of Identity*, Standard 68A, International Dairy Federation, Brussels.

IDF, (1996) *Oxidized Sterols*. Bulletin **315**, 52–58, International Dairy Federation, Brussels, Belgium.

Illingworth, D. (2002) Fractionation of fats. *Physical Properties of Lipids*, (eds. A. G. Marangoni & S. S. Narine), pp. 411–447, Marcel Dekker, New York.

Illingworth, D. & Bissell, T. G. (1994) Anhydrous milkfat production and applications in recombination. *Fats in Food Products*, (eds. D. P. J. Moran & K. K. Rajah), pp. 111–154, Blackie Academic & Professional, Glasgow.

Illingworth, D. & Hartel, R. W. (1999) Crystallisation kinetics studies on anhydrous milkfat. Paper presented at *23rd World Congress*, International Society for Fat Research, Brighton.

Illingworth, D., MacGibbon, A. K. H. & van der Does, Y. E. (1990) Pilot-scale multi-stage fractionation of milkfat from the melt. Paper presented at *81st Annual Meeting of the American Oil Chemists' Society*, April 22–25, Baltimore.

Jacobson, M. S. (1987) Cholesterol oxides in Indian ghee: possible cause of unexplained high risk of atherosclerosis in Indian immigrant populations. *Lancet (British Edition)*, **ii**(8560), 656–658.

Jebson, R. S. (1970) Fractionation of milkfat into high and low melting point components. *XVIII International Dairy Congress*, **IE**, pp. 239.

Joshi, S. (1998) *A Comparative Pharmaco-clinical Study of Churna, Ghrita and Sharkara of Yashti-madhu in Parinama Shoola with Special Reference to Duodenal Ulcer*, MD Thesis, Gujarat Ayurved University, Jamnagar, India.

Joshi, N. S. & Thakar, P. N. (1994) Methods to evaluate deterioration of milk fat–a critical review. *Journal of Food Science and Technology*, **31**, 181–196.

Joshi, C. H. & Vyas, S. H. (1976a) Studies on buffalo ghee: I. Seasonal variation in fatty acid composition and other properties of buffalo ghee. *Indian Journal of Dairy Science*, **29**, 7–12.

Joshi, C. H. & Vyas, S. H. (1976b) Studies on buffalo ghee: II. Various conditions affecting the granulation of ghee. *Indian Journal of Dairy Science*, **29**, 13–17.

Joshi, C. H. & Vyas, S. H. (1977) Seasonal variation in phospholipids content of buffalo ghee. *Gujarat Agricultural University Research Journal*, **3**, 26–28.

Kaur, N., Sukhija, P. S. & Bhatia, I. S. (1982) A comparison of seed phosphatides and synthetic compounds as antioxidants for cow and buffalo ghee (butter fat). *Journal of the Science of Food and Agriculture*, **33**, 576–578.

Kaylegian, K. E., Hartel, R. W. & Lindsay, R. C. (1993) Applications of modified milk fat in food products. *Journal of Dairy Science*, **76**, 1782–1796.

Kaylegian, K. E. & Lindsay, R. C. (1994) *Handbook of Milkfat Fractionation Technology and Applications*, American Oil Chemists' Society Press, Champaign.

Kehar, N. D., Krishnan, T. S. & Chanda, R. (1956) *Studies on Fats, Oils and Vanaspati*, pp. 85–90, Manager of Publications, Veterinary Research Institute, Izatnagar, India.; Cited from Pandya, N. C. & Kanawjia, S. K. (2002) Ghee: a traditional neutraceutical. *Indian Dairyman*, **54**, 67–75.

Kellens, M. (2000) Oil modification processes. *Edible Oil Processing*, (eds. W. Hamm & R. J. Hamilton), pp. 129–173, Sheffield Academic Press, Sheffield.

Kempanna, C. & Unnikrishnan, V. (1985) The role of carotene in the oxidative stability of ghee. *Indian Journal of Dairy Science*, **38**, 329–331.

Kuchroo, T. K. & Narayanan, K. M. (1972) Effect of addition of antioxidants on the oxidative stability of buffalo ghee. *Indian Journal of Dairy Science*, **25**, 228–232.

Kuchroo, T. K. & Narayanan, K. M. (1973) Preservation of ghee. *Indian Dairyman*, **25**, 405–407.

Kuchroo, T. & Narayanan, K. M. (1977) Distribution and composition of phospholipids in ghee. *Indian Journal of Animal Sciences*, **47**, 16–18.

Kumar, M. V., Sambaiah, K. & Lokesh, B. R. (1999a) Effect of dietary ghee–the anhydrous milk fat, on blood and liver lipids in rats. *Journal of Nutritional Biochemistry*, **10**, 96–104.

Kumar, M. V., Sambaiah, K & Lokesh, B. R. (2000) Hypocholesterolemic effect of anhydrous milk fat ghee is mediated by increasing the secretion of biliary lipids. *Journal of Nutritional Biochemistry*, **11**, 69–75.

Kumar, M. V., Sambaiah, K., Mangalgi, S. G., Murthy, N. A. & Lokesh, B. R. (1999b) Effect of medicated ghee on serum lipid levels in psoriasis patients. *Indian Journal of Dairy and Biosciences*, **10**, 20–23.

Kumar, N. & Singhal, O. P. (1992) Effect of processing conditions on the oxidation of cholesterol in ghee. *Journal of the Science of Food and Agriculture*, **58**, 267–273.

Kurup, M. P. G. (2002) Smallholder dairy production and marketing in India: Constraints and opportunities. Rangnekar D. and Thorpe W. (eds), *Smallholder dairy production and marketing – opportunities and constraints: Proceedings of a South – South workshop held at NDDB, Anand, India, 13–16 March 2001*. NDDB (National Dairy Development Board), Anand, India, and ILRI (International Livestock Research Institute), Nairobi, Kenya. pp. 65–87.

Lal, D., Rai, T., Santha, I. M. & Narayanan, K. M. (1984) Standardization of a method for transfer of phospholipids from ghee-residue to ghee. *Indian Journal of Animal Sciences*, **54**, 29–32.

MacGibbon, A. K. H (1988) Thermal analysis of milkfat and butter. *Chemistry in New Zealand*, **52**(3), 59.

MacGibbon, A. K. H. & McLennan, W. D. (1987) Hardness of New Zealand patted butter: seasonal and regional variation. *New Zealand Journal of Dairy Science and Technology*, **22**, 143–156.

Maes, P. J. & Dijkstra, A. J. (1985) *Process for separating solids from oils*. **United States Patent Application**, 4542036.

Mehta, S. R. & Wadhwa, B. K. (1999) Chemical quality of ghee prepared by microwave process. *Indian Journal of Dairy Science*, **52**, 134–141.

Merai, M., Boghra, V. R. & Sharma, R. S. (2003) Extraction of antioxygenic principles from *Tulsi* leaves and their effects on oxidative stability of ghee. *Journal of Food Science and Technology*, **40**, 52–57.

Munro, D. S., Cant, P. A. E., Mac Gibbon, A. K. H., Illingworth, D., Kennett, A. & Main, A. J. (1992) Concentrated milk fat products. *The Technology of Dairy Products*, (ed. R. Early), pp. 117–145, Blackie and Sons Ltd., Glasgow.

Munro, D. S., Cant, P. A. E., MacGibbon, A. K. H., Illingworth, D. & Nicholas, P. (1998) Concentrated milkfat products. *The Technology of Dairy Products*, (ed. R. Early), 2nd edn., pp. 198–227, Blackie Academic & Professional, Glasgow.

Munro, D. S. & Illingworth, D. (1986) Milkfat based food ingredients: present and potential products. *Food Technology in Australia*, **38**, 335–337.

Narayanan, K. M., Ramamurthy, M. K. & Bhalerao, V. R. (1966) Effect of processing on the phospholipid content of milk fat. *XVII International Dairy Congress*, **C**, 215–218.

Nath, B. S. & Ramamurthy, M. K. (1988a) Effect of browning compounds on the autoxidative stability of ghee (milk fat). *Indian Journal of Dairy Science*, **41**, 116–119.

Nath, B. S. & Ramamurthy, M. K. (1988b) Cholesterol in Indian ghee. *Lancet (British Edition)*, **2**(8601), 39.

Nath, B. S., Usha, M. A. & Ramamurthy, M. K. (1996) Effect of deep-frying on cholesterol oxidation in ghee. *Journal of Food Science and Technology*, **33**, 425–426.

Nhavi, N. G. & Patwardhan, V. N. (1946) The absorption of fats from the human intestine. *Indian Journal Medical Research*, **34**, 49–58.

Nicolosi, R. J., Courtemanche, K V., Laitinen, L., Scimeca, J. A. & Huth, P. J. (1993) Effect of feeding diets enriched in conjugated linoleic acid on lipoproteins and aortic atherogenesis in hamsters. *Circulation*, **88**, 1–457.; Cited from Parodi, P. W. (1996) Milk fat components – possible chemopreventive agents for cancer and other diseases. *Australian Journal of Dairy Technology*, **51**, 24–32.

Nielsen, J. H., Olsen, C. E., Jensen, C. & Skribsted, L. H. (1996) Cholesterol oxidation in butter and dairy spread during storage. *Journal of Dairy Research*, **63**, 159–167.

Niranjan, T. G. & Krishnakantha, T. P. (2000a) Membrane changes in rat erythrocyte ghosts on ghee feeding. *Molecular and Cellular Biochemistry*, **204**, 57–63.

Niranjan, T. G. & Krishnakantha, T. P. (2000b) Effect of ghee feeding on rat platelets. *Nutrition Research*, **20**, 1125–1138.

Niranjan, T. G. & Krishnakantha, T. P. (2001) Effect of dietary ghee – the anhydrous milk fat on lymphocytes in rats. *Molecular and Cellular Biochemistry*, **226**, 39–47.

Norris, R. (1976) Fractionating fats: butter spreadable over wide temperature range. **New Zealand Patent Application**, 172 101.

Pal, M. & Rajorhia, G. S. (1975) Technology of ghee I-Effect of multiple separation of cream on the phospholipids content of ghee. *Indian Journal of Dairy Science*, **28**, 8–11.

Pandya, T. N. (1996) Ghrit. *Ayurveda Research Journal*, **17**(9), 1–4.

Pandya, N. C. & Kanawjia, S. K. (2002) Ghee: a traditional neutraceutical. *Indian Dairyman*, **54**, 67–75.

Pantulu, P. C. & Ramamurthy, M. K. (1988) Keeping quality and texture of ghee packed in aluminum pouches and glass bottles. *Indian Journal of Dairy Science*, **41**, 509–510.

Parmar, S. S. & Sharma, R. S. (1990) Effect of mango (*Mangifera indica L.*) seed kernels pre-extract on the oxidative stability of ghee. *Food Chemistry*, **35**, 99–107.

Parodi, P. W. (1994) Conjugated linoleic acid: an anticarcinogenic fatty acid present in milk fat. *The Australian Journal of Dairy Technology*, **49**, 93–97.

Patel, R. D. (1979) *Studies on the Fatty Acids and Glycerides Composition of Milk Fat Under Different Dietary Conditions*, Report on PL-480 Scheme, Sardar Patel University, Gujarat, India.

Patel, R. S., Mathur, R. K., Sharma, P., Sheth, S. A. & Patel, N. N. (2006) Industrial method of ghee making at Panchmahal dairy, Godhra. *Indian Dairyman*, **58**, 49–55.

Patel, R. S. & Rajorhia, G. S. (1979) Antioxidative role of curry (*Murraya koenigi*) and betel (*Piper betel*) leaves in ghee. *Journal of Food Science and Technology*, **16**, 158–160.

Podmore, J. (1994) Fats in bakery and kitchen products. *Fats in Food Products*, (eds. D. P. J. Moran & K. K. Rajah), pp. 213–253, Blackie Academic and Professional, London.

Prasad, R. & Pandita, N. N. (1987) Variations in the cholesterol content of butter fat. *Indian Journal of Dairy Science*, **40**, 55–57.

Prasad, C. R. & Subramanian, R. (1992) Qualitative and comparative studies of cholesterol oxides in commercial and home-made Indian ghee. *Food Chemistry*, **45**, 71–73.

Pruthi, T. D. (1980) Phospholipid content of ghee prepared at higher temperatures. *Indian Journal of Dairy Science*, **33**, 265–267.

Pruthi, T. D. (1984) Distribution of phospholipids between the solid and liquid portions of ghee. *Indian Journal of Dairy Science*, **37**, 175–176.

Pruthi, T. D., Kapoor, C. M. & Pal, R. N. (1972a) Phospholipid content of ghee prepared by direct clarification and pre-stratification methods. *Indian Journal of Dairy Science*, **25**, 233–235.

Pruthi, T. D., Kapoor, C. M. & Pal, R. N. (1972b) Browning potential of different milk phospholipids from butterfat. *Milchwissenschaft*, **27**, 698–699.

Pruthi, T. D., Raval, N. P. & Yadav, P. L. (1973) A method for recovering phospholipids from ghee residue. *Indian Journal of Dairy Science*, **26**, 151–154.

Pruthi, T. D. & Yadav, P. L. (1998) Preparation of phospholipids rich ghee. *Indian Journal of Dairy Science*, **51**, 63–65.

Punjrath, J. S. (1974) New development in ghee making. *Indian Dairyman*, **26**, 275–278.

Puravankara, D., Boghra, V. & Sharma, R. S. (2000) Effect of antioxidant principles isolated from mango (*Mangifera indica* L) seed kernels on oxidative stability of buffalo ghee (butter fat). *Journal of the Science of Food and Agriculture*, **80**, 522–526.

Rajorhia, G. S. (1980) Advances in the preservation of of ghee and regional preferences for quality. *Indian Dairyman*, **32**, 745–750.

Rajorhia, G. S. (1993) Ghee. *Encyclopaedia of Food Science, Food Technology and Nutrition*, (eds. R. Macrae, R. K. Robinson & M. J. Sadler), Vol. 4, pp. 2186–2192, Academic Press, London.

Rajput, D. S & Narayanan, K. M. (1968) Effect of ripening of cream and storage of butter on phospholipids content of ghee. *Indian Journal of Dairy Science*, **21**, 112–116.

Ramamurthy, M. K. (1980) Factors affecting the composition, flavour and textural properties of ghee. *Indian Dairyman*, **32**, 765–768.

Ramamurthy, M. K., Narayan, K. M. & Bhalerao, V. R. (1968) Effect of phospholipids on keeping quality of ghee. *Indian Journal of Dairy Science*, **21**, 62–63.

Ramamurthy, M. K., Narayan, K. M. & Bhalerao, V. R. (1969) The role of ghee residue as antioxidants in ghee. *Indian Journal of Dairy Science*, **22**, 57–58.

Ramamurthy, M. K. & Narayanan, K. M. (1966) A method for estimation of phospholipids in milk and milk products. *Indian Journal of Dairy Science*, **19**, 45–47.

Ramamurthy, M. K. & Narayanan, K. M. (1974) Partial glycerides of buffalo and cow milk fats. *Milchwissenschaft*, **29**, 151–154.

Ramamurthy, M. K. & Narayanan, K. M. (1975) Glyceride composition of buffalo and cow milk fats. *Indian Journal of Dairy Science*, **28**, 163–170.

Ramesh, B. & Bindal, M. P. (1987) Influence of fatty acid composition on softening point and melting point of cow, buffalo and goat ghee. *Indian Journal of Dairy Science*, **40**, 94–97.

Rao, K. S. H. (1981) Packaging of ghee. *Indian Dairyman*, **33**, 149–152.

Rao, C. N., Rao, B. V. R., Rao, T. J. & Rao, G. R. R. M. (1985) shelf-life of buffalo ghee prepared by different methods by addition of permitted antioxidants. *Asian Journal of Dairy Research*, **3**, 127–130.

Rao, C. N., Rao, T. J. & Reddy, K. K. (1987) Studies on the shelf-life of ghee prepared by different methods. *Journal of Research APAU*, **15**, 136–138.

Rao, D. V. & Singh, H. (1990) Use of antioxidants in ghee – a review. *Indian Journal of Dairy Science*, **43**, 359–363.

Ray, S. C. & Srinivasan, M. R. (1976) Pre-stratification method of ghee making. *ICAR Research Series*, **8**, 14.

Richardson, T. & Korycka-Dahl, M. (1988) Lipid oxidation. *Developments in Dairy Chemistry-2*, (ed. P. F. Fox), pp. 241–263, Applied Science Publishers, London.

Rodenburg, H. (1973) Use of subsidised EEC concentrated butter (as butter oil or dried butter) in bakery and confectionery. *Kakao und Zucker*, **25**, 496–499.

Santha, I. M. & Narayanan, K. M. (1978) Antioxidant properties of ghee residue as affected by temperature of clarification and method of preparation of ghee. *Indian Journal of Animal Sciences*, **48**, 266–271.

Santha, I. M. & Narayanan, K. M. (1979) Studies on the constituents responsible for the antioxidant property of ghee residue. *Indian Journal of Animal Sciences*, **49**, 37–41.

Sarkar, S., Kuila, R. K. & Misra, A. K. (1993) Production of ghee with enhanced flavour by the dairy industry. *Indian Dairyman*, **45**, 337–342.

Saxena, R. B. & Daswani, M. T. (1996) Study of dairy Ghrita. *Ayurveda Research Journal*, **17**, 8–10.

Semwal, A. D., Sharma, G. K. & Arya, S. S. (1997) Antioxidant activity of turmeric (*Curcuma longa*) in sunflower oil and ghee. *Journal of Food Science and Technology*, **34**, 67–69.

Sethna, K. & Bhatt, J. V. (1950) Optimum temperature for ghee making. *Indian Journal of Dairy Science*, **3**, 39–41.

Shekar, S. & Bhat, G. S. (1983) Effect of dissolved oxygen level on keeping quality of milk and milk fat (ghee). *Indian Journal of Dairy Science*, **36**, 147–150.

Singh, R. B., Naiz, M. A., Ghosh, S., Beegom, R., Rastogi, V., Sharma, J. P. & Dube, G. K. (1996) Association of trans fatty acids (vegetable ghee) and clarified butter (Indian ghee) intake with higher risk of coronary artery disease in rural and urban populations with low fat consumption. *International Journal of Cardiology*, **56**, 289–298.

Singh, S. & Ram, B. P. (1978) Effect of ripening of cream, manufacturing temperature and packaging materials on flavour and keeping quality of ghee. *Journal of food Science and Technology*, **15**, 142–145.

Singh, S., Ram, B. P. & Mittal, S. K. (1979) Effect of phospholipids and method of manufacture on flavour and keeping quality of ghee. *Indian Journal of Dairy Science*, **32**, 161–167.

Singhal, O. P., Gangili, N. C. & Dastur, N. N. (1973) Physico-chemical properties of different layer of ghee (clarified butterfat). *Milchwissenschaft*, **28**, 508–511.

Sripad, S., Kempanna, C. & Bhat, G. S. (1994) Distribution of browning compounds in ghee and ghee-residue and their influence on keeping quality of ghee. *Indian Journal of Dairy and Biosciences*, **5**, 16–19.

Sripad, S., Kempanna, C. & Bhat, G. S. (1996) Effect of alcohol extract of defatted ghee residue on the shelf-life of ghee. *Indian Journal of Dairy and Biosciences*, **7**, 82–84.

Sserunjogi, M. L., Abrahamsen, R. K. & Narvhus, J. (1998) Current knowledge of ghee and related products. *International Dairy Journal*, **8**, 677–688.

Stark, W., Urbach, G. & Hamilton, J. S. (1976) Volatile compounds in butteroil: V. The quantitative estimation of phenol, *o*-methoxyphenol, *m*- and *p*-cresol, indole, skatole by cold-finger molecular distillation. *Journal of Dairy Research*, **43**, 479–489.

Steggerda, F. R. & Mitchell, H. H. (1951) The calcium balance of adult human subjects on high and low fat (Butter) diets. *Journal of Nutrition*, **45**, 201–11.

Tandon, R. N. (1977) Effect of feeding cottonseed to milch animals on the opacity pattern of ghee and changes in its physico-chemical constants on storage. *Indian Journal of Dairy Science*, **30**, 341–343.

Tanzia, N. & Prakash, J. (2000) Storage stability of ghee with added spices. *Indian Journal of Nutrition and Dietetics*, **37**, 20–27.

Tirtiaux, F (1976) Le fractionnement industrial de corps gras par cristallisation dirigee. *Procede Tirtiaux Oleagineux*, **31**, 279–285.

Unnikrishnan, V. & Rao, M. B. (1977) Copper and iron in ghee and their influence on oxidative deterioration. *Journal of Food Science and Technology*, **14**, 164–169.

Urbach, G. & Gordon, M. H. (1994) Flavours derived from fats. *Fats in Food Products*, (eds. D. P. J. Moran & K. K. Rajah), pp. 347–405, Blackie Academic and Professional, London.

Urbach, G., Stark, W. & Forss, D. A. (1972) Volatile compounds in butteroil: II. Flavour and flavour thresholds for lactones, fatty acids, phenols, indole and skatole in deodorized synthetic butter. *Journal of Dairy Research*, **39**, 35–47.

Wadhwa, B. K. & Bindal, M. P. (1995) Ghee residue: a promise for simulating flavours in vanaspati (hydrogenated edible vegetable oils) and butteroil. *Indian Journal of Dairy Science*, **48**, 469–472.

Wadhwa, B. K., Bindal, M. P. & Jain, M. K. (1979) Comparative study on the keeping quality of butteroil, flavour induced butteroil and ghee. *Indian Journal of Dairy Science*, **32**, 227–230.

Wadhwa, B., Bindal, M. P. & Jain, M. K. (1980) A comparative study in the lactone profile of fresh and stored ghee prepared from cow and buffalo milk. *Indian Journal of Dairy Science*, **33**, 21–23.

Wadhwa, B. K. & Jain, M. K. (1984) Studies on lactone profile of ghee. IV. Variations due to temperature of clarification. *Indian Journal of Dairy Science*, **37**, 334–342.

Wadhwa, B. K. & Jain, M. K. (1985a) Simulation of ghee flavour in butter oil with synthetic flavouring compounds. *Journal of Food Science and Technology*, **22**, 24–27.

Wadhwa, B. K. & Jain, M. K. (1985b) Studies on lactone profile of ghee. III. Variations due to method of preparation. *Indian Journal of Dairy Science*, **38**, 31–35.

Wadhwa, B. K. & Jain, M. K. (1990) Chemistry of ghee flavour – a review. *Indian Journal of Dairy Science*, **43**, 601–607.

Wadhwa, B. K. & Jain, M. K. (1991) Production of ghee from butter oil – a review. *Indian Journal of Dairy Science*, **44**, 372–374.

Wadhwa, B. K., Kaur, S. & Jain, M. K. (1991a) Enhancement in the shelf-life of flavoured butter oil by natural antioxidants. *Indian Journal of Dairy Science*, **44**, 119–121.

Wadhwa, B. K., Kaur, S. & Jain, M. K. (1991b) Enhancement of the shelf-life of flavored butter oils by synthetic antioxidants. *Journal of Food Quality*, **14**, 175–182.

Wadodakar, U. R., Murthi, T. N. & Punjrath, J. S. (1996) Isolation of ghee volatiles by vacuum degassing, their separation and identification using gas chromatography/mass spectrometry. *Indian Journal of Dairy Science*, **49**, 185–198.

Wadodkar, U. R., Punjrath, J. S. & Shah, A. C. (2002) Evaluation of volatile compounds in different types of ghee using direct injection with gas chromatography-mass spectrometry. *Journal of Dairy Research*, **69**, 163–171.

Warner, J. N. (1976) Ghee. *Principles of Dairy Processing*, Wiley Eastern Ltd., Delhi, India, pp 253–259.

Watt, G. N. (1982) *The Inversion of High Fat Creams*, Thesis in Partial Fulfilment of the Post-Graduate Diploma in Dairy Technology, Massey University, Palmerston North.

Willner, T., Sitzmann, W. & Munch, E.-W. (1990) Cocoa butter replacers produced by dry fractionation. Paper presented at *81st Annual Meeting of the American Oil Chemists' Society*, April 22–25, Baltimore.

Yadav, J. S. & Srinivasan, R. A. (1984) Qualitative and quantitative changes in flavour characteristics of ghee made from ripened cream. *Indian Journal of Dairy Science*, **37**, 350–356.

Yadav, J. S. & Srinivasan, R. A. (1985a) Mechanism of flavour enhancement in ghee by starter culture. *Indian Journal of Dairy Science*, **38**, 196–202.

Yadav, J. S. & Srinivasan, R. A. (1985b) Biotechnological parameters for preparing ghee (clarified butterfat) with improved flavour using a starter culture. *New Zealand Journal of Dairy Science and Technology*, **20**, 29–34.

Yadav, J. S. & Srinivasan, R. A. (1985c) Effect of ripening, cream with *Streptococcus lactis* subsp. *diacetylactis* on the flavour of ghee (clarified butterfat). *Journal of Dairy Research*, **52**, 547–553.

Yadav, J. S. & Srinivasan, R. A. (1992) Advances in ghee flavour research. *Indian Journal of Dairy Science*, **45**, 338–348.

7 Production of Yellow Fats and Spreads

B.K. Mortensen

7.1 Introduction

Milk fat has a unique flavour, a desirable mouth-feel, and a natural image, but ordinary butter has some shortcomings as the product does not always meet the consumer's expectations concerning spreadability and health.

According to Codex Alimentarius (FAO/WHO, 1971-revised in 1999 and amended in 2003 and 2006), butter is a fatty product derived exclusively from milk. It is also specified that butter must contain (g 100 g^{-1}) minimum 80 fat, maximum 16 water, and 2 milk solids-not-fat (SNF). National legislation in many countries used to have the same specifications, and the mixture of milk fat and fat of other origin was prohibited. Over the years, these restrictions almost completely prevented product development in the butter sector, but things have changed, and nowadays legislation in most countries permits the manufacturing and marketing of a multitude of butter-related products, varying in both fat content and fat composition.

How this has influenced the butter market in a number of countries was reviewed by Mann (1997). The main conclusion seems to be that, even though the sales of regular butter has declined in most countries, the total volume of sales of milk fat in butter and related products is more stable, although exact figures are normally not accessible.

A variety of butter-related products has been developed and marketed in recent years. One of the driving forces behind this is the dairy industry's attempt to maintain or even increase its market share by improving the functionality of butter products and by meeting the consumer's request for more healthy products. The main headlines in this development have been and still are lower fat content, modified fat composition and modified functionality.

7.2 Legislations

Codex Alimentarius, the joint FAO/WHO food standards programme, adopted a standard for dairy fat spreads in 2006 (FAO/WHO, 2006), applying it to products intended for use as spreads for direct consumption or for further processing. Dairy fat spreads are defined as milk products relatively rich in fat in the form of a spreadable emulsion principally of the water-in-oil type that remains in solid phase at a temperature of 20°C.

The raw materials, which should be milk and/or products obtained from milk including milk fat, may have been subjected to any appropriate processing (e.g. physical modifications including fractionation) before their use. The milk fat content (g 100 g^{-1}) shall be no less than 10 and less than 80, and shall represent at least two-thirds of the dry matter. The standard also comprises a long list of food additives, which may be used, and the list includes colours, emulsifiers, preservatives, stabilisers/thickeners, acidity regulators, antioxidants, anti-foaming agents and flavour enhancers.

In addition, it is stated in the standard that the name of the product shall be 'Dairy Fat Spread', although other names may be used if allowed by national legislation in the country of retail sale. Concerning declaration of the fat content, it is specified that the milk fat content shall be declared either as a percentage by mass, or in grams *per* serving.

The FAO/WHO (2007) also issued a standard for the production and labelling of fat spreads and blended spreads. The products, which are covered by the standard available, are foods that are plastic or fluid emulsions, principally of water and edible fats and oils of vegetable or animal (including milk) or marine origin. Fats and oils, which have been subjected to processes of physical or chemical modification including fractionation, inter-esterification or hydrogenation, are also included. It is stated in the standard that the total fat content (g 100 g^{-1}) in blends should be at least 80, and in blended fat spreads below 80. It is also stated that blended spreads are products in which milk fat is more than 3 g 100 g^{-1} of the total fat content. In fat spreads, the milk fat content must be no more than 3 g 100 g^{-1} of the total fat content; however, a higher minimum percentage of milk fat in fat spreads may be specified in accordance with the requirements of the country of the retail sale.

Within the European Union (EU), Council Regulation number 2991/94 (EU, 1994) provides standards regarding the trade of certain spreadable fats, such as butter, blends and spreads, which are allowed to be manufactured and marketed in member countries, and requirements for the composition of such products regarding milk fat, water and solids non-fat contents. The regulation concerning butter products states that these products are in the form of a solid, malleable emulsion, principally of the water-in-oil type, derived exclusively from milk. The requirements for composition and the authorised name of the products are shown in Table 7.1. For products with a fat content (g 100 g^{-1}) between 41 and 62, the term 'reduced fat' may be used as an additional description and, for products with a fat content of 41 g 100 g^{-1} or less, the term 'low-fat' or 'light' may be used.

Table 7.1 European Union (EU) regulations for butter products.

Product	Interval for milk fat (g 100 g^{-1})	Water (g 100 g^{-1})	Solids non-fat (g 100 g^{-1})
Butter (full-fat)	Min. 80	Max. 16	Max. 2
	Max. 90		
Butter (3/3-fat)	Min. 60	No requirements	No requirements
	Max. 62		
Butter(half-fat)	Min. 39	No requirements	No requirements
	Max. 41		
Dairy spread (x% butter)	Less 39	No requirements	No requirements
	More than 41 and less than 60	No requirements	No requirements
	More than 62 and less than 80	No requirements	No requirements

After EU (1994).

Table 7.2 European Union (EU) regulations for spreadable blended products.

Name	Range for fat content (g 100 g^{-1})
Blend	Min. 80 and max. 90
3/4 fat blend	Min. 60 and max. 62
1/2 fat blend	Min. 39 and max. 41
Blended spread x%	Less than 39; more than 41 and less than 60; more than 62 and less than 80

After EU (1994).

The same regulation lays down rules for the production and marketing of products composed of plant and/or animal products, so-called blends, and defined as products in the form of a solid, malleable emulsion principally of the water-in-oil type, derived from solid and/or liquid vegetable and/or animal fats suitable for human consumption, with a milk fat content between 10 and 80 g 100 g^{-1} of the fat content (EU, 1994). Table 7.2 shows the requirements for fat content in such products and the name to be used; there are no requirements regarding the content of water and solids non-fat. For blended spreads, additional descriptions, such as 'fat-reduced', 'extra fat-reduced' or 'light', may be used with the same rules as are valid for butter.

7.3 Dairy fat spreads

7.3.1 *Introduction*

Nutritional recommendations to reduce the amount of fat, in particular saturated fat, in the human diet have resulted in a decline in the consumption of full-fat butter products, and a growing interest in low-fat spreads. The idea of producing dairy spreads with a reduced fat content is by no means new. A pioneer work was carried out in the United States more than 50 years ago at the University of Wisconsin in Madison (Weckel, 1965), where a product, already at that time called 'Dairy Spread', was developed and intended for use on bread, crackers and sandwiches. This product contained 45 g fat 100 g^{-1} and no ingredients other than milk components. A mixture of milk fat in the form of cream or butter, milk solids and water, to which was added lactic acid, starter distillate, colour and salt was prepared; the mixture was heated, homogenised, packed while still warm and finally cooled. The product resembled cream cheese, had an oil-in-water emulsion structure, a rather tough and sticky consistency, although easily spreadable at low temperatures, and a limited shelf-life. Even though the product, on the basis of consumer tests, appeared to be acceptable, it failed to succeed in the marketplace.

Progress in the development of yellow spreads gathered momentum in the 1960s when scraped surface heat exchangers (SSHE) and stirred crystallisers, normally used in the production of margarine, were employed for the production of low-fat spreads, that is, water-in-oil emulsion products. SSHE were already at that time known in the dairy industry as they were used in the production of regular butter by the emulsification methods, such as the Creamery Package and the Cherry-Burrell Gold'n Flow, which were developed in the

United States (Munro, 1986). Till date, similar technology is used in the Ammix process that was developed in New Zealand (Truong & Munro, 1982; Munro, 1986).

At present, most low-fat spreads available on the market are of the water-in-oil emulsion type, but oil-in-water products are also known, and a considerable number of patents describing the manufacturing of these types of products have been granted over the years. However, oil-in-water products are rather different from butter in texture and consistency, and they have never really been successful as substitutes for butter.

Robinson & Rajah (2002) and Rajah (2004) reported extensive reviews on how the spread market has developed in a number of countries and described how the fat level in the products varied over the years. It started in the 1970s with spreads containing 40 g fat 100 g^{-1}, which created a lot of quality problems as it was difficult to combine 40 g fat 100 g^{-1} and the high water content into a stable product. Soon, 60 g fat 100 g^{-1} spreads were marketed, which was a much easier product to manufacture. However, further reductions in the fat content took place, and 20 g fat 100 g^{-1} spreads and almost non-fat spreads appeared in the late 1980s and early 1990s; mainly these spreads were formulated with protein-based fat substitutes.

After many years of discussions and considerations, the Codex Alimentarius (FAO/WHO, 2006) now defines dairy fat spreads as products exclusively obtained from milk. The product shall have a fat content (g 100 g^{-1}) no less than 10 and no more than 80. Furthermore, the fat phase shall represent at least two-thirds of the dry matter content of the products. The European Union (EU, 1994) has adopted a similar standard.

7.3.2 *Production technologies*

Many experiments concerning the manufacture of low-fat dairy spreads have been carried out through the years, and many methods of production have been patented claiming to give the desired spreadability, mouth-feel and flavour. The technologies available today are mainly based on utilising margarine technology, the inversion of cream and the direct increase of the water or air content in butter.

Margarine technology

The main principle in the margarine technology is that the fat and the water phases are prepared separately, mixed, emulsified, cooled and finally given an intensive mechanical working in a mixing device. This is the same process as used for production of recombined butter (Jebson, 1979).

As the interest in low-fat butter products emerged in Europe in the 1960s, many experiments were carried out using this technology. The fat phase, which is based on anhydrous milk fat (AMF), is melted in a tank where the temperature can be controlled. It is important that the temperature is high enough to avoid partial crystallisation, that is, the temperature should be at least 40°C. If the temperature is too low, the milk fat will start to crystallise, which will result in a grainy texture in the final product.

Spreads with 60 g fat 100 g^{-1} can normally be manufactured from regular AMF without the addition of emulsifiers. If a much lower fat content (e.g. 40 g 100 g^{-1} or even lower) is desired in the product, it will be necessary to use a mixture of AMF and a soft fat obtained by fractionation of milk fat because regular milk fat has too high an average melting point and it could otherwise be very difficult to obtain a sufficiently stable emulsion because the continuous liquid fat phase would be too small. Furthermore, the final product will be extremely firm and brittle at low temperatures and very soft at room temperature.

During the preparation of the fat phase, it will also be necessary to lower the interfacial tension between the fat and the water phase by using a suitable emulsifier in the formulation. Many different emulsifiers have been tested for creation of water-in-oil emulsions stable enough to be crystallised in an SSHE. Suitable emulsifiers like distilled monoglycerides or lecithin are added to the melted AMF. Normally, the emulsifiers are mixed into a small part of the milk fat, which is then added to the total quantity. The amount of emulsifiers added is often 0.5 to 1.25 g 100 g^{-1} of the fat phase. If a higher level of emulsifiers is used, a more stable emulsion is obtained, but then the product is difficult to destabilise in the mouth, which gives an unpleasant waxy mouth-feel. Flack (1997) reported the effect of types and blends of emulsifiers on the stability of a 20 and a 40 g 100 g^{-1} spread, respectively.

Fat substitutes, such as modified whey protein concentrate, are widely used in very low-fat spreads. It has also been reported that thermally stabilised microparticulated high amylose starch combined with maltodextrin is useful in oil-in-water spreads (Wursch & Tolstoguzow, 2000). The water phase has to be stabilised in low-fat spreads; otherwise phase separation and release of water during spreading of the product might occur. The water phase is prepared by mixing water with milk proteins, for example, buttermilk, skimmed milk powder, whey proteins or sodium caseinate. The added proteins have a dual function. First, they increase the viscosity of the aqueous phase and thereby help immobilise the system and second, they increase the flavour release by breaking the emulsion when the product is melted in the mouth. However, the addition of a high amount of proteins, that is, $\geq$12 g 100 g^{-1} of the aqueous phase, will result in undesirable off-flavours like bitterness and a chalky mouth-feel. Another disadvantage is the increased production costs caused by the high amount of expensive proteins. Burling *et al.* (2005) claimed that the protein concentration in spreads could be reduced to 2 to 5 g 100 g^{-1}, if a solution of heat denaturated whey protein concentrate with reduced calcium concentration is added. This low viscous water phase will gradually turn into a gel at low temperature (cold gelation) and it is, therefore, important to mix the solution into the fat phase before the gel formation is completed. The described procedure is claimed to stabilise both the oil-in-water and water-in-oil emulsions.

It will often be necessary to further immobilise the system by adding suitable stabilisers, such as starch, gelatine, alginate, pectin, inulin or carrageen. The choice of stabilisers is important as they may influence the taste of the products. Andersen and Hansen (2003) found, for instance, that the non-digestible and soluble hydrocolloid alginate contributes to the taste attributes of spreads, although it is normally claimed that alginate has a neutral taste. They also found that the typical butter flavour was reduced with increasing alginate concentration at the same time as a greasy mouth-feel evolved. As mentioned elsewhere, an important property of milk proteins in spreads is to destabilise the emulsion in the mouth, resulting in the desired meltdown profile. However, this ability to destabilise the emulsion is reduced when the protein is inactivated by the formation of an alginate–protein complex.

Salt, flavours (e.g. lactic acid) and preservatives (e.g. sodium, calcium, or potassium sorbate) might also be added to the water phase. It is important to note that lactic acid has the function of both flavour enhancement and microbial inhibition; it can be added directly in the form of a concentrate or indirectly in the form of a lactic acid starter culture. Charteris (1995b) discussed the possible additives, such as colouring material(s), flavourings and flavour enhancers, antioxidants, sweeteners and vitamins. These minor ingredients exhibit important functional properties and their use in the formulation should, therefore, be considered carefully.

After preparation of the fat and water phases, the two phases are mixed in a blending tank at a temperature above the melting point of the fat. It is important that the water phase is added slowly to the fat phase under intensive agitation so the two phases can be mixed into a homogeneous blend. It is also important that the two phases are at the same temperature and that no air is incorporated during the blending because this will destabilise the emulsion. After blending, the mixture is heated and cooled in an SSHE. The processed blend is then emulsified either by agitation or by pumping through a homogeniser at moderate pressure or through an emulsification pump where the creation of a water-in-oil emulsion takes place. A crucial point here is the stability of the emulsion. Proteins have, as mentioned earlier, a tendency to destabilise a water-in-oil emulsion and, if the emulsion is too unstable, it will be very difficult to avoid phase inversion into an oil-in-water emulsion in the following process. It is, therefore, vital that the composition and the properties of the water phase are carefully monitored and controlled so that the emulsion does not inverse during the crystallisation, which will spoil the texture in the final product.

The formed emulsion is pumped to an SSHE where, under the mechanical action of the equipment and fast cooling rate, it is crystallised in the form of a water-in-oil emulsion. The cooling rate and the in-line pressure in the cooler are most important, and should be controlled carefully as shearing forces that are too high might break the emulsion. An SSHE designed for heating or cooling of viscous products consists of a number of cooling tubes or cylinders linked in series or parallel, and the product is pumped through in a counter-current flow to the cooling medium, often ammonia, in the surrounding jacket. The cylinders are equipped with knives or blades mounted on a central rotor that continuously scrape the formed fat crystals from the cold inner surface of the tube and mix them to ensure uniform heat transfer (Anonymous, 2003). An example of an industrial scale SSHE is shown in Figure 7.1.

Between or after the cooling tubes, the emulsion is exposed to mechanical kneading in a special unit, which could be a sort of mixer or 'pinworker' (Figure 7.2). As a consequence, fat crystallisation takes place while the emulsion is intensively sheared by the central rotor that has a number of pins, which interact with static pins on the internal surface of the tube (Figure 7.3). The aim of the mechanical action is to further enhance crystallisation and break down the already formed crystal network to ensure a homogeneous product with optimal plasticity and spreadability. A detailed description of the crystallisation of milk fat in an SSHE, and the formation of fat crystals and crystal aggregates was reported by van Aken and Visser (2000).

As crystallisation takes time, the emulsion will contain a certain amount of super-cooled fat when leaving the SSHE. To leave time for further crystallisation, the partly crystallised water-in-oil emulsion could be lead through resting cylinders. These are large-diameter jacketed tubes cooled with water, which allow the product to crystallise under milder shear conditions often generated by internal sieve plates so that the amount of solid fat can increase and some inter-crystal bonds can develop. After the final cooling and working, the product is transferred to the packaging machine.

Fig. 7.1 An illustration of a scraped surface heat exchanger (SSHE). By permission of Gerstenberg Schröder A/S, Brøndby, Denmark.

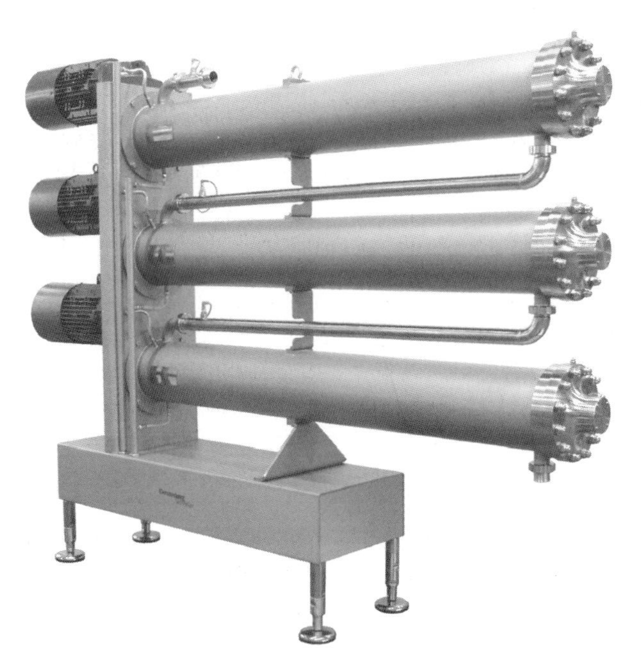

Fig. 7.2 Pin rotor machine. By permission of Gerstenberg Schröder A/S, Brøndby, Denmark.

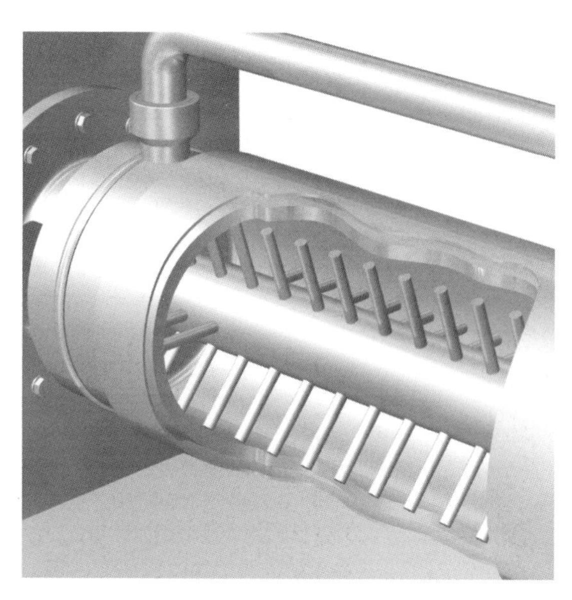

Fig. 7.3 A cross section view of a pin rotor. By permission of Gerstenberg Schröder A/S, Brøndby, Denmark.

Inversion of cream

Low-fat dairy spreads can also be produced directly from cream as described by Pedersen (1997) and the Gerstenberg Schröder Company (Anonymous, 2005). Cream with 40 g fat 100 g^{-1} is heated to 70°C, stabilised by 2 g 100 g^{-1} gelatine or carrageen and mixed with an emulsifier (e.g. 0.6 g 100 g^{-1} distilled monoglycerides). The emulsifier is melted together with a small portion of AMF and added to the cream. The mixture is then cooled to 18–20°C in an SSHE, and phase inverted in a high shear mixer. The formed water-in-oil emulsion is cooled, worked in a pin rotor mixer and finally cooled to 14°C and transferred to a packaging machine. The most important process parameters are the rotation speed of the rotor in the first chilling cylinder and the temperature of the cream when inverted.

Products made by this method are water-in-oil emulsions with a globular structure similar to that found in butter made by the churning method, which normally contains about 30–40% globular fat. This implies that the products' consistency and texture are closer to regular butter than products produced using the margarine technology. However, the product is not spreadable at refrigerator temperature and the addition of the soft fraction(s) of milk fat could, therefore, be an option to improve the spreadability of the product.

Pedersen (1997) also described the production of dairy spreads with 60 g fat 100 g^{-1}. Cream with 60 g fat 100 g^{-1}, at a temperature of 60°C, is cooled and inverted into a water-in-oil emulsion in an SSHE and worked in a pin rotor mixer before it is cooled to the packaging temperature. The texture of the product is very similar to the texture of regular butter, and it is claimed that no additives are needed to improve its spreadability.

Increased water content

In the traditional process for manufacturing of butter, either based on the batch churning method or on a continuous process the Fritz-method, a higher water content than 16 g

100 g^{-1} can be retained in the product by choosing a higher churning velocity, higher churning temperature, higher working intensity and/or increased dosage of water. However, there are limitations, as it will normally not be possible in this way to increase the water content to more than 25 to 30 g 100 g^{-1} and a proportional reduction in the fat content to about 70 g 100 g^{-1}. Alternatively, higher water content and a corresponding lower fat content can be obtained by adding a moisture phase with an increased viscosity. Nielsen (1993) described a process developed by Invensys APV where butter with a normal fat content (e.g. ~80 g 100 g^{-1}) is mixed with an aqueous solution of sodium caseinate, which is heated and chilled before it is added to the butter. The butter is pumped through a working section where it is softened by the mechanical action working under vacuum and transferred to a so-called butter homogeniser where a controlled amount of the aqueous solution is added. The blending is continued until the desired fat content and a homogeneous texture are reached. The product is then cooled to the packaging temperature in an SSHE and transferred to the packaging machine. The same author also reported that industrial plants based on this technology for production of spreads with 40 and 60 g fat 100 g^{-1} have been installed.

Increased air content

Regular butter contains 0.5 to 1.0 mL air 100 mL^{-1} if worked under vacuum. Increasing the air content will make the product more spreadable and, it could be argued, will also reduce fat consumption because the larger volume *per* gram implies that a smaller amount of butter is used for one serving.

Whipped butter is a special type of butter containing a considerable amount of air (e.g. up to 50 mL air 100 mL^{-1} or more) and, thereby, a lower fat content *per* volume. Whipped butter, which is made by incorporating air or an inert gas into the product, was originally developed in the United States and it has been widely used in North America for many years. The product can be manufactured in a rather uncomplicated batch process where the butter is worked in a batch mixer until it has reached the desired increase in volume. However, whipped butter can also be manufactured in a process where the butter is continuously softened in a mixing device like a pinworker or an inline mixer where a controlled amount of air is introduced. The product is then cooled in an SSHE and packed in plastic cups or tubs. Nitrogen is often used for the 'swelling' of the product to increase its oxidative stability. It is very important that the whipping temperature is not too high, as this will result in a final product with a very firm consistency and a crumbly texture. Fisker and Jansen (1970) found that the optimum whipping temperature was between 13 and 16°C.

The water dispersion of the product is similar to that found in ordinary butter, but the appearance of the product is very pale if no colour is added, and it has a solid, foam-like texture, which makes it easier to spread on bread than regular butter, although the product is rather brittle and not plastic like butter. Whipped butter is marketed in both salted and unsalted versions, and could also be produced with a lactic acid flavour. This product is used instead of traditional butter as a spread on bread or crackers, but can also be used as toppings on baked potatoes, pancakes and different hot dishes. The product is widely used in the catering sector for restaurants and airlines. However, as the density or weight of whipped butter is not the same as an equal measure of butter, it is not suited as a substitute for regular butter in baking recipes.

Whipped products can also be manufactured as low-fat products by weight. Lynch and O'Mahony (1999) described the production of a low-fat oil-in-water spread with a fat content between 10 and 45 g 100 g^{-1}, and a whipping overrun of at least 200%. The fat phase could be milk fat, but it is preferable to use a mixture of milk fat and vegetable oil with suitable emulsifiers and stabilisers. The emulsion is heated, homogenised and cooled, and the cold emulsion is whipped and packed. The product is claimed to be spreadable at all temperatures between 4 and 25°C.

The United States Department of Agriculture has issued an official standard for grades of whipped butter (Anonymous, 1994). In this standard, whipped butter is defined as a food product made by uniform incorporation of air or inert gas into butter made exclusively from milk, cream or a mixture of both, with or without salt, with or without additional colouring matter, and containing not less than 80 g fat 100 g^{-1}. The standard also describes typical quality defects, for example, acid, aged, coarse, free moisture, mealy or grainy and colour specks. Concerning the shelf-life of the product, it is stated that whipped butter should have a satisfactory quality after 7 days at 21°C.

7.3.3 *Quality aspects*

Dairy fat spreads of the water-in-oil emulsion type have a continuous fat phase formed as a matrix of solid and liquid fat in which microscopic droplets of the aqueous phase are distributed. The droplet size is most important for the product's ability to release flavour during melting in the mouth, and for its stability towards microbiological spoilage.

In butter, most of the water droplets have a diameter in the range of 1 to 5 μm, which results in a very good microbiological quality. However, in low-fat spreads with high water content, the water droplet size can increase dramatically because the dispersed aqueous phase becomes the major component and, as a consequence, the water droplets will be in much closer contact and tend to coalesce into larger droplets. Every stage of conversion of the emulsion between water-in-oil and oil-in-water states has been observed in low-fat spreads by Buchheim & Dejmek (1990). However, distribution of the water droplets in the range of 2 to 4 μm will give a very good stability against microbiological spoilage; it is still desirable for some of the water droplets to be larger (e.g. 10–20 μm) to give a better flavour release in the mouth. The overall quality perception of low-fat butter products seems to be that they still lack the texture, mouth-feel and flavour of full-fat products, although significant improvements have been obtained over the years by optimisation of the production technology.

One of the most important quality aspects related to low-fat dairy spreads is binding of the high water content. If the water is bound too tightly, the consistency becomes very tough and sticky, and the flavour release becomes very slow as the emulsion does not easily destabilise in the mouth. If the water is bound too loosely, syneresis might occur, meaning that water will leak out of the product. If low-fat spreads with a poor binding of water are spread on a slice of bread and the bread is left for a couple of hours, part of the water will be sucked into the bread, which then obtains an unpleasant pasty and spongy consistency while a thin and transparent film of fat is left on top of the bread. In addition, the distribution of the water droplets in the products is of vital importance as spreads containing high proportions of water are vulnerable to microbiological deterioration. This can be controlled by ensuring a finely dispersed aqueous phase, which is stabilised against coalescence by adding suitable emulsifiers and stabilisers. Charteris (1995a) concluded that physical entrapment of the aqueous phase in small discrete droplets restricts microbial growth to a small number of larger droplets. In practice, live bacteria are not detectable in droplets with a diameter less than 15 to 20 μm. Also, Charteris (1996) gave a comprehensive overview of the spoilage micro-organisms and pathogens that may develop in the aqueous phase of edible spreads and suggested a scheme for microbiological quality assurance of such products. van Zijl and Klapwijk (2000) reviewed the microbiological safety and quality of yellow fat products. They emphasised that the salt content and the pH of the composition of the aqueous phase are very important for obtaining an acceptable microbiological stability. However, the stability could be enhanced by adding preservatives like sorbic acid or its salts in concentrations of about 0.2 g 100 g^{-1} in spreads with a fat content below 60 g 100 g^{-1}. Typical spoilage organisms are moulds and lipolytic yeasts; no reports of food-borne illness caused by dairy spreads are known.

As already mentioned elsewhere, low-fat spreads are vulnerable products because of the high water content, and a high level of hygiene during processing is, therefore, highly recommended. A critical point is post-heat treatment contamination, and it is very important that the equipment used is properly cleaned and disinfected before the start of production. It is also vital that the product is protected against contamination during packaging; furthermore, the final product has to be stored at low temperatures.

7.4 Blends and blended spreads

7.4.1 *Introduction*

Because of its fatty acid composition, butter has poor spreadability at low temperatures and, although different softening technologies like temperature cycling of cream or work-softening have been developed, the product cannot compete with margarine produced from low-melting vegetable oils when spreadability is the main concern. Also from a health point of view, the high content of saturated fat in butter has been heavily criticised for many years, and at the same time vegetable oils with a high content of polyunsaturated fatty acids have been praised for their positive effect on cardiovascular diseases.

For these reasons, it was an obviously good idea to mix butter and vegetable oils with a high content of unsaturated fatty acids, and this approach has been debated for a very long time. However, legislation in most countries prohibited the production and marketing of such blends. In 1969, the product Bregott® containing 80 g fat 100 g^{-1}, of which 64 g 100 g^{-1} was milk fat and 16 g 100 g^{-1} soya bean oil, was launched in Sweden. Bregott® was well accepted by consumers and very quickly obtained a considerable market share; as a consequence, the dairy industry and farmers in many other countries realised that it was in their own interest to market such products and started lobbying for a change in legislation. Today, full-fat blends are produced and marketed with considerable success in many countries and, in several markets, blends have obtained a higher market share than regular butter.

As already discussed in Section 7.3.2, it is difficult to manufacture low-fat spreads with a satisfactory consistency if the fat phase is based exclusively on milk fat, and it was, therefore, obvious to develop a low-fat blended spread. Again, Sweden pioneered such developments and the first product of this type was launched in 1975 (Andersson, 1991). The product (lätt & lagom®) had a fat content of 40 g 100 g^{-1}, of which 24 g 100 g^{-1} was milk fat and 16 g 100 g^{-1} was soya bean oil, and it contained 48 g 100 g^{-1} water, 7.5 g 100 g^{-1} protein in the form of buttermilk concentrate, 1.2 g 100 g^{-1} salt and different additives.

The vegetable oils, which are used in most blended products today, are soya bean or rapeseed oil that replace 20 to 30 g 100 g^{-1} or even more of the milk fat. However, it is also possible to use other types of oil, for example, sunflower oil, olive oil or cotton seed oil. Rapeseed oil, which is low in erucic acid and high in oleic acid, is the preferred vegetable oil used in the production of blends and blended spreads in Northern Europe. The Codex Alimentarius (FAO/WHO, 2007) has issued a standard in which blends and blended spreads are defined as edible fats and oils of vegetable or animal (including milk) or marine origins. Blends shall have a fat content no less than 80 g 100 g^{-1}, and blended spreads shall have a fat content less than 80 g 100 g^{-1}. In the EU, Council Regulation No. 2991/94 (EU, 1994) required that the milk fat content in blends and blended spreads shall be between 10 and 80 g 100 g^{-1} of the total fat content. Over the years, a number of blended spreads have been developed and marketed in many countries and described in many reviews (Mann, 1990, 1997, 1998, 2000; Rajah, 2004).

7.4.2 *Production technologies*

The technologies available today for production of blends and blended spreads are mainly based on the churning method, the margarine making method and the inversion of cream method.

Churning method

If the production of blends is based on the batch churning of cream, vegetable oil could be added either in the cream storage tank or in the churn before churning. The mixture is then churned in the normal way but, since the blend is softer than regular butter, it is necessary to churn the mixture of cream and oil at a temperature lower than is normally used. It is also important to keep the temperature low by spraying the churn with cold water. During the churning process, the vegetable oil is emulsified into the cream forming small fat globules and, when the formation of butter grains takes place, these small fat globules are incorporated. The grains formed in this way are very soft, and it is necessary to lower the temperature by washing the grains with cold water after draining of the buttermilk; otherwise, it will be very difficult to squeeze sufficient water out of the blended product during working. The working time has to be shorter and less intensive than working of normal butter because overworking will result in a very greasy product.

If the production is based on a continuous churning process, vegetable oil might be added either in the cream storage tank or injected directly into the pipeline transferring the cream from the storage tank to the continuous churn. To obtain a low churning temperature, it is necessary to cool the cream as well as the oil to about 5 to 7°C before churning. However, it is vital that the churning temperature is chosen according to the type of vegetable oil, as it is most important that the oil does not start to crystallise at the churning temperature.

Churning of cream at 5 to 7°C requires a high energy input, which will generate undesired heat due to friction. To compensate for that, the fat content in the cream and oil mixture could be increased to 42 to 44 g 100 g^{-1}, which will speed up the phase inversion. If the continuous churn is equipped with facilities for cooling the butter granules by spraying with either cold water or chilled buttermilk, this cooling stage should be used. The mechanical working of the blend must be sufficient to obtain a homogeneous structure in the product. However, it is important that the working is not overdone as this will result in a greasy consistency in the final product.

Addition of vegetable oil to the cream before churning has one big disadvantage as residues of the oil will inevitable end up in the buttermilk. This will greatly limit the utilisation of the buttermilk in other products like cream cheese or in the manufacture of milk powder. To overcome this problem, Invensys APV has developed a procedure where the vegetable oil is added downstream from the separation section in the continuous butter machine after the buttermilk has been drained. Nielsen (1993) and Berntsen (1999) described this method in detail, which implies that the vegetable oil is added in the first working section together with other additives, such as lactic acid concentrate, salt and starter culture. To keep the temperature of the product low during working of the grains, the working sections should be cooled be circulating ice water in the jacket of this section of the machine.

When the product leaves the second working section, the temperature will typically be 13 to 16°C. It is then cooled to 10 to 12°C in an SSHE, which makes the product firmer and easier to handle during packaging. Before packaging, the product passes through a couple of in-line mixers ensuring that the oil is mixed homogeneously into the butter. It is important that the intensity of the working in the mixers and the temperature of the product when it leaves the SSHE are adjusted to avoid overworking of the product. Furthermore, this method of processing can also be used for the production of blended spreads with reduced fat content, if a higher amount of aqueous phase with increased viscosity is incorporated simultaneously with the vegetable oil. The increased viscosity in the aqueous phase could be obtained by adding proteins, for example, sodium caseinate (Nielsen, 1993).

A similar technology for the production of blended spreads was developed by Gerstenberg Schröder A/S (Ljunggren & Hansen, 2006). Butter, vegetable oil and an aqueous phase are pumped into a high shear mixer or emulsifying equipment where they are blended. The vegetable oil added can be liquid, semi-solid or solid. When a homogeneous mix is obtained, it is cooled in an SSHE before packaging. A significant temperature increase will occur in the mixer, and it is important to keep the temperature low during the processing stages. In addition, a comparable technology is described in a recent patent application (Nielsen *et al.*, 2005). It is claimed that the product obtained by adding solidified vegetable oil has good spreadability and improved stand-up properties without oiling-off at increasing temperature.

Margarine making method

The process used for the production of blends by the margarine technology is similar to the one described earlier. Skimmed milk powder or another protein source is dissolved in water and followed by heating the reconstituted powder. Melted AMF is mixed with vegetable oil, heated to ~40°C and blended with a suitable emulsifier. The two phases are mixed, emulsified and crystallised in an SSHE. The product, which still contains a lot of super-cooled fat, is then pumped through a pinworker where the super-cooled fat is crystallised under mechanical action.

Dungey *et al.* (1996) investigated the effect of different processing parameters on the texture of blends made by the margarine making method. They studied the effect of selected process variables like flow rate, coolant temperature, rotation speed in the cooler, rotational speed of the pinworker and the effect of the oil injection point and the proportion of oil injected. They found that the oil injection position was of great importance as partial or full injection of vegetable oil after the first cooling section gave a softer blend compared to incorporating the oil in the emulsion before processing. They also found that for each additional 1 g 100 g^{-1} of oil in the formula, the blend is softened by about 2%, whereas the type of vegetable oil did not significantly influence the firmness of the final product.

The margarine process can also be used for the production of blended spreads with reduced fat content. The preparation of the fat phase is in this case straight forward as it is not necessary to add fractionated milk fat because the addition of vegetable oil will secure a satisfactory consistency in the final product, at least, if the amount and the melting point of the vegetable oil are optimised. However, the preparation of the water phase is more complicated. Initially, when the first development of low-fat spreads started in the 1960s, the most common approach was to prepare a protein-free aqueous phase stabilised with hydrocolloids and combined with high amounts of emulsifiers, which resulted in products with an unpleasant waxy mouth-feel. To overcome this problem, low-fat spreads, such as the Swedish lätt & lagom®, should contain a high protein content to stabilise the aqueous phase, and this approach was developed in the 1970s. However, when the price of milk proteins increased during the 1980s and 1990s, spreads containing low- and medium-levels of protein were developed at the same time as the use of thickeners and stabilisers changed from gelatine to alginate, and later changed to starch (Robinson & Rajah, 2002).

In conclusion, mixing of the fat and water phases, creation of the water-in-oil emulsion, crystallisation of the emulsion in an SSHE and the final mechanical working of the product is similar to that described for the production of dairy fat spreads by the margarine making method.

Inversion of cream

Blends can also be produced in a process based on the inversion of cream as described by Anonymous (2003). Cooled, heat-treated cream with a fat content of ~40 g 100 g^{-1} is heated to 60°C in a plate heat exchanger (PHE), and concentrated to a fat content of ~80 g 100 g^{-1} in a special separator capable of concentrating the fat content of the cream without shattering the fat globules. The high-fat cream is then cooled to 20°C and pumped to a holding and pre-crystallisation tank where it is aged overnight.

When production starts, the cream is blended with an appropriate amount of vegetable oil, salt, starter culture and lactic acid concentrate, and the mixture is pumped by a high pressure pump through a battery of scraped surface coolers connected in series in two stages where the oil-in-water emulsion is turned into a water-in-oil emulsion. Between the two cooling stages a pinworker is placed, giving the emulsion a short but intensive mechanical working. The product, which is still partly liquid, is then pumped from the second cooling step into a butter silo where it hardens to the final consistency before packaging.

By this method of processing, even blended spreads with reduced fat content can be produced. In this case, the concentrated cream is diluted with water before further processing. This will result in a lower content of proteins and lactose, which is claimed to improve the flavour of the spread. It is also claimed that a further advantage of using concentrated cream as a base for production of blended spreads with reduced fat content is that no extra emulsifiers are needed, as the level of natural emulsifiers found in the cream is sufficient.

The technology, which is described earlier for the production of dairy fat spreads by inversion of the cream, can also by used for production of blended spreads (Pedersen, 1997). In this case, the vegetable oil is added to pre-crystallised cream before the phase inversion and crystallisation or added in the middle of the crystallisation process in the SSHE.

7.4.3 *Quality aspects*

If the composition of the fat phase is optimised by mixing butter or AMF and vegetable oil, low-fat blended spreads can be manufactured with acceptable consistency, and it is also easier to obtain a sufficient binding of the aqueous phase compared to a situation where the origin of the fat phase is exclusively from the milk fat. As far as butter flavour is concerned, at least 20–30 g 100 g^{-1} milk fat has to be included in the formulation of the spread to give a distinct buttery flavour (Andersson, 1991).

However, unsaturated lipids may undergo spontaneous autoxidation during processing and distribution and the addition of vegetable oils will, therefore, decrease the oxidative stability of the product. Addition of antioxidants will decrease the reaction rate and prolong the induction of the oxidation, but it cannot completely prevent autoxidation.

It has been reported (Youcheff & Hensler, 2001) that the oxidative stability of milk fat is enhanced by blending with vegetable oils. Mortensen and Danmark (1981) investigated the oxidative stability of blends produced from milk fat and soya bean oil and found that blends, as expected, were more sensitive towards oxidation than regular butter. It was confirmed that addition of soya bean oil had an antioxidative effect in the beginning of the storage period as the natural content of tocopherols in soya bean oil delayed the fat oxidation. However, when the antioxidative effect was exhausted, the oxidation accelerated as a consequence of the high content of unsaturated fat.

In this connection, cholesterol oxidation products (oxysterols) are of particular interest as they are known to initiate atherosclerosis and to have mutagenic effect. The accumulation of oxysterols in blends during storage was investigated by Nielsen *et al.* (1996), and they found that the accumulation in a product with a fat phase consisting of 75 g 100 g^{-1} milk fat and 25 g 100 g^{-1} rapeseed oil was higher than in butter. However, the concentration of oxysterols was quite low, but highly dependent on the storage time and temperature. In a review of the formation of oxidised sterols in dairy products, Appelqvist (1996) concluded that fresh butter has very low levels, but confirmed that accumulation took place during storage at a rate dependent on the storage time and temperature. He also stated that there is a substantial increase in cholesterol oxidation products in butter on heating under normal kitchen practices.

These results emphasise the importance of a close monitoring of the quality of the vegetable oil used and of the conditions under which it is stored. Also, blended spreads are just as vulnerable towards microbiological spoilage as dairy spreads, so the statements in Section 7.3.3 are also valid for these products.

7.5 Products with modified functionality

7.5.1 *Introduction*

The functional properties of milk fat are not suitable for all applications. One of the main limitations is the rheological properties, which are influenced by the chemical composition of the fat. The easiest way to overcome this shortcoming is to blend butterfat with fat of other origin as described in Section 7.4, but this is not an option for all applications. It might, therefore, be necessary to use other methods for tailoring milk fat for specific applications. This can be done by changing the feeding of the dairy cows, by fractionation of milk fat, and by inter-esterification of milk fat with vegetable oil.

7.5.2 *Production technologies*

Feeding process

It has been known for many years that major changes in milk fat composition can be achieved by supplying dietary fat to dairy cows. To control such an approach, it is important to distinguish between the origins of the different fatty acids found in milk fat. Short-chain fatty acids with less than 12 carbons in the chain are produced in the udder by *de novo* synthesis using the very volatile fatty acids produced in the rumen. Even numbered fatty acids with 12 to 16 carbons in the chain are partly synthesised *de novo* and partly transferred from dietary fat *via* circulating blood lipids. Odd numbered fatty acids with 15 and 17 carbons in the chain are synthesised by micro-organisms in the rumen, and fatty acids with more than 17 carbons in the chain are entirely supplied through the diet. This means that the fatty acid composition of a typical dairy cow diet is very different from the composition of the milk fat because fatty acids synthesised by ruminal micro-organisms and by the *de novo* synthesis in the mammary tissue dilute dietary fatty acids before incorporation into milk fat. Furthermore, hydrogenation of unsaturated fatty acids in the rumen makes fatty acids in milk fat more saturated than fatty acids in the diet (Grummer, 1991).

Hermansen (1995) studied the possibility of predicting the milk fatty acid composition from the composition of the dietary fat; he showed that feeding cows with oil/oilseeds could increase the content of C18:1 to $\sim$40 g 100 g^{-1} and the content of polyunsaturated fatty acids to 7 g 100 g^{-1}. If the fatty acids are protected against hydrogenation, they will pass the rumen and be released and absorbed in the abomasum. A protection technique was originally developed in Australia (Scott *et al.*, 1970) in which vegetable oil with a high content of polyunsaturated fat was encapsulated in formaldehyde-treated casein. It has since then been shown in several investigations that feeding cattle with rumen-protected soya bean oil or rapeseed oil high in linoleic acids provides an opportunity to produce softer butter, which might even be spreadable at low temperatures. However, the keeping quality of the butter is reduced substantially, which is caused by oxidative instability (Kristensen *et al.*, 1974), and there are also concerns about possible transfer of formaldehyde to the milk.

Several studies have shown that production of milk with high concentrations of mono- and poly-unsaturated fatty acids without a corresponding increase in antioxidants decreases the oxidative stability of the milk as a result of an imbalance between pro- and anti-oxidants in milk. Vitamin E is often used as a supplement in animal feed even though it seems that oxidised flavour and concentration of vitamin E are not very well related (Nicholson, 1993). On the other hand, Stapelfeldt *et al.* (1999) confirmed that α-tocopherol protects milk fat from oxidation and found that the concentration of free radicals was negatively correlated with the concentration of α-tocopherol.

Omega-3 (ω-3) fatty acids are of interest, and Børsting and Weisbjerg (1989) studied the effect of supplementing the diet of dairy cows with protected vegetable oil, saturated fatty acids and protected fish oil. They found that protected fish oil increased the content of ω-3 fatty acids in milk fat to 4 to 6 g 100 g^{-1}, but unfortunately this increase was accompanied by a severe off-flavour.

Dairy products are the major source in human diet of conjugated linolenic acid (CLA), which is claimed to exhibit a number of health benefits like anticarcinogenic and antiatherogenic effect and, in recent years, CLA has been much debated. CLA is formed as a result of incomplete bio-hydrogenation of polyunsaturated fatty acids in the rumen. Typical concentrations of CLA in milk fat are 3 to 6 mg g^{-1} fat, but the level can vary widely among herds and among individual cows. The level of CLA can be increased by feeding, for example, sunflower oil rich in linolenic acid. Most of the studies of the effect of CLA have been carried out on experimental animals, and the effects on humans are still not conclusive. A number of patents describe CLA-enriched milk fat and the use thereof, but so far the commercial success has not been overwhelming.

In conclusion, there are possibilities for manipulating the composition of milk fat through feeding of the dairy cows. However, special feeding has some limitations concerning the economy. One limitation is the increased price of the feeding, and another, even more important, is the cost of separate collection of the milk and separate processing at the dairy factory.

Fractionation process

Milk fat consists of a mixture of triglycerides with different physical properties, for example, melting point, and this can be used in a fractionation process where milk fat is split up in fractions with different melting points. Different methods of fractionation have been developed. One of the simplest and most efficient methods is crystallisation from a solvent, such as acetone, but the use of organic solvents in food production and the consequent loss of butter flavours have prevented commercial application of this method.

The most common fractionation process in commercial operation today is one in which liquid AMF is crystallised in a tank where it is taken through a controlled temperature programme (Figure 7.4). During this process, triglycerides with high melting points will crystallise and, afterwards, the crystals are separated from the melt by vacuum or pressure filtration or by centrifugation, or by a combination of these methods. The cooling temperatures and the rate of crystallisation applied strongly influence the composition of the fat crystals, and the cooling rate influences the quantity of the crystallised fat and the size of the crystals. This so-called dry fractionation or crystallisation from the melt is relatively cheap and based on a physical process not involving solvents. Furthermore, this process has the least effect on flavour and is generally accepted by legislation.

The result of the fractionation is a high-melting fraction, the so-called stearin fraction, and a liquid fraction, the so-called olein fraction. These two fractions can be further fractionated by a multi-step fractionation process into even harder and softer fractions, respectively, as

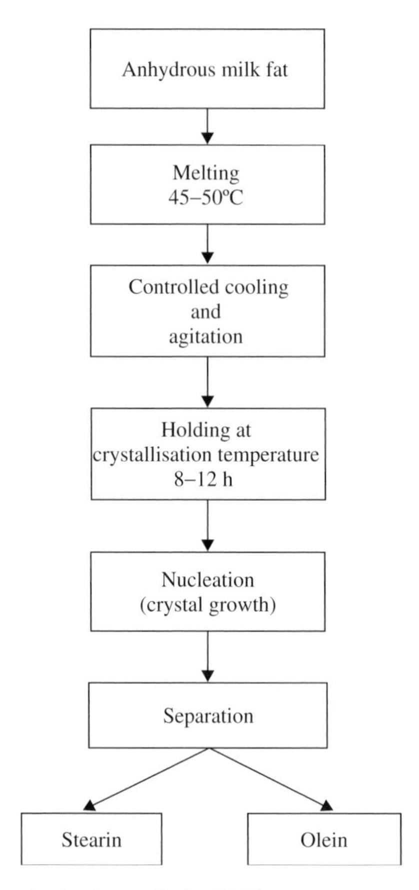

Fig. 7.4 Fractionation process of anhydrous milk fat (AMF).

described by Deffense (1993) and Deffense and Hartel (2003). It has also been reported by Kaufmann *et al.* (1982) that super-critical extraction can be used for fractionation of milk fat into a liquid and a solid fraction using carbon dioxide at 20 MPa and 80°C as a solvent. Commercialisation of this process has not been reported. A new trend in fractionation seems to be the use of membrane filtration, which is claimed to give a better separation between the soft and the solid fractions. No doubt, more research is needed before this technology can be put to industrial use.

Inter-esterification process

Inter-esterification of fats is a process in which the distribution of the fatty acids among the triglycerides is altered. The resultant product has the same total fatty acid composition as the starting material, but the composition of the triglycerides and the physical properties are changed. Inter-esterification can be carried out by both chemical and enzymatic means. Chemical inter-esterification is a concept that has been known for many years. Sodium methoxide is often used as catalyst and, if inter-esterification of milk fat with vegetable oil is carried out, randomisation of the fatty acid residues will be the result. Enzymatic inter-esterification with lipase has certain advantages over chemical inter-esterification, such as milder reaction conditions, which might be less harmful to the butter flavour. Depending on the lipase used, the result could be random, regiospecific or fatty acid specific. However, both chemical and enzymatic inter-esterification can give rise to off-flavours due to the release of short-chain fatty acids and to oxidation. In both cases, removal of off-flavours by deodorisation has to be done.

Several studies have shown that the spreadability of butter could be significantly improved by chemical and enzymatic inter-esterification of butter oil with rapeseed oil (Marangoni & Rousseau, 1998; Rousseau & Marangoni, 1999). However, losses in butter flavour and creation of off-flavours were evident, and it was concluded that butter flavour degradation has to be minimised before inter-esterification can be commercially used. Rønne *et al.* (2005) studied the batch and continuous processes for inter-esterification of milk fat and rapeseed oil based on the use of immobilised lipase enzyme. They found a tendency towards a reduction of the saturated fatty acids C14:0 and C16:0 and an increase of the long-chain saturated and unsaturated fatty acids C18:0 and C18:1 in the milk fat after the reaction. Even though they also found that the release of short-chain fatty acids during the process makes the sensory quality unacceptable for direct edible applications, they concluded that the process has a potential for industrial implementation. However, so far very little commercial interest in inter-esterification has emerged.

7.5.3 *Applications*

Milk fat fractions are often used as functional ingredients in other dairy products or in bakery, confectionery and chocolate products. A review of the applications of both the stearin and the olein fractions of milk fat was given by Deffense and Hartel (2003).

Dairy products

The most obvious application for low-melting fractions of milk fat is in butter and spreads to make the products more spreadable at low temperatures. It is also well known that addition of a high-melting fraction will make the butter more stable at high temperatures, which could be needed, for example, in tropical countries.

Possible uses of the fat fractions have been studied by Dolby (1970), who improved the spreadability of butter by recombining the fractions in different ratios. The fat fractions were emulsified with skimmed milk to form a cream with 40 g fat 100 g^{-1}, which was processed into butter in the normal way. Kankare and Antila (1974) simply added the liquid fat fraction to cream before churning and showed that the spreadability of the resulting butter at low temperature was markedly improved by adding the liquid fraction in amounts of up to 33 g 100 g^{-1} of the total fat content. Schäffer *et al.* (2000) later reported that addition of up to 25 g 100 g^{-1} of a milk fat fraction with a melting point of 10°C increased the spreadability of butter at low temperature considerably, while further addition had only a minor effect.

In butter and spreads, the amount of solid fat present is important for the rheological properties of these products over a broad temperature range. According to Lane (1998), a product with optimum spreadability from the refrigerator, good 'stand-up' properties and resistance to oiling-off at room temperature should have a solid fat content of 30 to 40 g 100 g^{-1} at 5°C, 10 to 20 g 100 g^{-1} at 20°C and below 3 g 100 g^{-1} at 35°C. To obtain a melting curve like this, it might be necessary to recombine high- and low-melting fractions, implying that the intermediate middle-melting fraction must be removed to obtain an ideal melting curve. From a technical point of view this is possible, but the increased productions cost might limit the utilisation of this procedure.

In ice cream, the balance between liquid and crystallised fat is important as too high a content of liquid fat results in an inferior quality and storage stability (Andreasen & Nielsen, 1998). The use of high-melting fractions of AMF in the mix has been suggested, but the advantage of this application compared to the use of regular milk fat or cream seems to be of minor importance.

Another application could be in low-fat cheese, where low-melting milk fat fractions might be able to improve the rheological properties. Very little knowledge is available in this area, but it might be worthwhile investigating.

Bakery products

Fat is an important ingredient in baked products as described by Flack (1997). Fat provides lubricant in the mouth and gives an easy breakdown and dispersion of the product in the mouth because fat interrupts the gluten network of the dough and prevents the formation of a hard cohesive texture. The main advantage of using butter in the bakery industry is the buttery flavour and the creamy mouth-feel it imparts. An overview of the use of milk fat in the bakery industry is shown in Table 7.3.

The main part of butter and hard fractions of milk fat used in bakeries is applied in products like puff pastry, croissants and Danish pastry. The function of the fat in these products, which consist of many thin alternate layers of dough, is to keep the layers from sticking together, which on baking gives a separation and an open network of crisp and flaky layers. It is important that the fat has a good sheetability (i.e. the fat should be able to form thin sheets) and a high tolerance to withstand the mechanical folding and rolling processes used in bakeries today, so little or no work-softening occurs. This requires butter with a rather firm and tough, but still sufficiently plastic, consistency. Such a product could be manufactured, if butter with a high content of high-melting trigycerides after storage at low temperatures is given a very powerful mixing to break down the crystal network, which has built up in the continuous fat phase. The product should have a finely dispersed water phase (the largest water droplets should not exceed 2–3 μm), and a very low air content, which could be

Table 7.3 Use of anhydrous milk fat (AMF) and milk fat fractions in bakery products.

Butter cookies, biscuits and short and sweet pastries	AMF + 20 g 100 g^{-1} olein fractions
Ice-cream cones and waffles	Olein fractions
Fermented pastries	AMF + stearin fractions
Puff pastries	AMF + stearin fractions
Cakes	AMF
Butter sponges	Olein fractions

Adapted from Barts (1991).

achieved by mixing the product under vacuum. An easier way to obtain the desired functional properties, that is, allowing the fat to move coherently with the dough layers, is to use a stearin fraction of milk fat with a higher melting point and a higher solid fat content than ordinary butter.

The main concern in butter cookies is flavour, tenderness and shortening properties like crunchiness, which can be obtained by using high levels of butter. Problems with fat blooming, which are pale stains on the surface of the cookies formed by high melting fat crystals, could be avoided by using olein fractions of milk fat with iodine values above 35.

The most commonly used fats for the production of white bread, rolls, and buns are bakery compounds, which are water-in-oil emulsions with water contents of 20 to 60 g 100 g^{-1}. When used in yeast-raised white bread, this product has a softening effect. Emulsifiers are often used to reduce the interfacial tension between the fat and water phases and to stabilise the emulsion and to impart fine stable water dispersion. It is important for handling bakery compounds that they appear as soft pasta easy to mix with flour. In a Danish study (Jansen, 1969), the applicability of butter in such mixtures was examined. It was concluded that a mixture (g 100 g^{-1}) consisting of 65 butterfat, 3 milk protein curd and 32 water had an advantageous influence on the structure, consistency and taste of bread, as it gave the bread a more dense body and structure and, at the same time as the taste of fresh bread was accentuated and preserved for a longer time. The use of milk fat fractions in bakery compounds seems to be unimportant.

Confectionery products

Milk fat is used as an important ingredient in a variety of confectionery products (see the reviews by Munksgaard & Ipsen (1989) and Shukla (1994)), despite the relatively high cost of milk fat compared to other fats and oils. Milk fat used in confectioneries could be in the form of butter, AMF, full-cream milk powder or chocolate crumb, which is a dried blend of sugar, milled cocoa beans and fresh or concentrated milk. Milk fat is primarily used for its buttery flavour, but another important property of milk fat is its ability to form part of the continuous fat phase in chocolate as a result of its compatibility with cocoa butter at most levels of addition. Milk fat is one of the few fats that are compatible with cocoa butter, although it should be noticed that the stearin fraction of milk fat is even more compatible with cocoa butter than regular milk fat.

Softening of chocolates, which is due to addition of milk fat, is widely recognised and is attributed to the mixed crystallisation effect, also called the eutectic effect. This means that the triglycerides of one fat co-crystallise with the triglycerides of the other fat, resulting in a reduction in the solid fat index of the mixture below the level of either fat. In dark chocolates, a small amount (2–3 g 100 g^{-1}) of milk fat is used to soften the product and control texture formation. In milk chocolates, a higher addition of milk fat is used, for example, up to approximately 20–30 g 100 g^{-1}. However, by using the stearin fraction of milk fat, which has a solid fat index closer to cocoa butter, it may be possible to further increase the proportion of milk fat in the formulation. Although it is desirable to increase the addition of milk fat because it is cheaper than cocoa butter and has a lower viscosity, there is a maximum level because too high an addition will retard cocoa butter crystallisation, which will result in too soft a chocolate and longer tempering time.

Milk fat is also known to inhibit fat bloom formation in chocolates, which is the formation of a white powdery surface on chocolate composed of large fat crystals caused by crystallisation of fat at low ambient temperature. This inhibitory effect is another reason for the addition of small quantities of milk fat to dark chocolate. Hartel (1996) described how the crystallisation rate of the cocoa butter is reduced when milk fat is added, so that the chocolate processing temperatures must be lowered to counteract this effect (Gordon, 1991).

7.6 Nutritionally modified products

7.6.1 *Introduction*

Over the years, discussion has been going on concerning the health aspects of butter and milk fat and, criticism related to high fat content, high content of saturated fat and high content of cholesterol has frequently been expressed both by professionals like nutritionists and dieticians and by health-conscious consumers. This has, of course, influenced product development concerning butter and spreads, and several nutritionally modified products have been developed and marketed based on either extraction of unwanted substances or enrichment with bioactive substances. Another example is spreads fermented with functional cultures, such as *Lactobacillus acidophilus* and *Bifidobacterium bifidum*.

7.6.2 *Production technologies*

Reduced cholesterol

When the link between blood cholesterol and coronary heart disease was established many years ago, it was recommended that people in the high-risk category avoid foods containing cholesterol. As milk fat contains $\sim$0.3 g 100 g^{-1} cholesterol, this was one of the types of fat that was discussed and still is, although the debate about the oral intake of cholesterol has almost faded away at least at a scientific level. Cholesterol in milk is bound to milk fat, but can be removed by biological, physical or chemical methods. Many methods have been tried for the removal of cholesterol from milk fat as described by Spreer (1998) and Mohamed *et al.* (2003), and most of them are based on extraction, distillation, adsorption, enzymatic conversion or combinations hereof. Cholesterol can be extracted with organic solvents like methanol in a single-stage or multiple-stage batch extraction process followed by de-aeration with steam under vacuum. As the use of organic solvents in food processing is problematic, these methods are of minor commercial interest.

When the super-critical extraction process was introduced in the food industry in the 1980s, many expected that this would be the method, which successfully could be used for removing cholesterol from milk fat. Bradley (1989) reported that up to 90 g 100 g^{-1} of the cholesterol can be extracted from milk fat by using carbon dioxide as a solvent under high pressure, and at moderate temperatures, in a multistage process. Super-critical extraction was detailed by Mohamed *et al.* (2003), and Bradley (1989) described a rather complicated super-critical fluid extraction plant and gave an estimation of the production costs. In spite of many attempts, real commercial success has not been reported mainly due to the very large investments and considerable operational costs.

Another method for the removal of cholesterol from milk fat is distillation. Cholesterol is a low volatile compound, but it is more volatile than most triglycerides found in milk fat, and it can, therefore, be removed by steam distillation and concentrated in the condensate. Different distillation techniques, like vacuum steam distillation and short-path molecular distillation, have been applied, and some of these methods seem to be rather effective. However, the problem is that the typical butter flavour is removed together with the cholesterol. The commercial interest in these complicated and expensive methods has been modest.

Cholesterol can also be removed by adsorption to activated carbon, diatomaceous earth or specially coated glass, ceramics or plastic. Melted AMF is brought into contact with the absorbent material either in a batch process or in a more or less continuous process using columns. In Europe, absorption methods, which are based on complex formation between cholesterol and cyclodextrin, have found some application on an industrial scale. In this method, melted AMF is blended with a 1 to 10 g 100 g^{-1} aqueous solution of 60 g 100 g^{-1} cyclodextrin and, when the complex with cholesterol is formed, it is removed from the milk fat by centrifugation (Spreer, 1998). It is reported that a more than 90 g 100 g^{-1} reduction in the cholesterol content can be obtained (Roderbourg *et al.*, 1994). The process is rather simple and does not require a large investment; the reduction in butter flavour is stated to be insignificant. However, large quantities of cyclodextrin must be used and the production costs are consequently high. Nevertheless, this process has been commercialised.

The enzyme cholesterol reductase can convert cholesterol into non-toxic forms that cannot be absorbed by humans. Several laboratory trials have been reported, but the process seems complicated, combined with high production costs. Reports on industrial use are lacking.

A large variety of cholesterol-reduced butters and spreads have been developed and marketed all over the world, but the commercial success is difficult to assess. Many products seem to have had a short life on the market place before they disappear again, but they are quickly replaced by new products in the category.

Added sterols

Phytosterols, also known as plant sterols, and their hydrogenated form, stanols, are structurally very similar to cholesterol, and they are found as a structural component of cell membranes of plants. They are present in low levels in grains, fruits, vegetables and vegetable oils, especially in soya bean oil and in tall oil, which is a by-product of paper manufacture from pine trees. Phytosterols have a limited solubility in both water and fat but, by esterification with long-chain fatty acids, the solubility in the fat is increased by 10-fold, making the ingredient suitable for addition to products like margarine or spreads without any particular technical problems.

It is well documented that phytosterol or phytostanol esters are absorbed only in trace amounts in the human gastrointestinal tract, and at the same time inhibit the absorption of intestinal cholesterol and, thereby, reduce the cholesterol level in blood even at a modest daily intake. According to a controlled study by Maki *et al.* (2001), reduced fat spreads enriched with plant sterol esters caused considerable reductions in serum levels of both total cholesterol and low-density lipoprotein (LDL) cholesterol. The study concluded that a reduced fat spread containing plant sterol esters provided a beneficial effect in the dietary management of hypercholesterolemia.

The first yellow fat spread with these ingredients, Benecol®, was launched in Finland by Raisio Oyi in 1995 with considerable market success, and it was later followed by similar products from Unilever under the brands Flora Pro-activ® and Becel Pro-activ®. The market development of this product category is reviewed by Watkins (2005). The addition of phytosterols to butter or low-fat spreads containing milk fat has been considered, but no such products seems to have reached the market place yet. One reason might be that the product category is tightly covered by patents limiting the possibilities for product development.

ω-3 Fatty acids

ω-3 Fatty acids comprise of α-linolenic acid (C18:3, ALA), eicosapentaenoic acid (C20:5, EPA) and docosahexaenoic acid (C22:6, DHA), which are polyunsaturated fatty acids, and are essential for humans. Health authorities normally recommend a higher intake of these fatty acids as many studies have shown that this may afford some degree of protection against coronary heart disease, although contradictory information can be found in the literature. Hooper *et al.* (2006) published a systematic review of studies from 2002 to 2006, and concluded that no evidence of a clear benefit of ω-3 fats on health was found.

The most widely available source of ω-3 fatty acids is cold water oily fish like salmon, herring and mackerel. Fish oil has to be refined before it can be used in spreads, and special precautions like addition of antioxidants and/or storage and shipment under modified atmosphere should be taken to prevent oxidation of the highly unsaturated oil. If this is done, fish oil can easily be added to butter and spreads in the same way as vegetable oil. A study by Kolanowski *et al.* (2001) established the sensory and nutritionally acceptable enrichment levels of ω-3 fats in fat spreads including blends with added unhydrogenated fish oil. They found that fat spreads might be enriched with up to 1 g 100 g^{-1} EPA and DHA without any significant influence on the sensory acceptability. The results also indicated that such enriched spreads could be stored for up to 6 weeks. Blends enriched with fish oil have been and still are marketed, but apparently with modest success probably because of the relatively high price due to the costly refining and shipment of the fish oil.

'Healthy' vegetable oils

In recent years, a number of blended spreads based on mixtures of milk fat and 'healthy' vegetable oils have been developed and marketed. The oils could be, for example, sunflower seed oil and olive oil high in monounsaturated fatty acids or red palm oil with a high content of carotenoids acting as antioxidants. Especially, olive oil seems to be popular probably because of the association with claimed health benefits of the 'Mediterranean diet', and a number of olive oil blends and spreads have been marketed. These oils could be used like any other oil in the production technologies already described.

7.7 Conclusions

Milk fat has a unique flavour, a desirable mouth-feel and a natural image, but ordinary butter does not always meet the consumer's expectations concerning spreadability and health. A variety of products have, therefore, been developed with lower fat content, modified fat composition and/or modified fat functionality.

The technologies for production of dairy fat spreads, that is, defined as milk fat products with less than 80 g fat 100 g^{-1}, are mainly based on the utilisation of margarine technology, inversion of cream and direct increase of the water or air content in butter.

One of the most important quality aspects related to low-fat dairy spreads is binding of the high water content. If the water is bound too tightly, the consistency becomes very tough and sticky and, if the water is bound too loosely syneresis might occur. The overall quality perception of low-fat dairy spreads seems to be that they still lack the texture, mouth-feel and flavour of full-fat products.

Owing to its fatty acid composition, butter has poor spreadability at low temperatures and, from a health point of view, the high content of saturated fat in butter has been heavily criticised for many years. It is, therefore, an obvious idea to mix butter and vegetable oils with a high content of unsaturated fatty acids. The technologies available today for the production of blends and blended spreads are mainly based on churning technology, margarine technology and inversion of cream. Addition of vegetable oils will decrease the oxidative stability of the product. This deterioration can be delayed, but not completely prevented by the addition of antioxidants.

The functional properties of milk fat can be modified by changing the feeding of the dairy cows, fractionation of the milk fat or by inter-esterification of the milk fat with vegetable oils. Modifying the composition of milk fat through feeding and inter-esterification is expensive and only limited commercial success has been reported. Fractionation is more commonly employed, and the milk fat fractions are often used as functional ingredients in products, such as in bakery goods, confectionery and chocolate.

The high content of cholesterol has frequently been criticised both by professionals, like nutritionists and dieticians, and by health-conscious consumers; several technologies have been developed for reducing the level of cholesterol. A large variety of cholesterol-reduced butters and spreads have been developed and marketed all over the world, but the commercial success is difficult to assess.

Other nutritionally modified products, such as spreads with an increased content of ω-3 fatty acids or products with an increased content of olive oil, have been developed and marketed with reasonable commercial success. The addition of phytosterols to butter or low-fat spreads containing milk fat has also been considered to reduce the cholesterol level in blood even at a modest daily intake. No such products containing milk fat seem to have reached the market place yet.

References

van Aken, G. A. & Visser, K. A. (2000) Firmness and crystallization of milk fat in relation to processing conditions. *Journal of Dairy Science*, **83**, 1919–1932.

Andersson, K. (1991) Modified "butters". *Utilizations of Milkfat*, Document No. 260, pp. 17–18, International Dairy Federation, Brussels.

Andersen, U. & Hansen, G. (2003) *Influence of Stabilized Water Phase on Meltdown Properties in Low Fat Spreads*, Unpublished Poster, Arla Foods Innovation Centre, Brabrand.

Andreasen, T. G. & Nielsen, H. (1998) Ice cream and aerated desserts. *The Technology of Dairy Products*, (ed. R. Early), 2nd edn., pp. 301–326, Blackie Academic & Professional, London.

Anonymous (1994) *United States Standards for Grades of Whipped Butter*, United States Department of Agriculture, Agricultural Marketing Service, Dairy Division, Washington, DC (www.ams.usda.gov/standards/Whipped.pdf).

Anonymous (2003) Butter and dairy spreads. *Dairy Processing Handbook*, 2nd and revised edn., pp. 289–291, Tetra Pak Processing Systems AB, Lund.

Anonymous (2005) *Butter Products Made by the Inversion of Cream*, Technical Information supplied by Gerstenberg Schröder A/S, Brøndby.

Appelqvist, L.-Å. (1996) *Oxidized Sterols*, Document No. 315, pp. 52–58, International Dairy Federation, Brussels.

Barts, R. (1991) The use of milkfat and milkfat fractions in the food industry. *Utilizations of Milkfat*, Document No. 260, pp. 19–20, International Dairy Federation, Brussels.

Berntsen, S. (1999) Dairy blend – new trend. *Scandinavian Dairy Information*, **13**(1), 24–26.

Bradley Jr., R. L. (1989) Removal of cholesterol from milk fat using supercritical carbon dioxide. *Journal of Dairy Science*, **72**, 2834–2840.

Buchheim, W. & Dejmek, P. (1990) Milk and dairy-type emulsions. *Food Emulsions*, 2nd edn., (eds. K. Larsson & S. E. Friberg), pp. 203–246, Dekker, New York.

Burling, H., Madsen, J. C. & Frederiksen, H. K. (2005) *Stabilisers Useful in Low Fat Spread Production*. **World Patent Application**, WO 041677 A1.

Børsting, C. F. & Weisbjerg, M. R. (1989) *Fatty Acid Metabolism in the Digestive Tract of Ruminants*, PhD Thesis, Danish Institute of Agricultural Sciences, Copenhagen.

Charteris, W. P. (1995a) Physicochemical aspects of the microbiology of edible table spreads. *Journal of the Society of Dairy Technology*, **48**, 87–96.

Charteris, W. P. (1995b) Minor ingredients of edible table spreads. *Journal of the Society of Dairy Technology*, **48**, 101–106.

Charteris, W. P. (1996) Microbiological quality assurance of edible table spreads in new product development. *Journal of the Society of Dairy Technology*, **49**, 87–98.

Deffense, E. (1993) Milk fat fractionation today: a review. *Journal of American Oil Chemical Society*, **70**, 1193–1201.

Deffense, E. & Hartel, R. W. (2003) Fractionation of milk fat. *Oils and Fats*, Volume 3, *Dairy Fats*, (ed. B. Rossell), pp. 158–186, Leatherhead Food International, Leatherhead.

Dolby, R. M. (1970) Properties of recombined butter made from fractionated fats. *XVIII International Dairy Congress*, **1E**, 243.

Dungey, S. G., Gladman, S., Bardsley, G., Iyer, M. & Versteeg, C. (1996) Effect of manufacturing variables on the texture of blends made with scraped surface heat exchangers. *Australian Journal of Dairy Technology*, **51**, 101–104.

EU (1994) Standards for spreadable fats, Council Regulation No. 2991/94 of 5 December 1994. *Official Journal of the European Communities*, **L 316**, 2–7.

FAO/WHO (1971) *Standard for Butter*, Codex Stan 279, Revised 1–1999, Amended 2003 and 2006, Food and Agriculture Organization of the United Nations, Rome.

FAO/WHO (2006) *Standard for Dairy Fat Spreads*, Codex Stan 253, Food and Agriculture Organization of the United Nations, Rome.

FAO/WHO (2007) *Standard for Fat Spreads and Blended Spreads*, CODEX STAN 256, Food and Agriculture Organization of the United Nations, Rome.

Fisker, A. N. & Jansen, K. (1970) Experiments on the manufacture of whipped butter. *XVIII International Dairy Congress*, **1E**, 249.

Flack, E. (1997) Butter, margarine, spreads, and baking fats. *Lipid Technologies and Applications*, (ed. F. D. Gunstone & F. B. Padley), pp. 305–327, Marcel Dekker, Inc., New York.

Gordon, M. (1991) The utilization in chocolate. *Utilisations of Milkfat*, Document No. 260, pp. 20–22, International Dairy Federation, Brussels.

Grummer, R. R. (1991) Effect of feed on the composition of milk fat. *Journal of Dairy Science*, **74**, 3244–3257.

Hartel, R. W. (1996) Applications of milk fat fractions in confectionery products. *Journal of American Oil Chemical Society*, **73**, 945–953.

Hermansen, J. E. (1995) Prediction of milk fatty acid profile in dairy cows fed dietary fat differing in fatty acid composition. *Journal of Dairy Research*, **78**, 872–879.

Hooper, L., Thompson, R. I., Harrison, R. A., Summerbell, C. D., Ness, A. R., Moore, H. J., Worthington, H. V., Durrington, P. N. Higgens, J. P. T., Capps, N. E., Riemersma, R. A., Ebrahim, S. B. J. & Smith, G. D. (2006) Risks and benefits of omega 3 fats for mortality, cardiovascular disease, and cancer: systematic review. *British Medical Journal*, **332**, 752–760. (doi:10.1136/bmj.38755.366331.2F).

Jansen, K. (1969) *Fremstilling af Bageriblandinger til Gærdej*, Report No. 181, Danish Government Research Institute for Dairy Industry, Hillerød.

Jebson, R. S. (1979) Recombined butter. *Recombination of Milk and Milk Products*, Document No. 116, pp. 30–32, International Dairy Federation, Brussels.

Kankare, V. & Antila, V. (1974) Use of low-melting milk fat fractions for improvement of the spreadability of butter. *XIX International Dairy Congress*, **1E**, 671.

Kaufmann, V. W., Biernoth, G., Frede, E., Merk, W., Precht, D. & Timmen, H. (1982) Fraktionierung von Butterfett durch Extraktion mit überkritischem CO_2. *Milchwissenschaft*, **37**, 92–96.

Kolanowski, W., Swiderski, F., Lis, E. & Berger, S. (2001) Enrichment of spreadable fats with polyunsaturated fatty acids omega-3 using fish oil. *International Journal of Food Sciences and Nutritrion*, **52**, 469–476.

Kristensen, V. F., Andresen, P. E., Fisker, A. N. & Mortensen, B. K. (1974) *Influence on Milk Yield, Fatty Acid Composition of Milk Fat and Butter by Feeding Dairy Cows Encapsulated Soybean Oil*, Joint report No. 1, Danish Government Research Institute for Dairy Industry and the Danish Institute of Agricultural Sciences, Hillerød.

Lane, R. (1998) Butter and mixed fat spreads. *The Technology of Dairy Product*, (ed. R. Early), 2nd edn., pp. 158–197, Blackie Academic & Professional, London.

Ljunggren, A. & Hansen, J. (2006) *Cold Mix Process for Blended Spreads*, http://www.gs-as.com.

Lynch, R. & O'Mahony, J. S. (1999) *Whipped Low Fat Spread*. **US Patent Application** 5 869 125.

Maki, K. C., Davidson, M. H., Umporowicz, D. M., Schaefer, E. J., Dicklin, M. R., Ingram, K. A., Chen, S., McNamara, J. R., Gebhart, B. W., Ribaya-Mercado, J. D., Perrone, G., Robins, S. J. & Franke, W. C. (2001) Lipid responses to plant-sterol-enriched reduced-fat spreads incorporated into a National Cholesterol Education Program. *American Journal of Clinical Nutrition*, **74**, 33–43.

Mann, E. J. (1990) Modified butters and spreads. *Dairy Industries International*, **56**(8), 9–11.

Mann, E. J. (1997) Fat spreads. *Dairy Industries International*, **63**(1), 13–14.

Mann, E. J. (1998) Fat spreads. *Dairy Industries International*, **64**(1), 19–20.

Mann, E. J. (2000) Butter-related spreads. *Dairy Industries International*, **66**(11), 20–21.

Marangoni, A. G. & Rousseau, D. (1998) Chemical and enzymatic modification of butterfat and butterfat-canola oil blends. *Food Research International*. **31**, 595–599.

Mohamed, R. S., Saldana, M. D. A., de Azevedo, A. B. A. & Kopcak, U. (2003) Removal of cholesterol from food products using supercritical fluids. *Extraction Optimisation in Food Engineering*, (ed. C. Tzia & G. Liadakis), pp. 369–389, Marcel Dekker Inc., New York.

Mortensen, B. K. & Danmark, H. (1981) *Holdbarhed af Blandingsprodukter Indeholdende Smørfedt og Sojaolie*, Appendix to Report No. 247, The Danish Government Research Institute for Dairy Industry, Hillerød.

Munksgaard, L. & Ipsen, R. (1989) *Confectionery Products*, Newsletter No. 109, International Dairy Federation, Brussels.

Munro, D. S. (1986) Alternative processes. *Continuous Butter Manufacture*, Document No. 204, pp. 17–20, International Dairy Federation, Brussels.

Nicholson, J. W. G. (1993) *Spontaneous Oxidized Flavour in Cow's Milk* and *Catalogue of Tests for the Detection of Post-Pasteurization Contamination of Milk*, Document No. 281, pp. 1–12, International Dairy Federation, Brussels.

Nielsen, W. K. (1993) Processing plants for low-fat butter and dairy blends. *Scandinavian Dairy Information*, **7**(2), 28–29.

Nielsen, J. H., Olsen, C. E., Jensen, C. & Skibsted, L. H. (1996) Cholesterol oxidation in butter and dairy spread during storage. *Journal of Dairy Research*, **63**, 159–167.

Nielsen, M., Olsen, P. B., Rokkedahl, K. & Thorning, P. F. (2005) *A Butter-like Dairy Spread and Method for Production*. **European Patent Application**, 1 688 044 A1.

Pedersen, A. (1997) Inversion of creams for butter products. *Dairy Industries International*, **63**(7), 39–41.

Rajah, K. (2004) Innovations in the spreads market. *Oils & Fats International*, **May**, 18–20.

Robinson, D. J. & Rajah, K. K. (2002) Spreadable products. *Fats in Food Technology*, (ed. K. K. Rajah), pp. 192–227, Sheffield Academic Press, Sheffield.

Roderbourg, H., Dalemanns, D. & Bouhon, R. (1994) Fundamentals of processing with supercritical fluids. *Supercritical Fluid Processing of Food and Biomaterials*, (ed. S. S. H. Rizv), pp. 1–26, Blackie Academic & Professional, London.

Rousseau, D. & Marangoni, A. G. (1999) The effects of interesterification on physical and sensory attributes of butterfat and butterfat-canola oil spreads. *Food Research International*, **31**, 381–388.

Rønne, T. H., Yang, T., Mu, H., Jacobsen, C. & Xu, X. (2005) Enzymatic interesterification of butterfat with rapeseed oil in a continuous packed bed reactor. *Journal of Agriculture and Food Chemistry*, **53**, 5617–5624.

Schäffer, B., Szakály, S., Lõrinczy, D. & Belágyi, J. (2000) Structure of butter – IV: effect of modification of cream ripening and fatty acid composition on the consistency of butter. *Milchwissenschaft*, **55**, 132–135.

Scott, T. W., Cook, L. J. Ferguson, K. A., McDonald, I. W., Buchanan, R. A. & Hills, G. L. (1970) Production of polyunsaturated milk fat in domestic ruminants. *Australian Journal of Science*, **32**, 291–293.

Shukla, V. K. S. (1994) Milk fat in sugar and chocolate confectionery. *Fat in Foods Products*, (ed. D. P. J. Moran & K. K. Rajah), pp. 255–276, Blackie Academic & Professional, Glasgow.

Spreer, E. (1998) Butter manufacture. *Milk and Dairy Products Technology*, pp. 205–242, Marcel Dekker, Inc., New York.

Stapelfeldt, H., Nielsen, K. N., Jensen, S. K. & Skibsted, L. H. (1999) Free radical formation in freeze-dried raw milk in relation to its α-tocopherol level. *Journal of Dairy Research*, **66**, 461–466.

Truong, H. T. & Munro, D. S. (1982) The quality of butter produced by the Ammix process. *XXI International Dairy Congress*, **1**, 341–342.

Watkins, C. (2005) The spread of phytosterols. *Food Technology*, **16**(6), 344–345.

Weckel, K. G. (1965) Dairy spreads. *Manufacture of Milk Products Journal*, **56**(7), 5–6.

Wursch, P. & Tolstoguzow, V. B. (2000) *Low Fat Spreadable Food Product*, **US Patent Application** 6 013 301.

Youcheff, G. G. & Hensler, A. P. (2001) *Anhydrous Milk Fat/Vegetable Fat Food Ingredients*. **US Patent Application** 6 265 007 B1.

van Zijl, M. M. & Klapwijk, P. M. (2000) Yellow fat products (butter, margarine, dairy and non-dairy spreads). *The Microbiological Safety and Quality of Food*, (eds. B. Lund, T. Blair-Parket & G. Gould), Vol. 2, pp. 784–806, Springer-Verlag, Heidelberg.

8 Cream Cheese and Related Products

T.P. Guinee and M. Hickey

8.1 Introduction

Cream cheese is a cheese of the semi-soft, fresh acid curd variety, obtained by slow quiescent acid gelation of milk. Compositionally, it is quite similar to unripened Neufchatel (Neufchâtel), Boursin, Petit Suisse, Mascarpone and Kajmak, all of which are characterised by relatively high levels of fat in dry matter (FDM, $\sim\geq$ 60 g 100 g^{-1}), low protein ($\sim\leq$ 10 g 100 g^{-1}), low calcium level ($<$100 mg g^{-1}) and low pH ($<$5.0) (Table 8.1). Compared to rennet–curd cheese varieties, such as Cheddar, Gouda and Emmental, these products generally have milder flavours and a softer texture.

However, cream cheese and related varieties as a group differ significantly (Davis, 1976; Kosikowski & Mistry, 1997a, 1997b; Robinson & Wilbey, 1998; Guinee *et al.*, 1993; Schulz-Collins & Senge, 2004) in terms of gelation mode, make procedure, flavour and texture and geographic region of manufacture and compositional standards. Gelation is induced by

- slow quiescent acidification as for cream cheese, Boursin and North American-style Neufchatel;
- quiescent acidification and rennet treatment in the case of French Neufchâtel or Petit-Suisse;
- heat plus acid in Mascarpone; or
- heating/dehydration and cooling in the case of Kajmak.

Cream cheese, Mascarpone and North American-style Neufchatel (designated Neufchâtel in Canada) have a clean, fresh, butter-like lactic flavour and a soft/semi-soft smooth creamy consistency with a texture that ranges from spreadable to short and brittle, depending on composition (e.g. fat content), manufacturing conditions (e.g. degree of heating and shear during curd treatment), and/or the type and level of added hydrocolloids. During the manufacture of these products, the curd is heated to $\sim$80°C, hot packed and cooled to $<$8°C to allow fat crystallisation and product structuring. Owing to the high temperature, microbial and enzymatic activities are low. Consequently, an ageing process (in the conventional sense of cheese ripening) is not viable, even though a period of $\sim$1 to 2 days is required for fat crystallisation and product structuring to occur. These cheeses may be classified as unripened varieties, being ready for consumption within a few days after manufacture (FAO/WHO, 2007e).

In contrast, French-style Neufchâtel is a ripened white-rinded, grainy-textured cheese, which develops a sharp, salty, mushroom-like flavour on ageing (Harbutt, 1998). Kajmak has a taste and texture similar to cream cheese when it is fresh, but on ageing develops a buttery consistency and a more rancid sharp flavour (Pudja *et al.*, 2008). The latter cheeses thus require an ageing period to enable the necessary biochemical changes (e.g. glycolysis,

Table 8.1 Composition of some cream cheeses and related products.

Cheese	Fat ($g\ 100\ g^{-1}$)	Protein ($g\ 100\ g^{-1}$)	Moisture ($g\ 100\ g^{-1}$)	FDM[a] ($g\ 100\ g^{-1}$)	Calcium ($mg\ 100\ g^{-1}$)	Phosphorous ($mg\ 100\ g^{-1}$)	Salt ($g\ 100\ g^{-1}$)	pH
American-style (blocks)[b]	28.8	5.6	62.4	76.6	69	84	0.7	5.1
American-style (tubs)[b]	23.8	5.9	65.8	69.5	97	93	1.0	4.9
Rahmfrischkäse[b]	24.5	5.8	66.2	72.6	79	94	0.7	4.9
Petit-Suisse[c]	NR[d]	NR	70–77	40–60	NR	NR	NR	NR
American Neufchatel[c]	20–33	10.0	≤ 65	NR	NR	NR	0.75	NR
Mascarpone[b]	54.1	2.6	38.9	88.6	29	NR	0.2	5.8
Kajmak (fresh)[e]	40–55	5–10	30–40	65–80	NR	NR	0.5–2.0	NR
(ripened)[e]	50–70	2–7	15–35	75–90	NR	NR	1.0–3.5	NR

[a] FDM, fat in dry matter.
[b] Data based on the analysis of retail samples available on the Irish and UK markets.
[c] Based on specified compositional standards (cf. Tables 8.1–8.12) and data from USDA (1976).
[d] NR, not reported.
[e] After Pudja *et al.*, (2008).

proteolysis and lipolysis) and the development of the attendant sensory characteristics. Hence, these cheeses are classified as ripened cheeses by the Codex Alimentarius Commission (CAC) (FAO/WHO, 2008a).

Overall, cream cheese and North American-style Neufchatel are the largest varieties, especially in the United States, where the per capita consumption in 2006 was 1.15 kg (US Department of Agriculture (USDA), 2008), and estimated production is 450 000 tonnes. On the other hand, the production volume of Kajmek, being a home-made product largely confined to the Balkan Peninsula, is probably very low.

8.2 Background and development

While cheese making has been used to preserve the nutritional value of milk from prehistoric times, following the emergence of larger cheese factories in Europe and the United States during the last decades of the nineteenth century, efforts were initiated to try to extend the shelf-life of cheese to facilitate access to more distant markets. Given the long history of cheese, it is not easy to identify the source of any particular cheese type. Many individual varieties have geographical names, towns, districts, regions, monasteries and so on that indicate their origins, for example, Cheddar, Emmental, Roquefort, Port Salut and Neufchâtel. However, this is not the case with cream cheese, where the name is more generic or descriptive in nature.

It is likely that cheeses produced from milk enriched with cream have evolved independently in many regions of the world, but the origins of the most common form in North America have been identified. The US production of Neufchatel cheese started in Orange County, New York, in 1862 (Kosikowski, 1977; Kosikowski & Mistry, 1997a). In 1872, the first cream cheese was made in New York by a dairyman, William Lawrence. Then in

1880, a New York cheese distributor, A. L. Reynolds, first began distributing a cream cheese packed in tin-foil wrappers, calling it Philadelphia. This name was used because at that time the name of that city was associated with the best-quality foods. In 1903, the Phenix Cheese Company bought the Philadelphia brand and in 1928, the Phenix Cheese Company merged with Kraft Foods, which was established by James L. Kraft, a cheese wholesaler in Chicago in 1903, and opened its first cheese plants in 1914 (Kraft Foods, 2008). The original cream cheese and Neufchatel cheese, which had a higher moisture and lower fat content, as produced in the United States, was a ripened cheese, was packed cold, had a relatively short shelf-life of less than 2 weeks and was liable to spoilage by yeast, mould and the development of fruity flavours (Tuckey, 1967; Kosikowski, 1977). However, cream cheese as produced today is quite different from the product of the early days. In 1920s, cream cheese, as produced by the cold-pack method, was improved to give a shelf-life of 3 to 4 weeks. The modifications included pasteurising or heating the cream cheese mix at 68.5°C and holding for 30 min, cooling to 49°C and subjecting to single-stage homogenisation, followed by cooling to 23.5°C, adding mixed lactic starter and incubating to pH 4.4 to 4.5, cooking the resulting curd with gentle mixing to 65°C and holding for 15 min, cooling to 26.5°C, adding ~1 g salt 100 g^{-1} in warm solution, cooling to 8 to 10°C, pumping into drainage bags and allowing to drain in a cold room to reduce the mix to about one-third the original volume, packaging the cheese immediately and keeping refrigerated. A hot-pack method was also developed at Cornell University in the late 1920s; while this had some similarities to the cold-pack method, it eliminated the use of the muslin drainage bags and gave a product with a reputed shelf-life of 2 to 3 months. The use of stabilisers, such as gums, gelatine, alginates, carrageenan or agar, was common in the hot-pack products to prevent syneresis (Dahlberg, 1927; Lundstedt, 1954; Tuckey, 1967; Kosikowski, 1977; Guinee *et al.*, 1993; Fox *et al.*, 2000).

In 1945, there was a major advance when a separator was developed and patented by Oscar Link, which allowed the continuous removal of whey from the hot cream–cheese mix. This patent, which has now expired, was assigned to the Kraft Cheese Company (Link, 1942). When combined with the hot-pack method, this further improved shelf-life. By the mid-1970s, all major US manufacturers of cream cheese were using centrifugation techniques or other extrusion methods of production (Kosikowski, 1977). More recent developments include the use of ultrafiltration (UF) techniques, in conjunction with, or replacing, centrifugation techniques (Kosikowski, 1974; Guinee *et al.*, 1993; Fox *et al.*, 2000). In the 1980s, a softer cream cheese, suitable for spreading, was developed by Kraft and other manufacturers. This product is particularly suitable for spreading on bagels and biscuits and has expanded the uses of cream cheese (Kosikowski & Mistry, 1997a).

In the United Kingdom, the origins of cream cheese are less certain, but it was probably manufactured by farmhouse production for some centuries. Different varieties were produced from mixtures of milk and cream. Manufacture was by the natural souring of cream, with or without the use of rennet and the curd was allowed to drain in a cloth bag. From about the 1920s, manufacture became more standardised and two main types evolved: first, single cream cheese, made from cream with a fat content of somewhere between 15 and 30 g 100 g^{-1}, where rennet could be added to facilitate thickening, and second, double cream cheese (DCC), made from cream of about 49 to 60 g fat 100 g^{-1}, where rennet was not needed (Davis, 1976; Scott, 1981). In later years, where larger quantities of cheese were manufactured from ripened cream, using a special separator, 9 to 11 g 100 g^{-1} fat cream could be used (Scott, 1981). The principles and compositional standards of these types of cream cheese were included in the UK legislation, which was developed in mid-1960s (Davis, 1966). A consequence of this is that cream cheese produced in the United Kingdom had higher fat levels than the similarly named cheeses produced in other countries. One further development in recent years is, given the focus on fat levels in the diet, that lower fat variants of cream cheese have been developed. Table 8.2 outlines compositional data of some cream cheese-like products on the UK and Irish markets. It will be noted that none of the products are actually designated as cream cheese. The reasons for this will be discussed further in the subsequent sections.

Table 8.2 Designations and nutritional information (g 100 g^{-1}) on some branded and own-label cream cheese-like products on the Irish and UK markets.

Brand name	Product designation	Fat	Protein	Carbohydrate	Salt (equivalent)
Philadelphia (block wrapped in foil)	Full-fat soft cheese	31.0	6.8	2.6	0.5[a]
Philadelphia (tub)	Full-fat soft cheese	24.0	5.9	3.2	1.0
Philadelphia Light - Reduced-Fat	Medium-fat soft cheese	11.5	8.7	4.0	1.0
Philadelphia Light - Chives	Medium-fat soft cheese	12.0	8.4	4.2	1.3
Philadelphia Extra Light	5 g 100 g^{-1} Fat soft cheese	4.7	12.0	5.0	1.0
Kerry Low Low - Soft Cream Cheese	Medium-fat soft cheese	16.5	8.7	3.0	0.8[a]
Pic Frisch Classic	Full-fat soft cheese	23.5	6.0	2.8	NR[b]
Boursin Ail & Fines Herbes	Full-fat soft cheese with garlic and fine herbs	41.0	7.0	2.0	NR
Boursin Light (32 g 100 g^{-1} fat in dry matter – FDM)	Reduced-fat soft cheese with garlic and fine herbs	1.8	2.4	0.5	0.3[a]
Crefeé 60 g 100 g^{-1} Fidm[c] Pasteurised	Full-fat soft cheese	23.0	NR	NR	NR
Linessa Vital and Active 1 g 100 g^{-1} Fidm	Low-fat soft cheese	0.2	11.0	3.9	0.8[a]
Castelli Mascarpone	Mascarpone (80 g 100 g^{-1} FDM)	35.5	NR	NR	NR
Tesco Soft and Creamy	Full-fat soft cheese	26.0	5	3.2	0.8
Tesco Light Choices	Reduced-fat soft cheese	6.1	13.5	4.1	0.7
Tesco Mascarpone	Full-fat soft cheese	44.0	6.0	3.5	0.0

[a] The labels of these products give sodium content only – the value stated is multiplied by 2.61 to calculate the salt equivalent in g 100 g^{-1}.
[b] NR, not reported.
[c] Fidm, fat in dry matter in French.

Cream-type cheeses are also produced in many other countries, for example, Petit-Suisse and Fromage frais à la crème in France, Rahmfrischkäse (fresh cream cheese) and Dopplerahmfrischkäse (fresh DCC) in Germany, Mascarpone in Italy, Flødeost in Denmark and Kajmak in the Balkans. The compositional and other requirements for these cheeses, while having certain similarities, differ from those in the United States and United Kingdom. On the other hand, Neufchâtel cheese (Appellation d'Origine Contrôlée – AOC) as produced in France – the cheese mentioned earlier as a likely origin of cream cheese in the United States – and Roomkaas (literally translated as cream cheese) in the Netherlands are both ripened cheeses and are outside the scope of this chapter.

8.3 Definitions and standards of identity

8.3.1 *Background and evolution*

The impetus and demands for food standards result frequently from demands of manufacturers who wish to protect their products and brands against similar products that may be regarded as of inferior quality and also cheaper, particularly those owing to differences in the composition of the competitor products. Standards and requirements thus developed tend to reflect the manufacturing processes, ingredients and chemical composition at the particular point in time. Hence, there is a need for regular revision of such standards to take into account process and product innovations, emergence of new functional ingredients and additives, evolving developments in hygiene and food safety and changing consumer and societal demands. Furthermore, a significant proportion of food law and regulation requires scientific or technological interpretation.

To understand these points it is worth considering the origins of standards and legislation governing food production and composition that date back to the Middle Ages. Many will have heard of the Reinheitsgebot (literally, the Purity Order), which concerned the purity of beer and originated in Bavaria in 1516 (Eden, 1993; Rieck, 2008). This listed the only permitted ingredients for beer as water, barley and hops. While the original order also set the price of beer at a mere 2 Pfennig per Maß, of course this did not last! It will be noticed that the list of ingredients did not include yeast; it would be more than three centuries before the role of micro-organisms in food fermentations was recognised. Nonetheless, it was only by 1987 that the requirements of the Reinheitsgebot were fully lifted in Germany as the result of a decision of the European Court of Justice (ECJ) (ECJ, 1987). Even to this day, certain German beers claim that they comply with the Reinheitsgebot.

By the middle of the nineteenth century, there were concerns about the adulteration, purity and wholesomeness of foods that led to the development of food legislation in many different jurisdictions. Nowadays the basis for food legislation is given as food safety, consumer protection and fair trade. The words may differ, but the fundamentals have not really changed.

Because of the diversity of cheese varieties and types, with local and regional variations, even among the more generic varieties, the development of international standards for cheese has proved to be a challenge. The concept of the protection of the authenticity and diversity of certain traditional food products may be traced to the Convention of Paris for the Protection of Industrial Property in March 1883, and in particular Art 6 quinquies thereof (Bertozzi

& Panari, 1993; McSweeney *et al.*, 2004). An amended version of this convention can be accessed (Anonymous, 2008b). This formed the basis of the agreement on Trade-Related Aspects of Intellectual Property Rights (TRIPS), which is now administered by the World Trade Organisation (WTO), and which led to the development of the AOC system in the legislation of France with the enactment of the Law for the Protection of the Place of Origin in May 1919. This system specified the place (region or department) in which certain products must be manufactured. Roquefort cheese, for example, was the first cheese granted an AOC in the law of 26 July 1925 (Anonymous, 2001; Anonymous, 2008a). Today, France has 45 AOC cheeses, one of which is Neufchâtel.

The concept behind the AOC system also spread to other European Union (EU) countries. An international convention was signed at Stresa on 1 June 1951, ratified by France, Italy, Switzerland, Austria, Scandinavia and Holland, on the use of designations of origin and names of cheeses. This was a series of multi-lateral agreements and is often referred to as The Stresa Convention (Anonymous, 1952; Les Autorités Fédérales de la Confédération Suisse, 2008). This stated that only *'cheese manufactured or matured in traditional regions, by virtue of local, loyal and uninterrupted usages'* may benefit from designations of origin governed by national legislation. Article 1 of the Convention prohibits the use of descriptions that contravene that principle. Further details of The Stresa Convention and some other bilateral agreements may also be accessed (Rieke, 2004).

Ultimately, these systems and conventions led to the European Regulation 2081/92 on the protection of geographical indications (PGI) and protection of designations of origin (PDO) (EU, 1992b); the European Regulation 1107/96 on the registration of products laid down in Article 17 of the former regulation, which lists protected products in the annex of the latter regulation as amended (EU, 1996b); the European Regulation 2082/92 on certificates of specific character (also referred to as Traditional Speciality Guaranteed TSG) (EU, 1992c) and the European Regulation 2301/1997 on entry of certain names in the 'register of certificates of specific character' provided for in the former regulation, which has three registered cheeses – Mozzarella (Italy), Boerenkaas (Netherlands) and Hushållsost (Sweden) (EU, 1997).

The CAC that implements the joint FAO/WHO foods standards programme was established in 1962 and developed international standards for 35 different cheese varieties between 1963 and 1978. A review of these standards was commenced in the early 1990s. The revision was lengthy and difficult; some of the original standards were dropped along the way; but in 2007, 16 revised standards and a new standard for Mozzarella were adopted.

Many individual countries also developed their own national legislation and standards over the years. Such legislations on cream cheese and certain cream cheese-like products shall be discussed, with particular reference to the legislation in the EU, United Kingdom, Ireland, Denmark, Italy Germany, United States, Canada and Australia as well as the international standards developed within the Codex Alimentarius.

8.3.2 *European legislation*

There is no specific European legislation on cheese or individual cheese varieties, except insofar as the regulations on the PGI and PDO and on certificates of specific character apply to certain cheese names. For this reason, a European cheese could not be designated as Neufchâtel, unless produced as prescribed by French legislation, as this is a protected name (Anonymous, 2006), but a cheese such as Cheddar or Emmental could be so designated, under European law, unless it used the name of a protected designation of such varieties, for example, West Country Farmhouse Cheddar or Emmental de Savoie. Of course, many individual Member States have their own legislation, and care must be taken as regards the implications of such national legislation in the countries of retail sale. However, the requirements of European legislation on hygiene, labelling, additives and packaging materials apply to all foodstuffs, including cheese.

Access to European legislation

Following its adoption, the EU legislation is published in the L-Series of the *Official Journal of the European Union*. It may also be accessed using the EUR-Lex website http://eur-lex.europa.eu/RECH_naturel.do, and the use of this website is facilitated greatly by knowing the type (directive, regulation, decision or COM-final), the year and the number of the relevant legislation. In this chapter, the necessary information will be given when referring to specific legislation. The most recent legislation is usually available electronically in portable document format (PDF), but earlier legislation may be accessible in hyper text markup language (HTML) format only. A copy of the original official journal document may be requested by e-mail. Amendments to legislation are also published in the official journal; however, these normally have merely the text that is being changed. Consolidated texts of most legislation can be accessed electronically, but such texts come with a warning that they are not official texts. Nonetheless such consolidated legislation, incorporating the amendments into the original text, can be very useful, as they facilitate use of the documents.

European Hygiene Legislation

The early legislative work in the EU was largely taken up with market regulation, and it was not until 1985 that the first Community hygiene measure for milk was adopted; this was Directive 85/397 (EU, 1985). This initiated a process of harmonising hygiene standards within the Community to facilitate intra-Community trade without compromising existing Member State hygiene rules. It covered all aspects of the production, transport and processing of milk from farm to the final consumer.

This was followed in 1992 by a new Milk Hygiene Directive 92/46 (EU, 1992a), which became effective from 1 January 1994. This Directive contained animal health requirements for raw milk, hygiene requirements for registered holdings, hygiene requirements in milking, collection and transport of milk to collection centres, standardisation centres, treatment establishments and processing establishments. For the first time, uniform EU-wide hygiene standards were created as the earlier Directive 85/397 (EU, 1985b) applied to intra-Community trade only. Directive 92/46 (EU, 1992a) laid down minimum compositional standards for milk, and also standards for the maximum plate count and somatic cell count for raw milk at collection from dairy farms intended for the production of certain milk-based products, including cheese.

A major review was carried out on the EU Hygiene Directives, following a recommendation in the EU White Paper on Food Safety (EU, 2000b). Before this review, there were a total of 16 commodity-specific EU Directives and one directive on general food hygiene, which had been gradually developed in the period from 1964 and had given a high level of protection to the consumer. However, they were comprised of a mixture of different disciplines (hygiene, animal health, official controls) and were detailed and complex. It was decided to overhaul the legislation to improve, simplify and modernise it and separate aspects of food hygiene from animal health and food control issues. The review aimed for a more consistent and clear approach throughout the food production chain from 'farm to fork'.

A package of new hygiene rules was adopted in April 2004 by the European Parliament and the Council. They became applicable from 1 January 2006 and, in the case of milk and milk products, replaced Directive 92/46 (EU, 1992a). The new rules are Regulations and not Directives, making them binding in Member States without the necessity of national legislation to be enacted to implement their provisions. Instead of all the hygiene requirements being embodied in a single piece of legislation, however, the hygiene requirements for the dairy sector are now contained across at least six different regulations. The three main Regulations are (a) Regulation 852/2004 on the hygiene of foodstuffs (EU, 2004d), (b) Regulation 853/2004 laying down specific hygiene rules for food of animal origin; Annex III Section XI thereof contains specific requirements for raw milk and dairy products (EU, 2004e) and (c) Regulation 854/2004 laying down specific rules for the organisation of official controls on products of animal origin intended for human consumption (EU, 2004f).

Then, in early December 2005, two important additional regulations were published, that is, Regulation 2074/2005 (EU, 2005b) and Regulation 2076/2005 (EU, 2005c). In addition to laying down implementing measures and transitional measures, they also contain important amendments and derogations to the original regulations. In 2006, a further Regulation 1662/2006 (EU, 2006a) was published, amending Regulation 853/2004 (EU, 2004e), which contained a replacement to the Complete Section XI of Regulation 853/2004. Fortunately, a consolidated version of Regulation 853/2004 (EU, 2004e), incorporating all the amendments up to 17 November 2007, is available on the EUR-Lex website. Microbiological criteria for foodstuffs are laid down in Regulation 2073/2005 (EU, 2005a) and those for cheeses made from pasteurised milk and additional requirements for unripened soft cheeses are laid down in Regulation 2073/2005 (EU, 2005a) and are outlined in Table 8.3. Guidance documents have also been developed by the Commission on Regulation 852/2004 (EU, 2005e; EU, 2005f; EU, 2006c) and Regulation 853/2004 (EU, 2005d).

European additive legislation

From the 1960s through to the mid-1970s, the EU established a series of basic directives addressing the use of colours, preservatives, antioxidants, emulsifiers, stabilisers and thickeners. Amendments were made to these over the years. Specific additive provisions were included in vertical legislation, and in other cases, authorisation for their use was left to Member States. Inevitably, this led to differences between the legislative provisions of Member States and this hindered the free movement of foodstuffs within the open market; thus harmonisation of this area became a major priority. With the move to horizontal legislation, as proposed in the White Paper on the Completion of the Internal Market in 1985 (EU, 1985a), moves were initiated to address additives in a horizontal and more comprehensive manner. First, the use of additives in foods throughout the EU was addressed

Table 8.3 Microbiological criteria for cheese and unripened soft cheeses (fresh cheeses) made from milk or whey that has undergone pasteurisation or a more severe heat treatment in EU legislation[a].

	Sampling	Plan	Limits	
Micro-organisms	n^{b}	c^{b}	m^{b}	M^{b}
Escherichia coli[c]	5	2	100 cfu[d] g^{-1}	1000 cfu g^{-1}
Coagulase-positive staphylococci[e]	5	2	10 cfu g^{-1}	100 cfu g^{-1}

[a] However, the limits do not apply to products intended for further processing in the food industry.
[b] *n*, number of units comprising the sample; c, number of sample units between m and M; m, the acceptable microbiological level in a sample unit; and M, maximum level for any sample unit which, when exceeded in one or more samples, would cause the lot to be rejected.
[c] *Escherichia coli* is used as an indicator of hygiene.
[d] cfu, colony-forming units.
[e] This requirement applies to unripened soft cheeses and excludes cheeses where the manufacturer can demonstrate, to the satisfaction of the competent authorities, that the product does not pose a risk of staphylococcal enterotoxins.
Data compiled from EU (2005a).

under the additive Framework Directive 89/107/EEC (EU, 1989) and the flavourings Directive 88/388 (EU, 1988), and, in 1994 and 1995, specific additive directives addressing the following were adopted:

- Colours by Directive 94/36 (EU, 1994b)
- Sweeteners by Directive 94/35 (EU, 1994a) as amended
- Additives other than colours and sweeteners by Directive 95/2 (EU, 1995c), as amended

Furthermore, food additives must at all times comply with the approved criteria of purity. These criteria are outlined in three Commission directives:

- Sweeteners by Directive 95/31/EC (EU, 1995a) as amended
- Colours by Directive 95/45/EC (EU, 1995b), as amended and
- Additives other than colours and sweeteners by Directive 96/77/EC (EU, 1996a) as amended

While these directives met the requirements of harmonising legislation in the EU and covered a more comprehensive list of additives than heretofore, they are not necessarily easy to address within the scope of this chapter. One major difficulty is that the EU does not have the equivalent of the Codex Food Categorisation System (FCS), which will be discussed later. Consequently, the appendices of the Directives contain references to foodstuffs that are not defined (or categorised) at Community level. Some references are specific and clear (e.g. partially dehydrated and dehydrated milk as defined in Directive 76/118/EEC-EU, 1976); others name particular, but undefined, products at European level (e.g. ripened cheese, whey cheese, Mozzarella); but others are either very general (e.g. dried powdered foodstuffs) or not clear (e.g. fine bakery wares). Within these constraints it may be worth looking at the specific additive directives that address additives in general, colours and additives other than colours and sweeteners.

The Framework Additive Directive 89/107/EEC (EU, 1989) has the following definitions. An additive is *'any substance not normally consumed as a food in itself and not normally used as a characteristic ingredient of food whether or not it has nutritive value, the intentional addition of which to food for a technological purpose in the manufacture, processing, preparation, treatment, packaging, transport or storage of such food results, or may be reasonably expected to result, in it or its by-products becoming directly or indirectly a component of such foods'* (Article 2).

A processing aid is *'any substance not consumed as a food ingredient by itself, intentionally used in the processing of raw materials, foods or their ingredients, to fulfil a certain technological purpose during treatment or processing and which may result in the unintentional but technically unavoidable presence of residues of the substance or its derivatives in the final product, provided that these residues do not present any health risk and do not have any technological effect on the finished product'* (Footnote to Article 1.3(a)).

This directive excludes flavourings and substances added as nutrients from the scope of the additives governed by its provisions; this refers to substances such as vitamins, minerals and trace elements. Annex II lays down three basic principles for the approval of use of additives, which may be summarised as follows:

- A technological need can be demonstrated and that need cannot be achieved by other means that are economical or technologically practical.
- Their use does not present a hazard to human health at the levels of use proposed on the basis of scientific evidence available.
- Their use does not mislead consumers.

Annex I of the directive lists 25 additive functional categories (Table 8.4); the definitions of these categories are given in the specific directives on colours, sweeteners and additives other than colours and sweeteners.

Table 8.4 The 25 additive functional classes listed in the EU framework Additive Directive 89/107/EEC, as amended.

Colour	Flavour enhancer	Glazing agent
Preservative	Acid	Flour treatment agent
Anti-oxidant	Acidity regulator	Firming agent
Emulsifier	Anti-caking agent	Humectant
Emulsifying salt	Modified starch	Sequestrant
Thickener	Sweetener	Enzyme
Gelling agent	Raising agent	Bulking agent
Stabiliser	Anti-foaming agent	Propellent gas and packaging gas

Note: Two additional additive functional classes (carriers and foaming agents) are defined in Directive 95/2, as amended.
After EU (1989a).

The Colours Directive 94/36 (EU, 1994b) has five annexes, listing 43 permitted food colours and the provisions for their use. Annex II lists foods that may not contain added colours except where specifically provided for in Annexes III, IV or V – this list includes *ripened and unripened cheese (unflavoured)*. None of the annexes have specific provisions on the use of permitted colours in unripened cheese; therefore, the use of colours is not permitted in unripened cheese.

Directive 95/2 on additives other than colours and sweeteners (EU, 1995c) is quite long, complex and detailed. The provisions may be summarised as follows:

- The latest consolidated text contains definitions of 24 additive functions (Article 1.3 (a) – (w) and Article 1.4). Included are definitions of two functions, carriers and foaming agents, not listed in Annex I of the Framework Directive. Additives are listed in the annexes without specified functions. It is up to food manufacturers to assign the principle or main additive function to each additive in product labelling, recognising that an additive may have more than one function in a food.
- The functional ingredients edible gelatine, certain starches, casein and caseinates are not considered as food additives.
- Article 2.3 lists foods where the permission to use the additives listed in Annex I does not apply unless specifically allowed for. This article does not list unripened cheese.
- Annex I lists 114 E numbers and specific food additives that are generally permitted for use in foods not referred to in Art 2.3 or Annex II. The use of the additives in this annex, which includes gums, alginates, agar, carrageenans, celluloses and modified starches, would be permitted in unripened cheese, at least in theory, where their use meets the basic principles outlined in the discussion on Directive 89/107/EEC (EU, 1989).
- The *quantum satis* principle is often misinterpreted as meaning that one may use as one likes; however, this is incorrect, and the term is defined in Article 2.8 as, *'that no maximum level is specified. However, additives shall be used in accordance with good manufacturing practice, at a level not higher than is necessary to achieve the intended purpose and provided that they do not mislead the consumer'*.
- Annex II lists foods where a limited number of Annex I additives may be used. This annex includes specific provisions for ripened cheese, whey cheese and Mozzarella. Although the latter is regarded as a specific variety of unripened cheese, the limitations do not apply to other unripened cheeses, such as cream cheese and related types.
- Annex III addresses conditionally permitted preservatives and anti-oxidants. It has four parts: (a) Part A deals with sorbates, benzoates and *p*-hydroxybenzoates; (b) Part B addresses sulphur dioxide and sulphites; (c) Part C other preservatives and (d) Part D other anti-oxidants. Note that Part A permits the use of sorbates at a maximum level of 1000 mg kg^{-1} in unripened cheese. Part C permits the use of nisin in Mascarpone, but not other soft unripened cheeses, at a maximum level of 10 mg kg^{-1}. Potassium nitrate and sodium nitrate are permitted in hard, semi-hard and semi-soft cheeses, but not in soft cheeses. Propionates are permitted for surface treatment of cheese at *quantum satis* level.
- Annex IV deals with 'Other Permitted Additives' and is the most complicated of the annexes to comprehend. It is often best to look at each additive of interest and then look up the annex to see if it is permitted for a particular product of interest. Phosphates

are addressed as a group and the specified maximum levels apply to their use singly or in combination, expressed as phosphate ion (P_2O_5). For unripened cheese, except Mozzarella, phosphates are allowed at a maximum level so expressed of 2 g kg^{-1}.

- Annex V addresses permitted carriers and carrier solvents, and shall not be discussed further.
- Annex VI deals with additives for use in foods for infants and young children, and is not relevant as regards cheese.

Forthcoming changes as regards additives

In 2006, the European Commission adopted a package of legislative proposals that aimed to upgrade rules for additives (EU, 2007c) and flavourings (EU, 2007b) and to introduce harmonised legislation on food enzymes (EU, 2007d). It also proposed the creation of a common authorisation procedure for food additives, flavourings and enzymes, based on scientific opinions from the European Food Safety Authority (EFSA) (EU, 2007a). Three of these proposals were adopted as common positions and published as three separate communications in early March 2008; they may be accessed on the EUR-Lex website as follows:

- For additives (EU, 2008b)
- For enzymes (EU, 2008c)
- For the common authorisation procedure (EU, 2008a).

The next step is for these proposals to be debated at the European Parliament.

European legislation on the labelling of foods

Horizontal European labelling requirements for foods are contained in the EU Labelling Directive 2000/13 (EU, 2000a), as amended. It should be noted that the scope of this directive applies to the labelling of foodstuffs to be delivered as such to the ultimate consumer, or to mass caterers (defined as restaurants, hospitals, canteens and other similar mass caterers). Hence, they may not necessarily apply directly to products in so far as they may be intended for further manufacture. In such instances, the products are normally traded to meet detailed specifications between the purchaser and vendor. Compliance with such specifications, especially where written and signed by both parties, would be governed by contract law.

The horizontal food labelling requirements include the following provisions:

- *Name of the food* – a hierarchy exists as regards the names/designations used. For example, when the product has a legal name specified in EU legislation, then that name should be used. Where there is no EU legal name, the name under which a product is sold shall be the name provided for in the legislation and administrative provisions applicable in the Member State in which the product is sold to the final consumer or to mass caterers. Where neither of the above provisions apply, the name under which a product is sold shall be the name customary in the Member State in which it is sold to the final consumer or to mass caterers *or* a description of the foodstuff, and if necessary of its use, which is clear enough to let the purchaser know its true nature and distinguish it from other products with which it might be confused. The use in the Member State of marketing of the sales name under which the product is legally manufactured and marketed in the Member State of production shall also be allowed. However, this has qualifications: (a) where the other labelling requirements would not enable consumers in the Member State of marketing to know the true nature of the foodstuff and to distinguish it from foodstuffs with which they could confuse it, in which case, the sales name shall be accompanied by other descriptive information which shall appear in proximity to the sales name and (b) this name cannot be used if the product so named is so different in the Member State of sale as regards its composition or manufacture from the foodstuff known there under that name that the provisions mentioned here are not sufficient to ensure correct information for consumers.

- *List of ingredients* (including additives) – if vitamins or minerals are added, these should be indicated. Cheese does not require such a list provided that no ingredient has been added other than lactic products, enzymes and micro-organism cultures essential to manufacture. However, an indication of the use of salt is required when it is used for the manufacture of fresh cheese. This can be taken to include unripened cheese.
- *Weight* – an indication of the net quantity.
- *The date of minimum durability* – this information should be indicated.
- *Special conditions of storage and use* that would affect the minimum durability.
- *Name and address of manufacturer or seller* – this information should be given in addition to the identification mark required by the hygiene regulations outlined earlier.

In 2006, Regulation (EC) No. 1924/2006 (EU, 2006b), which has been in force since the 1 July 2007, is aimed at ensuring that nutrition and health claims made on foodstuffs comply with specified requirements. A list of permitted nutrition claims and their specific conditions of use is included in the annex to Regulation (EC) No. 1924/2006 (EU, 2006b). The requirements for fat content claims, such as fat-free, low-fat, reduced-fat and light, are outlined in Table 8.5. Furthermore, nutrition or health claims may not be made that are inconsistent with generally accepted nutrition and health principles or if it encourages

Table 8.5 Requirements for fat-related nutrient claims on solid foodstuffs in the EU under Regulation No. 1924/2006.

Claim	Requirement for fat content	Additional requirements
Fat free[a]	$\leq$0.5 g 100 g^{-1}	–
Low-fat	$\leq$3g 100 g^{-1}	–
Reduced-fat	25% less fat than the reference standard product	–
Light or lite	Same as for 'reduced' claim, i.e. 25% less fat than the reference standard product	Claim shall be accompanied by an indication of the characteristic that makes the product 'light', i.e. fat

[a]Regulation 1924/2006 specifically prohibits nutrient claims of the form 'X% fat-free'.
After EU (2006b).

excessive consumption of a particular food or is not consistent with a good diet. In this regard nutrient profiles are being developed to address this principle. These profiles have not been finalised yet. This may have implications for claims on cheeses, including cream cheese and related products. In December 2007 guidance on the implementation of Regulation 1924/2006 (EU, 2006b) has been published by the EU Commission (EU, 2007e).

European legislation on food packaging materials

Cream cheese and cream cheese-like products may be packed in differing packaging formats, for example, foil wrappers, plastic tubs, etc. These packaging materials should comply with the general requirements of Regulation 1935/2004 (EU, 2004c) and the particular requirements such as those contained in Directive 2002/72 (EU, 2003), as amended by Directive 2004/19 (EU, 2004b) and Directive 2004/1 (EU, 2004a). It is normal for processors to specify to their packaging suppliers that their products comply with the requirements of these directives.

8.3.3 *UK legislation*

Legislative basis

Up to recent times, England and Wales had common legislation, signed by the appropriate minister of the UK government and the Secretary of State for Wales. Separate but similar legislation is enacted for Scotland and Northern Ireland; however, some differences could and sometimes did occur. Then from 2000, with the establishment of the Welsh Assembly, separate but a similar legislation for Wales was enacted.

The primary source of legislation is by Acts of Parliament, primarily the Westminster Parliament; secondary legislation in the form of Statutory Instruments (SIs), Statutory Regulations and Orders (SROs) are enacted under specified sections of the enabling act or acts. Details of the current legislation in the United Kingdom as well as the separate legislation applicable to Scotland, Wales and Northern Ireland may be found via the relevant link on the (UK) Foods Standards Agency website (http://www.foodstandards.gov.uk) or that of the Office of Public Sector Information (OPSI; http://www.opsi.gov.uk/legislation/uk.htm). When searching the OPSI website, it is necessary to know the year and number of the Act or SI of interest.

Background

In the 1850s there was increasing concern on the issues of food purity and food adulteration based on the identification of such issues by analysts and medical doctors. This led to the adoption of three separate pieces of legislation, addressing food adulteration; one such was the Adulteration of Food and Drugs Act 1860 (HMSO, 1860). However, this was ineffective but it paved the way for the enactment of the Sale of Food and Drugs Act 1875 (HMSO, 1875). The main requirements of this Act were

- that nothing that would be injurious to health should be added to food for sale;
- that sale of food that was not of the proper nature, substance or quality was prohibited;

- that analysts were appointed;
- that purchasers of a food were empowered to have it analysed;
- officers entitled to obtain samples for submission to an analyst be named.

Although it was not without its critics, the act, with subsequent amendments, enlargement and consolidation, remained in force for the next 60 years (Monier-Williams, 1951). In the early 1930s, a Departmental Committee on the Composition and Description of Foods was established to look into the whole area of definitions, standards, labelling and advertising. This committee was in favour of a limited number of standards, the main aim of which would be to inform consumers of what they were purchasing (Monier-Williams, 1951). The report of this committee in 1934 resulted in a new consolidated Sale of Food and Drugs Act (HMSO, 1938). The 1938 Act remained in place until it was replaced by the Sale of Food and Drugs Act 1955 (HMSO, 1955). Subsequently, as the result of a number of food scares in the 1980s, with causes such as Salmonella, Listeria and bovine spongiform encephalopathy (BSE), the Food Safety Act 1990 (HMSO, 1990) was enacted. This was a broad measure that created a more systematic structure of UK food law and tightened up on offences, enforcement powers and penalties.

Among the first legislation that set standards for dairy products were the Butter and Margarine Act 1907 (HMSO, 1907; French & Phillips, 2000), the Condensed Milk Regulations 1923 (HMSO, 1923a) and the Dried Milk Regulations 1923 (HMSO, 1923b). However, the first cheese regulations were not adopted until the Cheese Regulations 1965 (Davis, 1966; HMSO, 1965).

On hygienic aspects, in 1912 or thereabouts, the quality and purity of milk supply began to receive increased attention. World War I interfered with the enforcement of the Milk and Dairies (Consolidation) Act 1915 (HMSO, 1915), but by 1918 a tentative system of milk grading was in operation. The production of clean and safe milk was first seriously addressed under the Milk and Dairies Amendment Act 1922 (HMSO, 1922a) and the Milk (Special Designations) Order 1922 (HMSO, 1922b), which incorporated different grades for raw milk.

When the United Kingdom joined the then European Economic Community on 1 January 1973, European legislation began to have a major role in shaping the evolving national legislation. However, as outlined in the discussion of European legislation earlier, harmonisation of legislation on vertical legislation did not start until the 1970s and on hygienic aspects of milk production until the mid-1980s.

European directives have to be enacted into the laws of Member States, while Community regulations are binding in their entirety on Member States. In the latter case, the relevant SIs reference the requirements contained therein and outline particular elements, such as interpretations/definitions, specify the competent authority, address administration, detail offences, defences and penalties and specify certain schedules. Where the European regulations specify general provisions, the UK SI may lay down more specific requirements and may address national provisions where discretion or optional provisions are delegated to Member States.

UK legislation on cheese

As mentioned earlier, the first UK cheese legislation was contained in the Cheese Regulations 1965 (HMSO, 1965; Davis, 1966). These applied to England and Wales, and Scotland and

Northern Ireland had separate, but identical legislation. It had a definition of cheese and specified the requirements for hard, soft, whey cheese, processed cheese and cheese spread; it subdivided soft cheese into full-fat, medium-fat, low-fat and skimmed milk cheese, on the basis of their composition. It also defined cream cheese (minimum 45 g fat 100 g^{-1}) and DCC (minimum 65 g fat 100 g^{-1}) – these fat levels were absolute levels, independent of moisture content. These standards were amended on a number of occasions, but in 1970, were replaced by the Cream Cheese Regulations 1970 (HMSO, 1970); they were re-enacted, with amendments; the 1970 Regulations thus contained many of the earlier provisions.

The equivalent Scottish Regulations were the Cheese (Scotland) Regulations 1970 (The Stationery Office, 1970a); those for Northern Ireland were the Cheese Regulations (Northern Ireland) 1970 (The Stationery Office, 1970b).

In the 1970 regulations, the requirements for soft cheeses of different fat and moisture compositions, including cream cheese and DCC, were the same as in the 1965 regulations and are outlined in Table 8.6. Specific ingredients were allowed in cheese in general and in soft cheese in particular. These included certain acidity regulators, colours, emulsifiers and stabilisers. The list included alginates, carrageenan and the major gums. The acidity regulators included lactic, acetic, citric and phosphoric acids and glucono-δ-lactone (GDL). Gelatine and starch were not permitted. Of course the European additive legislation, as discussed earlier, now supersedes this list; its interest at this time is that it may be taken as an indication of the additives in these categories that were in use at the time the legislation was in force and indeed today. This theory is supported by additives such as citric acid, carrageenan, locust bean gum (LBG) and carob bean gum being listed in the ingredient lists of some of the products contained in Table 8.2.

Over the years, probably due to their specified requirements for high fat levels, products designated cream cheese and DCC almost disappeared from the UK market. Furthermore, foil-wrapped Kraft Philadelphia was launched in the UK market in 1960; although this complied with the US requirements for cream cheese, as we shall see later, it had a lower fat level than that required for cream cheese in the United Kingdom, hence it was designated as a full-fat soft cheese. Its popularity led to other similar full-fat soft cheese products being developed. In the mid-1980s, light variants were launched to meet increasing demand

Table 8.6 Soft cheese and cream cheese compositional (g 100 g^{-1}) requirements under the Cheese Regulations 1970, as amended, now repealed[a,b].

Product designation	Milk fat	Moisture
Double cream cheese	≥ 65	≤ 60
Cream cheese	≥ 45	≤ 60
Full-fat soft cheese	≥ 20	≤ 60
Medium-fat soft cheese	≥ 10 and < 20	≤ 70
Low-fat soft cheese	≥ 2 and < 10	≤ 80
Skimmed milk soft cheese	< 2	≤ 80

[a]These regulations applied to England and Wales only.
[b]Repealed by The Cheese and Cream Regulations 1995.
After HMSO (1970), as amended.

for lower fat products; eventually, at the turn of the century, extra light products were developed (Kraft Foods, 2008). Although these are seen as cream cheese type products, in all probability due to the earlier compositional requirements, the products are designated as full-fat soft cheese, medium-fat soft cheese, reduced-fat soft cheese and X g fat 100 g^{-1} soft cheese (see Table 8.2). Use of terms, such as low-fat, is now strictly regulated by European legislation on nutritional and health claims.

The 1970 Regulations were amended on a number of occasions (HMSO, 1974; HMSO, 1984; The Stationery Office, 1984; The Stationery Office, 1995), but were repealed by the short-lived Cream and Cheese Regulations 1995, which applied to England, Wales and Scotland (HMSO, 1995). Northern Ireland had its own, but separate similar regulations (The Stationery Office, 1996). The 1995 regulations had a new definition of cheese, the old definitions of processed cheese and cheese spread and retained compositional requirements for Cheddar, Blue Stilton and 10 other UK territorial cheeses; the compositional requirements for cream cheese, DCC and soft cheeses were not retained. These regulations were repealed and replaced by the Food Labelling Regulations 1996, which retained the definition of cheese and the compositional requirements of the cheeses in the 1995 regulations, but did not retain the definitions of processed cheese and cheese spread (HMSO, 1996).

The definition of cheese in the Food Labelling Regulations 1996 is as follows: *'Cheese means the fresh or matured product intended for sale for human consumption, which is obtained as follows:*

In the case of any cheese other than whey cheese, by combining, by coagulation or by any technique involving coagulation, of any of the following substances, namely milk, cream, skimmed milk, partly skimmed milk, concentrated skimmed milk, reconstituted dried milk, butter milk, materials obtained from milk, other ingredients necessary for the manufacture of cheese provided that those are not used for replacing, in whole or in part, any milk constituent, with or without partially draining the whey resulting from coagulation. In the case of whey cheese – (a) by concentrating whey with or without the addition of milk and milk fat, and moulding such concentrated whey, or (b) by coagulating whey with or without the addition of milk and milk fat'.

This is a broader definition than that in the 1970 regulations, listing materials obtained from milk as an additional source of raw materials. Coagulation or 'techniques involving coagulation' are required, but not defined, and the compulsory removal of at least part of the resultant whey is no longer required. There is nothing specific regarding the requirements for cream cheese and related products.

What then are the present requirements as regards cream cheese? Do the earlier detailed provisions, especially as regards chemical composition, in the 1970 cheese regulations still apply, even though the legislation was repealed more than 10 years ago, in 1995? The position is somewhat unclear. It is likely that the status of the names of cheeses, which were defined in the repealed regulations, have become 'customary names' in the United Kingdom (i.e. name customary in the Member State in which it is sold, as discussed in Section 8.3.2). Probably the best advice was contained in the 1997 UK Local Authority Coordination Body on Trading Standards (LACOTS) opinion on cheese names that indicated that customary names could be used if they did not depart from the original compositional profile. They also indicated that use of the terms full-fat, medium-fat, hard and soft, in relation to cheese, would indicate the true nature of the food as required by Regulation 8 of the Food Labelling Regulations

1996. It must be stressed that the LACOTS advice contains the warning that only the courts have the authority to interpret statute law, that their advice was based on the information provided and might be revised in the light of further information, and not intended to be a definitive guide, or substitute for the relevant law (LACOTS, 1997). LACOTS is now known as LACORS (the Local Authorities Coordinators of Regulatory Services). Their advice is now intended for guidance for their local council colleagues only, and access to much of their advice is restricted.

Therefore, the answer to the question of why the products on the UK and Irish markets are designated as full-fat soft cheese is probably that their composition does not meet the compositional requirements of the repealed UK Cheese Regulations 1970. Hence, to use the names cream cheese or DCC, for products which do not meet those compositional requirements, would risk possible legal challenge. It should also be recognised that the present products are regarded, and indeed often referred to, as cream cheese by the general public. Thus, the products do not risk legal challenge and are accepted by the general public with their present designations.

8.3.4 *Irish legislation*

Introduction

The Irish Parliament (Oireachtas) consists of the President and two Houses of the Oireachtas (Dáil Éireann and Seanad Éireann), which are responsible for enacting new legislation in Ireland. A proposal for new legislation is published as a bill, which becomes an act and is declared law when it is agreed by both houses of the Oireachtas and signed by the President. Acts frequently give powers to specific ministers of governments (particularly those with responsibility for Health and Agriculture and Food) to make secondary laws known as SIs. These laws can be written in the form of regulations or orders and they detail specific rules and give enforcement powers to a particular authority. The similarity to the UK legislative process, on which it was based, will be apparent. Indeed, from the foundation of the state in 1922 until mid-century, reference was often made to acts of the UK Parliament, such as the Food and Drugs Act 1875, as discussed earlier, which had application in the whole of Ireland at the time of its enactment. For instance, such references are contained in the Dairy Produce Act 1924 (The Stationery Office Dublin, 1924) and the Sale of Food and Drugs (Milk) Act 1935 (The Stationery Office Dublin, 1935).

Until its joining the then European Economic Community in 1973, Ireland had quite limited legislation governing the composition of foods. Indeed, it may be argued that such legislation was not really required, because, as an exporter of the majority of its food products, Irish manufacturers frequently adhered to the compositional standards and requirements of its main export market, the United Kingdom, and production for the home market reflected the same requirements.

Irish legislation on cheese

There is no, and has never been, specific Irish legislation on cheese. Cheeses produced and labelled to meet UK legislation have not had problems when sold on the Irish market. This facilitated both producers and consumers. The size of the home market would create problems if this were not the case. However, in the case of reduced-fat variants of common cheese varieties, where there are specific compositional requirements in UK legislation, these can be designated reduced fat with the variety name on the Irish market, but not in United Kingdom (e.g. reduced-fat Cheddar is acceptable in Ireland, provided it meets the EU legislation on nutrition and health claims as regards a reduced-fat claim (EU, 2006b), but could not be so designated in the United Kingdom, where a name such as medium-fat hard cheese would be required). The requirements for claims related to fat content are outlined in Table 8.5.

Another possible approach for products intended for the Irish market alone would be to manufacture in accordance with the Codex Standard for cream cheese, which shall be discussed later. In such instances, it would be wise to seek prior approval of the competent Irish Authority, which is currently the Department of Agriculture, Fisheries and Food, as regards its manufacture, composition and proposed product labelling. A possible problem could arise from having similar products with different product names on the market, which could be seen to be misleading or confusing to consumers.

All horizontal European legislation applying to all foodstuffs, as discussed in Section 8.2.2, would apply in any event. Some cream cheese-like products on the Irish market at this time are included in Table 8.2.

8.3.5 *US legislation and standards*

Introduction and background

As mentioned at the outset of this chapter, cream cheese is a major cheese on the US market, thus it is worthwhile to look at US legislation and in particular that related to cheese, in some detail. It should be pointed out that this section shall address US federal legislation. Up to 1900, there was little federal legislation addressing food standards and the individual states controlled domestically produced and distributed foods; however, this control was markedly inconsistent from state to state. From the early 1880s, the USDA's Division of Chemistry (renamed the Bureau of Chemistry in 1901), under Harvey Wiley who had been appointed its chief chemist in 1883, began researching the adulteration and misbranding of food (and drugs) on the market. Their efforts coincided with a general trend for increased federal regulations in all matters pertinent to safeguarding public health. State laws provided varying degrees of protection against practices, such as misrepresenting the ingredients of food products or medicines (Swann, 2008). It should also be added that in the early 1900s the food industry strongly supported national food legislation to obtain national uniformity in regulatory requirements and to build credibility for the food supply (Porter & Earl, 1992).

Despite considerable debate on the issue of constitutionality surrounding States' rights, Congress enacted the Food and Drugs Act 1906 (Pub. L. No. 59–384 34 STAT. 768 (1906)), sometimes called the 'Wiley Act' in honour of its chief advocate (FDA, 1906). This act was aimed at '*preventing the manufacture, sale, or transportation of adulterated or misbranded or poisonous or deleterious foods, drugs, medicines, and liquors, and for regulating traffic therein, and for other purposes*.' Congressional acts identify and grant broad authority to federal agencies to interpret their provisions into the United States Code (USC), with the relevant enforcement agencies identified.

Under the Food and Drugs Act 1906, responsibility for administration and their examination for 'adulteration' or 'misbranding' was granted to the Wiley's US Department of Agriculture Bureau of Chemistry (USDABOC). Over the years, the name of this body has changed to the more familiar Food and Drug Administration (FDA). The evolution and development of this organisation shall be discussed a little later.

Over the next 25 years, amendments were made to the original act, but in the early 1930s, because of ongoing problems, the federal regulators, consumer groups and the media again pressed for a new act with more powers and scope. It took 5 years to be passed but the Federal Food, Drug, and Cosmetic (FDC) Act of 1938 was finally adopted (FDA, 1938). This act is sometimes referred to as Title 21, Chapter 9 of the United States Code (21 USC 9). As well as extending the scope of the earlier act to cover cosmetics and therapeutic devices, this new act repealed the Food and Drugs Act 1906 and contained the following new provisions of relevance to food:

- Allowed that safe tolerances be set for unavoidable poisonous substances
- Permitted standards of identity, quality and fill-of-container to be set
- Authorised the inspection of manufacturing premises
- Added the use of court injunctions to the previous penalties of seizures and prosecutions

This law, although it has been subject to frequent amendments in the intervening years, remains the basis for federal regulation by the FDA to the present day. One amending act of relevance to the nutritional claims on dairy products is the Nutrition Labelling and Education Act (NLEA) of 1990 (Pub. L 101–535).

Evolution and development of the FDA

When the Food and Drugs Act 1906 was passed into law, responsibility for its administration and for the examination of food for 'adulteration' or 'misbranding' was granted to the USDABOC. Over time, this body has changed its name to the Food and Drugs Administration and has transferred the department of government under which it operates on a number of occasions, according to the following time-line:

- **1927** – The non-regulatory duties of the USDABOC were transferred with the USDA, and its name was changed to the Food, Drug, and Insecticide Administration.
- **1930** – The name was shortened to the more familiar FDA.
- **1940** – It was transferred to the Federal Security Agency.
- **1953** – The agency was transferred again, this time to the Department of Health, Education, and Welfare (HEW).
- **1968** – The agency became part of the Public Health Service within HEW.
- **1980** – The education function was removed from HEW to create the Department of Health and Human Services (DHSS) – the FDA's current home.

The FDA is now responsible for about 80% of the US food supply. The exceptions are as regards the safety, wholesomeness, labelling and packaging of meat, poultry and certain egg products, which are the responsibility of the USDA (Swann, 2008).

The Code of Federal Regulations

The Code of Federal Regulations (CFR) is the consolidated source of the general and permanent rules developed by the relevant US government departments and/or their administrative agencies and also published in the Federal Register (US National Archives and Records Administration, 2008). It is divided into 50 titles that represent broad areas that are subject to Federal regulation. For instance, Title 7 covers agriculture, administered by the USDA, and Title 21 deals with food and drugs, administered by the FDA. Each volume of the CFR is updated once each year and is issued on a quarterly basis. It is published by the Office of the Federal Register, an agency of the National Archives and Records Administration:

- Titles 1 to 16 are updated as of 1 January.
- Titles 17 to 27 are updated as of 1 April.
- Titles 28 to 41 are updated as of 1 July.
- Titles 42 to 50 are updated as of 1 October.

Each title is divided into chapters; each chapter is further subdivided into parts that cover specific regulatory areas.

Specific US cheese legislation

Large parts of the CFR may be subdivided into subparts. All parts are organised in sections and most citations in the CFR are provided at the section level. Part 133 of the CFR addresses standards of identity for cheese; currently, there are 76 standards of identity for different cheeses, of which 11 are for pasteurised processed cheese, pasteurised cheese spreads or pasteurised blended cheeses. The full format of CFR citations are, for example, Cheddar Cheese 21 CFR Part 133 Subpart B §133.113. However, an abbreviated form comprising just the CFR part and subpart letter, such as 21 CFR §133.113, is also quite common. The latest CFR is also available online at http://www.access.gpo.gov/nara/cfr/cfr-table-search.html#page1. The online documents are available as ASCII text and PDF files.

As outlined in the introduction, both cream cheese and American Neufchatel cheese, a lower fat type of cream cheese, were developed in the United States. By the 1920s, a cream cheese containing 26 g fat 100 g^{-1} and 66 g moisture 100 g^{-1} had been made from 7 g fat 100 g^{-1} milk preparations for some years. The first US standard of identity for cream cheese was issued in 1921. This standard was advisory in nature and was not compulsory and stated that '*Cream cheese is the unripened cheese made by the Neufchatel process from whole milk enriched with cream. It contains in the water-free substance not less than 65 g 100 g^{-1} of milk fat*'. In the following 10 years, competition caused manufacturers to raise the fat content of cream cheeses to over 33 g 100 g^{-1}. It was believed that this produced a cream cheese that was far superior to the older type cheese and thus resulted in increased sales (Lundstedt, 1954). American Neufchatel cheese, a cream cheese product with lower fat content, was also the subject of a federal standard of identity.

From the late 1940s to the present day, compulsory federal standards of identity for cream cheese and American Neufchatel cheese have been promulgated. The present standards of identity for cream cheese and American Neufchatel cheese were first published in the CFR in 1989 and amended in 1993. The consolidated provisions for cream cheese may be found

Table 8.7 Cream cheese and Neufchatel/Neufchâtel cheese and cheese spreads compositional (g 100 g^{-1}) requirements under the USA Code of Federal Regulations and Canadian Food and Drug Regulations.

Product	USA		Canada	
	Milk fat	Moisture	Milk fat	Moisture
Cream cheese	≥ 33	≤ 55	≥ 30	≤ 55
Cream cheese with other foods	≥ 27	≤ 60	≥ 26	≤ 60
Neufchatel/Neufchâtel[a] cheese	≥ 20 and <33	≤ 65	≥ 20	≤ 60
Cream cheese spread	Not permitted	Not permitted	≥ 24	≤ 60
Pasteurised Neufchatel cheese spread with other foods	≥ 20	≤ 60	Not specified	Not specified

[a]This cheese is designated as Neufchatel in the USA and Neufchâtel in Canada.
After US National Archives and Records Administration (2008) and Department of Justice Canada (2008).

in §133.133, for cream cheese with other foods in §133.134, for (American) Neufchatel cheese in §133.162 and for pasteurised Neufchatel cheese spread with other foods in §133.178. The compositional requirements for these products are summarised in Table 8.7, where they are compared with the equivalent Canadian Standards, which shall be discussed in the next section. It will be noted that the US standards have lower fat requirements for cream cheese than the old UK legislation (minimum 33g fat 100 g^{-1} versus. minimum 45 g fat 100 g^{-1}). The following are some other requirements for cream cheese, as outlined in §133.133:

- The dairy ingredients are pasteurised.
- The permitted dairy ingredients for manufacture are milk, non-fat milk or cream, alone or in combination.
- The permitted dairy ingredients are subjected to the action of lactic acid producing bacterial culture.
- The above ingredients may be homogenised.
- Rennet or other clotting enzymes of animal, plant or microbial origin are added.
- The coagulated mass may be warmed and stirred and it is drained.
- The moisture content of the product may be adjusted by the addition of the optional ingredients cheese whey, concentrated cheese whey, dried cheese whey or reconstituted cheese whey.
- Stabilisers may be added to a maximum level of 0.5 g 100 g^{-1} of the finished product. The specific stabilisers are not listed, but the US legislation on additives shall be outlined briefly in the subsequent section.

The product shall be called cream cheese and the ingredients used shall be labelled in descending order of inclusion. Unlike in the European Union, the dairy ingredients, use of lactic cultures and rennet or milk clotting enzymes must be indicated; in the case of the latter indicating just enzymes will suffice.

Labelling of foods in the United States are addressed in CFR PART 101 Food Labelling. Unlike the situation on cheese in the United Kingdom, described earlier, it is permitted to use claims on nutrient content in conjunction with the names of standardised products, including cream cheese and similar products. The requirements for foods named by use of a nutrient content claim are given in §101.10. Such content claims are based on reference amounts customarily consumed per eating occasion; a list of such reference amounts can be found in CFR §101.12 – for cream cheese and similar products this reference amount is 30 g. Fat-related claims on nutrient content are contained in §101.62, with the requirements for light (or lite) addressed in §101.56. The nutrient claims – reduced and light – are recognised as comparative nutrient claims and the appropriate reference food must be specified. There are also requirements for labelling statements to be used on foods that make claims based on nutrient content; these will not be discussed further. Table 8.8 endeavours to provide a simple overview of fat content-related claims with special reference to cream cheese and similar products.

Food additives in US legislation

In federal standards of identity, in the case of cream cheese and cream cheese-type products as outlined earlier, certain food additive functional classes are listed as optional ingredients (e.g. stabilisers). A total of 32 such additive functional classes are defined in 21CFR§170.3(o). Specific individual additives are not listed in the standards of identity. Hence, it is worth considering briefly how additives are regulated in the United States. The definition of a food additive is given in the Food, Drug and Cosmetic Act 1938, as amended, in Section 201(s) and (t) and in the CFR (21CFR§170.3). Ingredients and substances (including food additives) that may be used are classified into four legal categories as follows:

- **New food additives** – for substances that have no proven track record of safety and must have prior review and approval as regards safety by the FDA before marketing; these additives may receive generally recognised as safe (GRAS) status.

Table 8.8 Fat-related relative claims on nutrient content in US product labelling with special reference to cream cheese and similar products (CFR §101.62 and 101.56).

Claim	Requirement	Reference amount (g) for cream cheese and related types[a]
Fat free	<0.5 g per reference amount	30
Low-fat	≤ 3 g per reference amount	30
Reduced-fat	25% less fat per reference amount than the reference product	30
Light/lite[b] where $\geq 50\%$ of calories come from fat	The number of calories is reduced by at least one-third compared to the reference product or contains 50% less fat per reference amount than the reference product	30
Light/lite[b] where $<50\%$ of calories come from fat	50% less fat per reference amount than the reference product	30

[a]Reference amounts for foods differ; the amounts for specific foods are given in CFR §101.12.
[b]A light claim may not be made for a product for which the reference food meets the definition of 'low-fat' and low-calorie.
After US National Archives and Records Administration (2008).

- **Colour additives** – for dyes that are used in foods, drugs, cosmetics and medical devices must be approved by the FDA before they can be marketed;
- **GRAS** – for substances whose use in food has a proven track record of safety based either on a history of use before 1958 or on published scientific evidence and that need not be approved by the FDA before being used; the various means of achieving this classification are given in 21CFR §170.30 and 21CFR §170.31.
- **Prior-sanctioned** – for substances that were assumed to be safe for use in a specific food by either the FDA or the USDA before 1958; an example here is the use of nitrate as a preservative in meat because it was sanctioned before 1958; however, it cannot be used on vegetables because they were not covered by the prior sanction.

There are four subsequent parts of the CFR that list and define the substances for food use:

- Part 181 – Prior-sanctioned Food Ingredients.
- Part 182 – Substances GRAS.
- Part 184 – Direct Food Substances Affirmed as GRAS.
- Part 186 – Indirect Food Substances Affirmed as GRAS.

A list of all the ingredients and substances is given at the start of each of these parts, giving the section reference for each compound. In addition, Part 189 lists substances prohibited from use in human food. A further useful reference point on substances for use in food in the United States is the Food Additive Status List on the FDA website (FDA, 2006). This lists substances alphabetically and outlines their status and limitations for use. For a brief overview on this topic, a useful document is a short publication of the International Food Information Council, prepared with the assistance of the FDA (International Food Information Council, 2005).

8.3.6 *Canadian legislation and standards*

Canadian food legislation that prescribes the standards of composition, strength, potency, purity, quality or other characteristics of certain foods is contained in the Canadian Food and Drug Regulations; a consolidated up-to-date version is accessible (Department of Justice Canada, 2008). Dairy products are addressed in Division 8 of the regulations, with detailed provisions for cream cheese in B.08.035 (1); for cream cheese with (naming the added ingredients) in B.08.037 (1); for cream cheese spread in B.08.038 (1); for cream cheese spread with (naming the added ingredients) in B.08.039 (1); and compositional requirements only for Canadian Neufchâtel cheese in Table 8.2 at the end of B.08.032 (1). The compositional requirements for these cheeses are summarised in Table 8.7, where they are compared with the equivalent US Standards, for similar cheeses as discussed in the previous section.

The following are some other requirements for cream cheese, as outlined in B.08.035:

- The product made by coagulating cream with the aid of bacteria to form a curd.
- Forming the curd into a homogeneous mass after draining the whey.
- Cream cheese may contain (a) cream added to adjust the milk fat content; (b) salt; (c) nitrogen to improve spreadability in an amount consistent with good manufacturing practice; (d) only emulsifying, gelling, preserving, stabilising and thickening agents

from lengthy specified lists; and (e) only specified enzymes, pepsin, rennet and chymosin A or chymosin B from specified microbial sources, at levels consistent with good manufacturing practice.

- Stabilisers may be added to a maximum level of 0.5 g 100 g^{-1} of the finished product; maximum levels of the preservatives (sorbates and propionates) are also specified, used singly or in combination.

Unlike in the European Union, the dairy ingredients, use of lactic cultures and rennet or milk-clotting enzymes must be indicated also in Canada; in the case of the latter indicating just enzymes will suffice. All cheeses should be labelled to show on the principal display panel a statement of the percentage of fat in the product followed by the words ‘milk fat’ (MF) or ‘butter fat’ (BF), and the percentage of moisture in the cheese followed by the word ‘moisture’ or ‘water’.

8.3.7 *German cheese legislation with particular reference to cream cheese-type products*

German national legislation as regards cheese is contained in the Käseverordnung (which can be translated as the Cheese Order). The original version of the present Käseverordnung was adopted in April 1986, but has been updated on a regular basis. Consolidated electronic and print versions are available; however, some electronic versions do not show all the tables in the Anlagen (Annexes) (Behr’s Verlag, 2007; Bundesministerium der Justiz, 2008).

Under German legislation, cream cheese is regulated under the names Rahmfrischkäse (which can be translated as cream fresh cheese) and Dopplerahmfrischkäse (which can be translated as double cream fresh cheese) can be regarded as variants of Frischkäse (fresh cheese), with specific fat content requirements. Table 8.9 outlines the compositional

Table 8.9 Frischkäse, Speisequark, Rahmfrishkäse and Dopplerahmfrishkäse compositional requirements (g 100 g^{-1}) under the German Käseverordnung (2007 update).

Product designation	Fat in dry matter (FDM)	Dry matter[a]	Protein[a]	Moisture in fat-free cheese (MFFC)
Dopplerahmfrischkäse	≥60 and ≤87	44	Not specified	>73
Rahmfrischkäse	≥50 and <60	39	Not specified	>73
Dopplerahmstufe Speisequark	≥60 and ≤87	30	6.8	>73
Rahmstufe Speisequark	≥50 and 60	27	8.0	>73
Volfettstufe Speisequark	≥45 and <50	25	8.2	>73
Fettstufe Speisequark	≥40 and <45	24	8.7	>73
Dreiviertelfettstufe Speisequark	≥30 and <40	22	9.7	>73
Halbfettstufe Speisequark	≥20 and <30	20	10.5	>73
Viertelstufe Speisequark	≥10 and <20	19	11.3	>73
Magerstufe	<10	18	12.0	>73

[a]Amounts quoted are minimum values.
After Behr’s Verlag (2007).

requirements for these products. It should be noted that the fat content is specified as a range of Fett in der Trockenmasse (Fett i.Tr.) (FDM) rather than in terms of absolute fat levels, as is the case in the United Kingdom and United States discussed earlier. For example, German Rahmfrischkäse with a dry matter (DM) of 40 g 100 g^{-1} and an FDM of 51 g 100 g^{-1} has a fat content of 20.4 g 100 g^{-1}; this is much lower than the minimum fat content prescribed for cream cheese in the US Code of Federal Regulations (33 g 100 g^{-1}) or the now-repealed UK Cheese Regulations 1970 (45 g 100 g^{-1}). The moisture on a fat-free basis (Wassergehalt in der fettfreien Käsemasse) is specified for Frischkäse at greater than 73 g 100 g^{-1}.

The raw materials for cream cheese are milk, cream and skimmed milk; however, milk products for fat adjustment are allowed, and for cream cheese that is heated, starch and gelatine are allowed to the level that is technologically justified. German cream cheese products with lower fat levels are designated with names such as Frischkäsezubereitung (which translates as fresh cheese preparations).

8.3.8 *Danish cheese legislation with particular reference to cream cheese-type products*

Danish legislation on cheese is detailed in the Mælkeproduktbekendtgørelsen (which can be translated as the Executive Order on Milk Products). The present version dates from 2004 and can be accessed electronically (Ministerialtidende Danmark, 2004). Therein, cheese is defined as '*the ripened or unripened (fresh), milk, prepared with or without rind, and in which the ratio of whey protein and casein does not exceed that in milk*', which is achieved by

- total or partial coagulation of milk protein in milk, skimmed milk, fat standardised milk, cream and buttermilk or a combination thereof by rennet or other appropriate coagulants and with partial drainage of whey or
- a manufacturing process that includes the full or partial coagulation of milk protein in [dairy] milk and other dairy products, which gives an end product that has similar physical, chemical and sensory characteristics as the product manufactured, listed under No. 1.

Cheese is further classified by texture, fat content and ripening type. The designation of *unripened soft cheese* may be replaced by names, such as fresh cheese, Quark, Hytteost or Cottage cheese, or Flødeost (cream cheese) subject to certain requirements. The compositional requirements for cheese, including Flødeost (cream cheese) are outlined in Table 8.10.

8.3.9 *French cheese legislation with reference to some cream cheese-type products*

As mentioned elsewhere, the earliest production of what we now regard as cream cheese was based on French Neufchâtel cheese. On the basis of an eleventh-century text, which mentions cheeses produced in the area around Neufchâtel-en-Bray in Normandy, some claim that this is the oldest cheese of Normandy. A more appropriate origin may be contained in the records of the l'abbaye de Saint-Amand in Rouen, where a cheese was reported as named Neufchâtel (Syndicat de Défense et de Qualité du Fromage Neufchâtel, 2008). In the Middle Ages, it had many shapes, but eventually six shapes and sizes came to be used: (a) bonde cylindrique, square (carré) and briquette of 100 g; (b) heart-shape (coeur) and double bonde of

Table 8.10 Compositional requirements (g 100 g^{-1}) for cheese, including cream cheese under Danish legislation wherein cheese is classified mainly by texture, fat content and ripening type.

Product designation	Fat in dry matter (FDM)	Moisture in fat-free cheese (MFFC)
Extra hard	–	<51
Hard	–	$>49\%$ and ≤ 56
Firm	–	$>54\%$ and ≤ 69
Soft	–	>67
Extra full-fat	≥ 60	–
Full-fat	≥ 45	–
Medium-fat (Middelfed)	≥ 30	–
Light (Mager)	≥ 10	–
Extra light (Ekstra mager) or skimmed milk cheese	<10	–
Cream cheese (Flødeost)	≥ 60	>67

After Ministerialtidende Danmark (2004).

200 g; and (c) large heart-shape (grand coeur) of 600 g. This ripened soft cheese is now protected as an AOC and it is governed by a special French Decree (Anonymous, 2006).

In general, the current French legislation on cheese is contained in the new French Cheese Decree of 27 April 2007 (Anonymous, 2007); this replaced an earlier Decree of 30 December 1988. The definition of cheese given therein is as follows: '*The name ''cheese'' is reserved for non-fermented or fermented, unripened or ripened products obtained from the following raw materials: milk, totally or partially skimmed milk and buttermilk used as such or in a mixture and totally or partially coagulated before draining or after partial elimination of the aqueous phase. The minimum dry matter content of the product corresponding to this definition should be at least 23 g for 100 g of cheese*' (this last criteria may vary for 'fresh cheeses', depending on their fat content) (Gillis, 2000).

One or several of the following ingredients may be used during cheese manufacture:

- Salt
- Spices and herbs
- Partially or totally dehydrated milk or buttermilk, dairy protein preparation where technologically justified; the initial protein content of the dairy raw material mixture used should not increase by more than 10 g L^{-1} (for 'defined' cheeses, this limit is reduced to 5 g L^{-1})
- The addition of fat and proteins of an origin other than milk is forbidden
- Spice extracts and natural flavour
- Sugars and other food products conferring the finished product with a specific flavour, to a maximum level of 30 g 100 g^{-1} of finished product
- Rennet, harmless lactic acid bacteria, yeasts and moulds cultures

- Other substances other categories of aroma compounds, subject to conditions and requirements of ministerial decrees

The denomination 'fromage blanc' (literally, white cheese) is reserved for an unripened cheese, which is produced by fermentation, mainly by a lactic fermentation, that is, where the adjective 'frais' (fresh) is used, as in fromage frais (fresh cheese); the cheese must contain living microbial flora at the time of sale to consumers. This means that a cream cheese that is mixed and heated to temperatures of about 68°C, as outlined in Section 8.1, could not be designated as 'fromage frais à la crème' under French cheese legislation. The word crème could be used for any cheeses, which have fat content of greater than or equal to 45 to less than 60 g 100 g^{-1} FDM. While the term *fromage frais à la crème* was used in the Marketing and Labelling section (Section 7) of old Codex Standard for Cream Cheese and mentioned in certain texts (Kosikowski, 1977; Kosikowski & Mistry, 1997a, 1997b; FAO/WHO, 2000), this has been replaced by the term *fromage à la crème* in the revised version of the Codex Standard for Cream Cheese (FAO/WHO, 2007g). The latter term is also used in the French language version of the Canadian Food and Drug Regulations (Ministère de la Justice Canada, 2008).

One could be forgiven for thinking, because of its name, that the cheese variety, Petit-Suisse, was a cheese of Switzerland. However, its origins were from Normandy, but the idea of the cheese came from Switzerland. In the mid-1800s, a Swiss employee in a dairy of Auvilliers (near Beauvais) suggested to the owner of the dairy, to add cream to the curd mass used to produce bondon-shaped cheese. However, the owner of the Gervais Company, which was well-known for its double cream, recognised its potential and this resulted in Petit-Suisse becoming one of the most important products of the company to this day. Originally the cheese was sold wrapped in a thin band of paper and placed six by six in a small wood crate (Courtois, 2008). It is now sold in small plastic cups, with foil lids, similar to fromage frais. The present French Cheese Decree contains compositional and shape requirements for this cheese and these are outlined in Table 8.11.

8.3.10 *Italian standard on Mascarpone*

Mascarpone originated around 1600 in Lombardy region of Northern Italy, where it is believed to have taken its name from *Mascherpa,* which is the local term for a sort of ricotta cheese. It is often referred to as a triple cream (triple-crème) cheese. The high fat content and smooth texture of Mascarpone cheese made it suitable as a substitute for cream or butter. Ingredient applications of Mascarpone cheese tend to focus on desserts. The most famous

Table 8.11 Requirements for Petit-Suisse cheese in French Cheese Decree No. 2007–628.

Parameter	Requirement
Size and shape	Cylindrical 30 or 60 g
Fat in dry matter (FDM) (g 100 g^{-1})	40–60
Dry matter (DM) (g 100 g^{-1})	23–30

After Anonymous (2007).

application of Mascarpone cheese is in the Italian dessert tiramisu. Its other uses include the following:

- A filling or topping for desserts such as tarts and cheesecake
- In pasta sauces, where the fat content of the cheese adds to the richness of the product
- A thickener to add richness to foods such as soups, risotto or dips where it adds viscosity and mild flavour
- As a spread in place of butter or margarine, its lower fat content compared to the butter and margarine being an advantage
- In ice-cream, as an alternative to cream in the mix formulation

There is no specific Italian legislation on Mascarpone. There is a voluntary standard, but this is not binding (Ente Nazionale Italiano di Unificazione, 1998). This standard describes Mascarpone as a soft, spreadable and unripened cheese. The manufacturing procedure is described in Section 8.7. The chemical and organoleptic characteristics for Mascarpone cheese as contained in the Italian standard Norma Italiana UNI 10710 are outlined in Table 8.12.

8.3.11 *Cheese legislation in Australia*

The Joint Food Standards Code Australia and New Zealand

Historically Australia and New Zealand had separate legislation and indeed that remains partly the situation to this day. However, in 1995 Australia and New Zealand signed a Joint Food Standards Setting Treaty that committed both countries to the development and implementation of a single set of food standards. The Food Standards Treaty provides for a bi-national agency, now called Food Standards Australia New Zealand (FSANZ), to undertake the relevant food standards development for both Australia and New Zealand. FSANZ has produced a Food Standards Code, which is regularly updated, that contains the joint standards developed to date (Food Standards Australia New Zealand, 2007). The present

Table 8.12 Chemical and organoleptic characteristics for Mascarpone cheese in Norma Italiana UNI 10710[a].

Parameter	Requirement
Dry matter (DM) (g 100 g^{-1})	48–60
Fat in dry matter (FDM) (g 100 g^{-1})	≥ 80
Protein (g 100 g^{-1})	2.8–6
pH	5.7–6.6
Appearance	Creamy mass with no rind
Body and texture	Soft, close and spreadable
Colour	From white to straw coloured
Flavour and taste	Delicate, creamy

[a]This is a voluntary standard, it is not legally binding.
After Ente Nazionale Italiano di Unificazione (1998).

Food Standards Code provides a definition of cheese that is similar to the 1999 Revision of the Codex General Standard for Cheese (FAO/WHO, 1998) and a list of ingredients permitted in cheese. Standard 4.2.4A of the Code, which applies to Australia only, sets out primary production and processing requirements for Gruyere, Sbrinz, Emmental and Roquefort cheese; references and requirements for other individual cheese varieties are not included.

The current definition of the cheese in Standard 2.5.4 of the code is as follows: '*Cheese means the ripened or unripened solid or semi-solid milk product which may be coated and is obtained by one or both of the following processes: (a) coagulating wholly or partly milk, and/or materials obtained from milk, through the action of rennet or other suitable coagulating agents, partially draining the whey which results from such coagulation; or (b) processing techniques involving concentration or coagulation of milk and/or materials obtained from milk which give an end-product with similar physical, chemical and organoleptic characteristics as the product described in paragraph (a)*'.

The ingredients permitted in cheese are

- Water
- Lactic-acid-producing micro-organisms
- Flavour-producing micro-organisms
- Gelatine
- Starch
- Vinegar and
- Salt.

The earlier, and now repealed, 1987 Food Standards Code, developed for Australia, but later adopted and updated by the Australia New Zealand Food Authority (ANZFA), was applicable up to the adoption of the 2000 Code. The consolidated text, updated to 1 August 1999, contained definitions of cream cheese and Neufchatel cheese in Standard H9 of the old code as follows: '*Cream cheese and Neufchatel cheese are products prepared by coagulating a mixture of cream and milk or skimmed milk through the action of protein coagulating enzymes, heat or acid or any two or all of them*'.

They were permitted to contain the following:

- Salt, potassium chloride or mixtures of both
- 'Modifying agents' as in Group III as set out in Standard A10
- 'Modifying agents' as in Group I as set out in Standard A10, to a maximum 5 g kg^{-1}
- Specified acids (phosphoric and hydrochloric) and the acidity regulator glucono-δ-lactone
- Starter cultures

The following compositional requirements of cream cheese and Neufchatel cheese were specified in a Schedule to Standard H9:

- Cream cheese – $\leq$55 g 100 g^{-1} moisture and $\geq$65 g 100 g^{-1} FDM
- Neufchatel cheese – $\leq$60g 100 g^{-1} moisture and $\geq$45 g 100 g^{-1} FDM

However, in the absence of specified requirements in the new Food Standards Code, the production of individual varieties of cheese, including cream cheese, that are in conformance with the revised Codex Standards, as discussed in the next section, should be acceptable in Australia and New Zealand.

8.3.12 *Codex Alimentarius – international standards for cheese and cream cheese*

Background

The CAC is an international intergovernmental body. Its membership is open to member nations and associate members of the FAO and/or the WHO. As of October 2007, it had 175 member countries and 1 member organisation (the European Community) (FAO/WHO, 2007f). Nowadays, the CAC meets annually, and the venue alternates between the FAO headquarters in Rome and WHO headquarters in Geneva.

Nominated senior officials represent member governments at Codex Alimentarius meetings. National delegations may also include representatives of the industry, consumers and academia. A significant number of other international governmental organisations, for example, the Office International des Epizooties (OIE), the WTO and recognised international nongovernmental organisations (NGOs), such as the International Dairy Federation (IDF), the Confederation of the Food and Drink Industries of the EU (i.e. Conféderation des Industries Agro-Alimentaires (CIAA)) and the International Special Dietary Foods Industries (ISDI) may also attend in an observer capacity. Observers are allowed to contribute to meetings at all stages except in final decisions; this is the exclusive prerogative of member governments.

The CAC has established two types of subsidiary committees: (a) Codex Committees and (b) Coordinating Committees. The former type of committee is subdivided into General Subject Committees (currently ten in number), and are so-called because of the horizontal nature of their work, and Commodity Committees (currently 21 in number, of which 16 are active), which develop the standards for specific foods or classes of foods. There are also five Regional Coordinating Committees whose role is to ensure that the CAC is responsive to regional interests, and particularly the needs of developing countries. The CAC also establishes *ad hoc* intergovernmental task forces given stated tasks on specific topics. Currently, there are three such task forces for foods derived from biotechnology, antimicrobial resistance and quick frozen foods. The Codex Alimentarius structure is shown in Figure 8.1.

The main aims of the Codex Alimentarius are as follows:

- The protection of consumer health
- Ensuring fair trading practices
- Facilitating international trade

Codex Alimentarius standards

Codex Alimentarius standards consist of 13 volumes, which contain general principles, general standards, commodity standards, definitions, codes, methods and recommendations, and as of July 2006, its content is shown in Table 8.13 (FAO/WHO, 2006). As may be seen, in addition to individual food commodity standards, it encompasses food labelling,

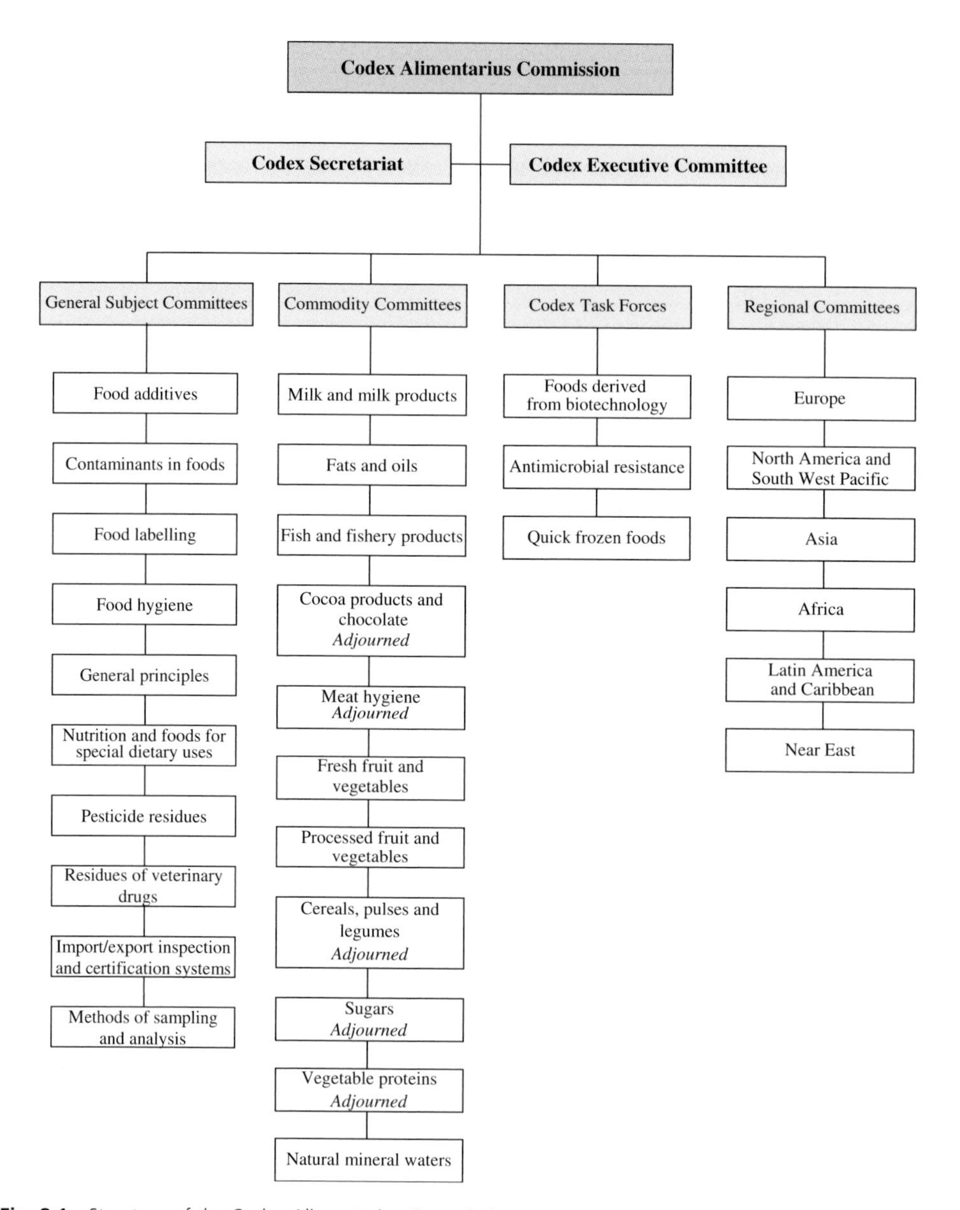

Fig. 8.1 Structure of the Codex Alimentarius Commission.

food additives, food hygiene, contaminants, nutrition and foods for special dietary uses and methods of analysis and sampling.

The CAC has established a number of principles on the scientific basis for it decision-making (Randell & Race, 1996). These principles ensure that the quality and food safety provisions shall be based on sound science and, that in establishing food standards, other legitimate factors that are relevant to consumer's health and the promotion of fair trade may be considered. The Codex Alimentarius standards and related texts are subject to revision, as and

Table 8.13 Content of the Codex Alimentarius (as of July 2006).

Category	Number of Codex Standards, guidelines and codes of practice
Food commodity standards	186
Food commodity related texts	46
Food labelling	9
Food hygiene	5
Food safety risk assessment	3
Sampling and analysis	15
Inspection and certification procedures	8
Animal food production	6
Contaminants in foods (maximum levels, detection and prevention)	12
Food additives provisions (covering 292 food additives)	1112
Food additives related texts	7
Maximum limits for pesticide residues (covering 218 pesticides)	2930
Maximum limits for veterinary drugs in foods (covering 49 veterinary drugs)	441
Regional guidelines	3

Data compiled from FAO/WHO (2006).

when deemed necessary by the CAC and its subsidiary bodies to ensure that they are consistent with and reflect current scientific knowledge. Any member of CAC may identify and present new scientific or other information that may warrant a revision to the relevant body.

The Uruguay round of multilateral trade negotiations held under the General Agreement on Tariffs and Trade (GATT), which took place between 1986 and 1994, led to the formation of the WTO on 1 January 1995. For the first time, GATT agreements included agriculture and food in its scope; however, the Marrakesh agreement of 1994 also included the agreements on sanitary and phytosanitary measures (commonly referred to as the SPS Agreement) and on technical barriers to trade (commonly referred to as the TBT Agreement). These agreements acknowledge the need for the harmonisation of international standards to minimise the risk of sanitary, phytosanitary and other technical standards becoming barriers to international trade. Thus, the SPS and TBT agreements gave formal recognition to international standards, guidelines and recommendations of international organisations, including the CAC, as reference points for facilitating international trade and resolving disputes. Hence, the role of Codex Alimentarius in this regard is now well recognised.

Codex Alimentarius cheese standards

The CAC, which implements the joint FAO/WHO foods standards programme, was established in 1962, and has developed, *inter alia*, international standards for cheese in general, unripened cheese, and 35 individual cheese varieties between 1963 and 1978. A review of these standards was commenced in the early 1990s. The present Codex General Standard for cheese was initially adopted in 1978, and was revised in 1999; the most recent amendment was adopted in 2006 (FAO/WHO, 2007c). This contains the following definition of cheese that is relevant to all cheeses: '*Cheese is the ripened or unripened soft, semi-hard, hard or extra hard product, which may be coated, and in which the whey protein/casein ratio does not exceed that of milk, obtained by*

- *Coagulating, wholly or partly the protein of milk, skimmed milk, partly skimmed milk, cream, whey cream or buttermilk or any combination of these materials, through the action of rennet or other suitable coagulating agents, and by partly draining the whey resulting from the coagulation, while respecting the principle that cheese-making results in the concentration of milk protein (in particular, the casein portion), and that consequently, the protein content of the cheese will be distinctly higher than the protein level of the blend of the above milk materials from which the cheese was made, and/or*
- *Processing techniques involving coagulation of the protein of milk and/or products obtained from milk which give an end-product with similar physical, chemical and organoleptic characteristics as the product defined under above*'.

This standard also defined unripened cheese, including fresh cheese, as *Cheese that is ready for consumption shortly after manufacture.* A separate Codex Standard for unripened cheese was adopted in 2001 (CODEX STAN 221–2001); this requires such cheese to conform to the general definition outlined above (FAO/WHO, 2007d).

As regards the 35 individual varieties of cheese standards, 19 were deleted, due to their limited, if any, involvement in international trade (FAO/WHO, 1996, 1999). Of the remaining 17, 16 revised standards were adopted in 2007, the 17th was retained, but not revised, and 1 new additional standard (Mozzarella) was also adopted in 2007 (FAO/WHO, 2007a; FAO/WHO, 2007b). Codex Standards were originally intended to be adopted by the member countries of the CAC, but this could only be encouraged and were not binding. However, the formal recognition of Codex Standards as reference points for facilitating international trade and resolving disputes in the WTO has increased their role and profile.

The original Codex Alimentarius standard for cream cheese

The original 1973 Codex Individual Standard for cream cheese names five depositing countries in Section 2 (United States, Denmark, Federal Republic of Germany, Australia, and Canada). The term depositing country indicates a country that contributed to, and had a special interest in, the development of the standard. The standard also recognised three variants based on fat content; (a) Type A – 'cream' cheese, with a minimum fat content of 33 g 100 g^{-1} and a maximum moisture content of 55 g 100 g^{-1}; (b) Type B – 'cream' cheese 28 g 100 g^{-1}, with a minimum fat of 28 g 100 g^{-1} and a maximum moisture content of 58 g 100 g^{-1}; and (c) Type C – 'cream' cheese 24 g 100 g^{-1}, with a minimum fat content of 24 g 100 g^{-1} and a maximum moisture content of 62 g 100 g^{-1}. In the labelling section, it was indicated that Type B should be designated as cream cheese 28 g 100 g^{-1} fat or cream cheese 28 g 100 g^{-1} and Type C should be designated as cream cheese 24 g 100 g^{-1} fat or cream cheese 24 g 100 g^{-1} (FAO/WHO, 2000). It is believed these fat levels reflected the fat levels common in the five depositing countries. The names ‘Rahmfrischkase’ and ‘Fromage frais a la crème’ were also recognised in the English version of this standard.

This old standard also listed as necessary ingredients lactic-acid- and aroma-producing bacterial cultures and sodium chloride (salt); optional additions, including rennet or other suitable coagulating enzymes, six vegetable gums; and other thickening agents (two alginates, pectins and gelatine). The total weight of these optional ingredients was limited to not more than 5 g 100 g^{-1} of the finished product.

The 2007 Revision of the Codex Alimentarius standard for cream cheese

The standard for cream cheese was included in the review of the existing Codex Standards for individual cheese varieties, begun in 1994. In the years from 1973, changes occurred in the technology used for manufacture, consumer tastes, demand for lower fat variants of standardised foods and nutritional guidelines. The emergence of reduced and low-fat variants of cream cheese in North America, Northern Europe and Australasia led to requests for inclusion of lower fat variants within the standard. Other countries were opposed to the concept of cream cheese with low-fat levels. Eventually, the revised standard was adopted in July 2007 (FAO/WHO, 2007e). In this standard, cream cheese is described as a soft, unripened and rindless cheese, has a near white to light yellow colour, a spreadable texture that is smooth to slightly flaky, without holes, and spreads and mixes readily with other foods. It must also comply with the Codex General Standard for Cheese and the Group Standard for Unripened Cheese including Fresh Cheese, as outlined earlier.

In the new standard, the requirement for fat content is expressed in terms of FDM, replacing the total fat content as in the old standard. A minimum level of 25 g 100 g^{-1} FDM, together with a reference fat level of 60 g 100 g^{-1}, with a range of 60 to 70 g 100 g^{-1} FDM, was eventually agreed. The reference level of fat was included for two purposes: *first*, to identify products that could be called cream cheese, without qualification, and *second*, the reference fat level could be used for making appropriate relevant nutrient claims, such as reduced fat. Unripened cheeses with FDM content below 25 g 100 g^{-1} cannot be designated as a cream cheese, even where a qualifying term is used in conjunction with the name. The compositional requirements for cream cheese contained in this standard are outlined in Table 8.14.

Furthermore, in the labelling of cream cheese, as specified in Section 7 of the standard, there are different requirements for products with FDM content between 40 and 60 g 100 g^{-1} or above 70 g 100 g^{-1} and those with FDM levels between 25 and 40 g 100 g^{-1}. In the former case, the name must be accompanied by an appropriate qualification describing

Table 8.14 Compositional requirements (g 100 g^{-1}) in the Codex Standard for cream cheese CODEX STAN 275–1973[a].

Constituent	Minimum	Maximum	Reference
Fat in dry matter (FDM)	25	Not restricted	60–70
Moisture in fat-free cheese (MFFC)	67	–	Not specified
Dry matter (DM)	22	Restricted by MFFC	Not specified

[a]Formerly CODEX STAN C-31-1973, adopted in 1973, Revision 2007.
After FAO/WHO (2007e).

the modification made or the fat content (as g 100 g^{-1} FDM or g 100 g^{-1} fat), as part of the product name or in a prominent place in the same field of vision as the name. In the case of the latter, the name must be accompanied by an appropriate qualification describing the modification made or the fat content (as g 100 g^{-1} FDM or g 100 g^{-1} fat) as part of the product name or in a prominent place in the same field of vision as the name, *or alternatively* the name specified in the national legislation of the country in which the product is manufactured and/or sold, or with a name existing by common usage, in either case provided that the designation used does not create an erroneous impression in the country of retail sale regarding the character and identity of the cheese. These provisions are quite complicated, but reflect the difficulties encountered in getting sufficient consensus in the relevant Codex committees.

The additive provisions are contained in Section 4 of the standard; they contain many more than the limited number in the old standard. The additive functional classes permitted are colours, acids, acidity regulators, stabilisers, thickeners, emulsifiers, antioxidants, preservatives and foaming agents (for whipped products only). Tables of specific additives in each of these functional classes are included in the standard.

In addition to individual product standards, Codex Alimentarius is also developing a General Standard for Food Additives (GSFA) and an associated Food Category System (FCS), which is contained in Annex B of the GSFA. The FCS is a means for assigning food additive uses in the Standard (FAO/WHO, 1995). All foods are included in the system, but the food category descriptors used in the FCS are not intended to be legal product designations nor are they intended for labelling purposes. The FCS is hierarchical, and an additive permitted in the general category is taken as permitted in all sub-categories, unless otherwise stated. Unripened cheese is assigned Food Category 01.6.1, and it includes fresh cheese, which is ready for consumption soon after manufacture. Among the examples listed is cream cheese (rahmfrischkase, an uncured, soft spreadable cheese).

At this time, additive provisions are included in all relevant adopted Codex product standards. However, it is the intent, within Codex Alimentarius, to have the GSFA as the single reference source for additives. This will mean that at some stage the additive provisions in individual standards will be moved to the GSFA. This will not be an easy task and will be carefully monitored by all interested sources. The Codex Committee on Food Additives is body that addresses the food safety aspects of additives and their use, while the relevant Commodity Committee addresses the technical justification for use in specific products. The completion of the GSFA is likely to take quite a few years more.

8.4 Cream cheese

Cream cheese is a cream-coloured, mild acid tasting – acid fresh cheese that is prepared by acid coagulation of mixtures of milk and cream. Its consistency ranges from brittle (especially DCC) to spreadable (e.g. single cream cheese). The product, which is most popular in North America, has a shelf-life of around 3 months at below 8°C.

8.4.1 *Principles of manufacture*

In milk, casein, which comprises ~80 g 100 g^{-1} of the total protein and is the main structural protein in acid-induced milk gels, exists in the form of spherical-shaped colloid particles

(~40–300 nm in diameter), casein micelles (Fox & Brodkorb, 2008; McMahon & Oommen, 2008). The micelles on a dry weight basis consist of ~7 g 100 g^{-1} ash (mainly Ca and P), 92 g 100 g^{-1} casein and 1 g 100 g^{-1} minor compounds including magnesium and other salts. The casein is heterogeneous, comprising of four main types: α_{s1}, α_{s2}, β and κ, present at a ratio of ~4:1:4:1.2. Model studies in dilute dispersions indicate that the individual caseins vary in the content and distribution of phosphate and consequently have different calcium-binding properties. Generally, α_{s1}-, α_{s2}-, β- caseins bind calcium strongly, and precipitate at relatively low calcium concentrations (e.g. ~0.005–0.1 M $CaCl_2$ solutions), inclusive of the calcium level in milk (30 mM); in contrast, κ-casein is not sensitive to these calcium concentrations and can stabilise up to 10 times the mass of the calcium-sensitive caseins. Owing to the differences in degree of phosphorylation, the arrangement of casein within the micelle is such that the core is occupied by the calcium-sensitive caseins and κ-casein is principally located at the surface, with the hydrophilic c-terminal regions oriented outwards towards the serum phase forming negatively charged hairs, which confer stability to the micelle by electrostatic repulsion, Brownian movement and steric effect.

The manufacture of cream cheese and all its variants involve the acid-induced gelation of the casein, removal of whey from the gel and curd formation and concentration, pasteurisation and homogenisation of the curd and cooling. The principles of the manufacture of cream cheese and other fresh acid curd cheeses have been reviewed in detail (Guinee *et al.*, 1993; Lucey & Singh, 1997; Lucey, 2004a, 2004b; Schulz-Collins & Zenge, 2004). During the manufacture of cream cheese, the milk undergoes slow quiescent acidification at temperatures of 20 to 30°C to pH values in the range 4.6 to 4.8 as a result of the *in situ* conversion of lactose to lactic acid by adding starter culture, or by the addition of acidogens, such as glucono-δ-lactone. This leads to a controlled destabilisation of the casein micelles in the form of a continuous gel netwok, which occludes water and fat; the main factor promoting gelation is charge neutralisation, which reduces the inter-micellar repulsive forces, the steric effect and degree of hydration. However, a closer examination of the acidification process indicates that acidification promotes two opposing sets of physico-chemical changes on the casein micelles:

- A tendency towards disaggregation of the casein micelles in milk into a more disordered system at: pH values >5.3 as a result of the following physico-chemical changes; (a) dissociation of colloidal calcium phosphate, which at 20°C is fully soluble at ~pH 5.2 to 5.3 (van Hooydonk *et al.*, 1986; Gastaldi *et al.*, 1994); (b) a concomitant increase in the dissociation of individual caseins from the micelle (especially β-casein), which becomes more pronounced as the temperature during gelation is reduced in the range 30 to 5°C (Dalgeish & Law, 1988; Singh *et al.*, 1996); (c) an increase in casein hydration between pH 6.0 and 5.3, where it is at a maximum (Creamer, 1985); and (d) a swelling of, and increase in the volume of, casein micelles at pH ~5.3 (Roefs *et al.*, 1985).
- A tendency towards casein aggregation owing to (a) an overall reduction in the net negative charge on the casein micelle, from ~8 to 20 mV at native pH (Schmidt & Poll, 1986; Banon & Hardy, 1992; Wade *et al.*, 1996) to ~0 at pH 4.6, owing to protonation of previously ionised amino acid residues such as glutamate and aspartate; (b) a re-adsorbtion of dissociated casein back onto the micelle as the pH of individual caseins approach their ioselectric pH; (c) a sharp reduction in casein hydration from ~3.5 g water g^{-1} at pH ~5.3 to ~2.5 at the isoelectric pH, 4.6, especially when rennet

is added (Creamer, 1985); and (d) a decrease in electrostatic repulsion between, and steric stabilisation of, the casein micelles and a simultaneous increase in the hydrophobic attractions (Lucey, 2004a).

At pH values greater than the onset of gelation (~5.1–5.4 at 20–30°C), depending *inter alia* on the heat treatment of the milk (Lucey *et al.*, 1998b), disaggregating forces predominate and a gel is not formed. However, at pH values lower than those at the onset of gelation, aggregation forces prevail and gel formation proceeds, resulting in the development of casein aggregates that initiate the formation of strands. Eventually, as the pH of milk approaches the isoelectric pH, dangling strands link together to form a three-dimensional particulate casein network that encloses the fat and moisture components of the milk. Gelation is accompanied by a marked increase in the elastic shear modulus, G′ (index of firmness), which increases with further pH reduction and casein aggregation (Lakemond & van Vliet, 2008). The acid-induced gel is dehydrated by cutting/straining, centrifugation or UF, and the resultant curd is further processed, as discussed subsequently.

8.4.2 *Manufacture stages*

Conventional manufacturing process

The Conventional manufacturing process of conventional cream cheese is described by Kosikowski & Mistry (1997a, 1997b) and by Guinee *et al.*, (1993). It involves the following steps, as summarised in Figure 8.2:

- Standardisation of the cheese milk, normally prepared by blending whole milk and cream
- Pasteurisation and homogenisation of the cheese milk
- Inoculation with a mesophilic starter culture and incubation at ~22°C
- Gelation by *in situ* conversion of lactose to lactic acid by the starter culture
- Curd formation and whey separation by subjecting the stirred gel to centrifugal separation of filtration
- Treatment of the curd, which includes the addition of salt and other materials, and pasteurisation and homogenisation
- Cooling and storage, during which time the product sets and acquires the final texture characteristics of final cream cheese

The individual steps are discussed in more detail in the following sections.

Standardisation

Standardisation involves the blending of milk and cream at ratios that give a product composition, which complies with the specification or standard of identity of the cheese being made, and gives the desired end product characteristics. For cream cheese, the specification as defined in the CFR is as follows: minimum milk fat content ≥33 g 100 g^{-1} and maximum moisture content 55 g 100 g^{-1}. From this definition and analysis of commercial cream cheeses, the following composition (g 100 g^{-1}) may be assumed: DM 45, fat 33, lactose/lactic acid 2.5 (assuming 4.5 lactose in cheese milk), NaCl 0.8, ash 0.4 (assuming 0.7 ash in milk) and protein 8.3. Standardisation also requires a number of assumptions

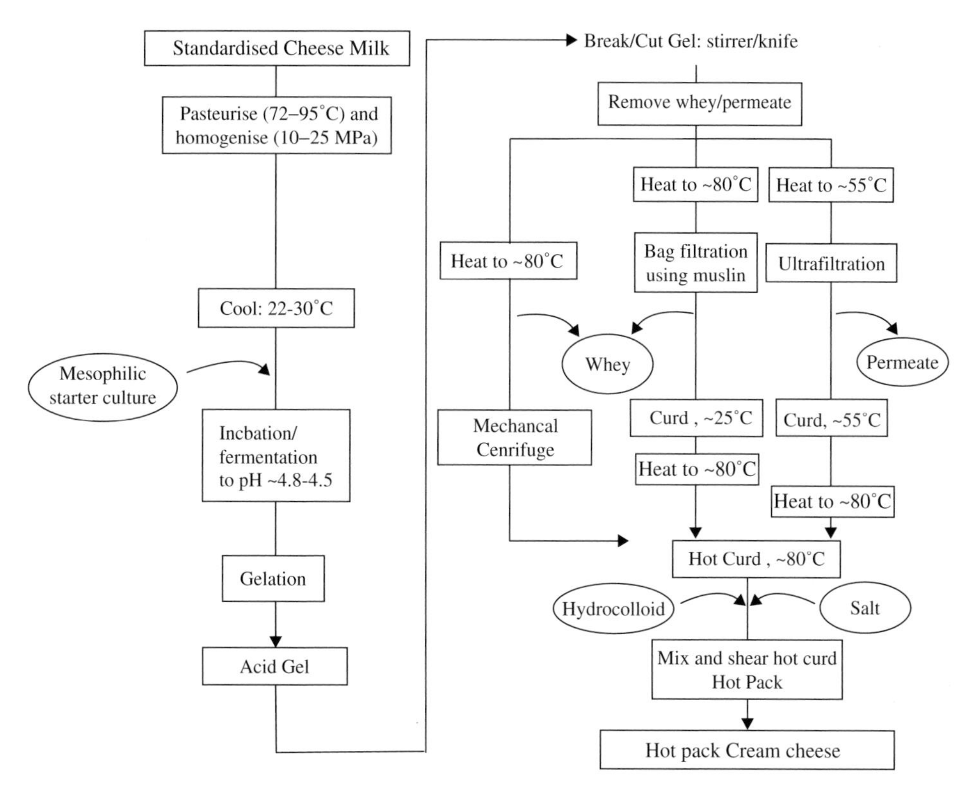

Fig. 8.2 Outline of manufacturing process for cream cheese.

on the percentages of milk fat and protein recovered during manufacture. These depend *inter alia* on

- The degree of whey protein denaturation and complexation with the casein, as affected by pasteurisation temperature of the standardised milk, and the temperature at which the gel is separated.
- The level of fat retention as affected by homogenisation of the cheese milk and the process used in separation.
- The degree of casein hydrolysis during the fermentation of the cheese milk.

Based on the typical heat treatments applied during pasteurisation (80°C for 1 min) and separation temperature (80°C), the level of whey protein denaturation in milk (as measured by insolubility at pH 4.6) and the yield increases for Quarg cheese when using the Thermoquarg process (Winwood, 1983; Guinee *et al.*, 1995; Rynne *et al.*, 2004), it is assumed that ~60 g 100 g^{-1} of the whey proteins are recovered during cream cheese manufacture. Similarly, it is assumed that fat recovery is high, in the order of ~95 g 100 g^{-1}, owing to the fact that standardised milk is subjected to homogenisation pressures in the range 15 to 25 MPa (Jana & Upadhyay, 1993) before fermentation. Given that ~68 g 100 g^{-1} of the solids of bulk starter culture prepared from skimmed milk are lost during cheese manufacture (Hicks *et al.*, 1985), it is estimated that up to 8 g 100 g^{-1} of the casein is hydrolysed during the manufacture of fresh acid curd cheese and lost to the cheese whey during subsequent whey drainage. Consequently, the total loss of protein (as non-protein nitrogen (NPN), hydrolysed casein and non-denatured whey protein) during the manufacture of cream cheese is estimated at ~17.5 g 100 g^{-1} of total. Using the aforementioned product composition, the required protein-to-fat ratio in the cheese milk (P_m/F_m) may be calculated by the following equation

$$\frac{P_m}{F_m} = \frac{\text{Protein in cheese/protein recovery factor}}{\text{Fat in cheese/fat recovery factor}} = \frac{5.3/0.825}{33/0.95} = 0.29$$

where the protein and fat recovery factors refer to fraction of these components recovered, and are estimated by deducing the losses during manufacture, as discussed earlier. Hence, for the cream cheese, the fat required in standardised milk containing 3.3 g 100 g^{-1} protein would be ~11.4 g 100 g^{-1} for DCC. For a light cream cheese with fat and protein contents of 14 and 12 g 100 g^{-1}, respectively, the fat content in standardised milk with 3.3 g 100 g^{-1} protein would then be 3.9 g 100 g^{-1}.

Pasteurisation and homogenisation, and pasteurisation of the standardised milk

The standardised milk is pasteurised, typically at temperatures of 72 to 75°C for 30 to 60 s, which leads to denaturation of 5 to 15 g 100 g^{-1} of the whey proteins and their interactions with the casein micelle through disulphide interaction (Rynne *et al.*, 2004). The heated milk is homogenised at first and second stage pressures, typically in the order of 15 to 25 MPa and 3 to 5 MPa, respectively. Homogenisation is important for the following reasons:

- It reduces/eliminates creaming of fat during the fermentation or acidification stage and therefore prevents compositional heterogeneity of the resultant gel.
- It reduces fat losses on subsequent whey separation.

Homogenisation of the standardised milk disrupts the fat globules and increases their fat surface area by a factor of 5 to 6. The reformed fat globules become coated with a protein layer consisting of casein micelles, sub-micelles and, to a lesser extent, whey protein (Walstra & Jenness, 1984). The membrane, referred to as the recombined fat globule membrane (RFGM) enables the fat globule to behave as a pseudo-protein particle, which can become an integral part of the gel network during acid gelation in the manufacture of products such as yoghurt and cream cheeses. Their participation in the formation of a composite gel thereby leads to a more uniform, stronger gel at the end of fermentation (van Vliet & Dentener-Kikkert, 1982; Ortega-Freitas *et al.*, 2000). In contrast, native milk fat globules do not participate in acid–milk gel matrix but rather become occluded by the casein/whey protein network. Consequently, homogenisation of the cheese milk is a key step in the manufacture of cream cheese, where it minimises creaming (fat flocculation) during incubation, increases gel firmness and leads to a more uniform, smoother and firmer gel and end product.

Fermentation and gelation

Following pasteurisation and homogenisation, the milk is cooled (20–30°C), pumped to the fermentation tank, inoculated with a mesophilic starter culture, generally comprised of strains of *Lactococcus lactis* subsp. *lactis* or *L. lactis* subsp. *cremoris* D-type starter culture and held

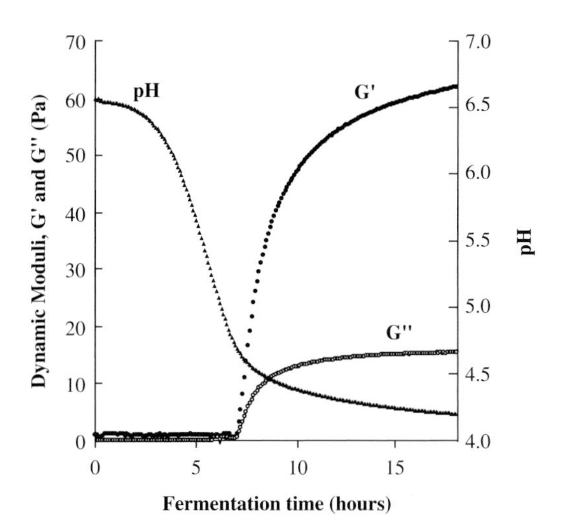

Fig. 8.3 Dynamic changes in pH (△) and rheology during the formation of an acid-induced skimmed milk gel: storage modulus, G′; loss modulus, G″. The skimmed milk (3.1 g protein 100 g^{-1}) was pasteurised at 72°C, cooled to 30°C, innocualted with 2 g 100 g^{-1} *Lactococcus lactis* subsp. *cremoris* and incubated for 18 h. Measurements were made using a CINAC continuous pH-monitoring system (Alliance instruments, 95740 Frepillon, France) and low-amplitude oscillation rheometry (CSL2 500 Carri-Med; TA Instruments, Inc., New Castle, DE, USA).

at this temperature until the desired pH (~4.5–4.8) is reached. The onset of gel formation in acid-induced milk gels occurs typically at pH 5.4 to 4.8, and the gel strength, as measured by development of elastic shear modulus (G′), increases continuously with further pH reduction to 4.6 (Figure 8.3). This may be important during subsequent stirring and separation, as a firmer gel would reduce the susceptibility of curd to shattering on exposure to the forces involved, thereby enabling more efficient separation and fewer losses in the whey stream.

The pH at the onset of gelation depends on extent of heat treatment of the milk before fermentation. High heat treatment (e.g. 95°C for 1–5 min) results in the onset of gelation at higher pH values (e.g. 5.4) and firmer gels compared to standard pasteurisation temperature (72°C for 30–60 s) (Davies *et al.*, 1978; Harwalkar & Kaláb, 1980; Lucey *et al.*, 1998b; Lucey 2004a; Figure 8.3). The higher pH at gelation onset in high heat-treated milk may be attributed to an increased hydrodynamic radius of the casein micelles in the heated milk as effected by the attachment of denatured whey protein to the k-casein at the micelle surface, resulting in the formation of a lining of filamentous appendages that protrude outwards from the micelle surface (Heertje *et al.*, 1985; Visser *et al.*, 1986). It may also be due to the higher isoelectric pH of the main whey protein, β-lactoglobulin, which would initiate isoelectric aggregation of the caseins at higher pH than their normal value, that is, 4.6 (Lucey *et al.*, 1998b). The firmer gel from high heat-treated milks is due to the incorporation of denatured whey proteins in the gel, and, hence, the higher effective concentration of gel-forming protein.

Curd formation and whey separation

After fermentation, the resulting gel is agitated gently, heated and concentrated by various methods; (a) continuous concentration using mechanical separator at 70 to 85°C; (b) UF at

50 to 55°C; or (c) draining through muslin bags at 60 to 90°C over 12 to 16 hours, as in the traditional batch method.

Mechanical separation of the curd is the most common method of manufacture, especially in large operations that would typically have a battery of large separators (e.g. 2000 L h^{-1}) operating simultaneously. The gel is heated batch-wise or continuously (e.g. in a wide gap plate heat exchanger) to a temperature of ~75 to 80°C and pumped to a centrifugal curd separator (e.g. Westfalia, cream cheese separator, model KNA), where the curd and whey are continuously separated into two different lines.

UF may also be used for separation of permeate and concentration of the gel into curd. Owing to the thick viscous consistency of cream cheese curd, concentration by UF necessitates a two-stage process so as to maintain satisfactory flux and DM level; (a) stage 1, standard modules with centrifugal or positive pumps; and (b) stage 2, high-flow modules with positive pumps. The gel is stirred, heated to ~55°C, pumped to the UF unit where it is concentrated by a factor of ~3, typically using ceramic or polysulphone-type membranes. In the manufacture of cream cheese by UF, whey proteins retained in the product (retentate) would be expected to undergo rapid denaturation and aggregation on heating the curd at the low pH (~4.6–4.8) (Harwalkar & Kaláb, 1985; Patoka *et al*., 1986). A recent study on the heating of β-lactoglobulin at pH 4.6 in dilute solutions (1 g 100 g^{-1}) with medium ionic strength show that heating at temperatures ~80°C is conducive to the formation of precipitable aggregates (O'Kennedy & Mounsey, 2008). This can lead to graininess of the curd. Consequently, the standardised milk is generally pasteurised at temperatures (e.g. 95°C for 1–5 min), which are higher than those used for pasteurisation (72–75°C for 30 s) when curd separation is achieved using a mechanical separator. The higher pasteurisation temperature ensures more extensive denaturation and complexation of the whey proteins with the casein before fermentation in the UF cream cheese process. This reduces the risk of graininess in the final product, which generally has a creamier and smoother consistency than the cream cheese produced using the separator method. High heating temperatures of the milk are not recommended in the manufacture of cream cheese by the separator method as they make both separation and the achieving of the desired level of DM in the finished cheese difficult.

The traditional batch method is hardly used today, except perhaps in some small farmhouse cheesemaking operations. In this case, it is a low-cost affordable system, as gel can be heated to the separation temperature by indirect steam injection in the fermentation tank, which is typically a jacketed vessel enabling steam heating and cooling. The gel is agitated gently and continuously while heating to ~80 to 90°C over a period of ~10 to 30 min. It is then fed by gravity flow into muslin 'bags' suspended from a stainless frame and the whey is allowed to drain over a period of 8 to 16 h. The drainage time, which is usually determined by the experience of the cheese operator, depends on pH and temperature of the gel at separation, the surface area of drainage and weight of heated gel per bag. The resultant curd is removed from the bags into a cooker. Alternatively, the curd may be retained in the bags and cooled to a temperature of 4 to 8°C until ready for processing; cooling involves placing the bags on trolley tables, which are placed at 4°C, or layering the bags with ice (Kosikowski & Mistrty, 1997a, 1997b).

Heating the curd and blending with other ingredients

In large plants, the hot curd (~80–85°C) from the mechanical separator is treated with salt and hydrocolloid, via online metering and mixing devices, and pumped directly to a mixing vessel where the temperature is maintained and further mixing takes place. Ribbon blenders are popular in large plants; these consist of a U-shaped horizontal trough and a specially fabricated ribbon agitator consisting of a set of inner and outer helical agitators, which move the curd both radially and laterally to ensure thorough mixing. These vessels can hold typically up to 20 000 L, and the curd holding time can range from 10 min to 1 h, depending on number of separators, number of blenders and plant design.

In modern plants using UF technology, the curd is typically cooked continuously using a scraped-surface heat exchanger to a temperature of ~75 to 85°C, and otherwise treated similar to curd produced using the separator method. In small operations, curd produced using the traditional, separator or UF processes, is heated to a similar temperature, typically by batch cooking/mixer to which salt and hydrocolloid are added. Process cheese-type cookers (e.g. Stephan, Damrow, Blentech, Limitech, Scanima) used employ various mixing devices (counter-rotating augers, high-speed rotating double knives, mixing valves), which can be operated at different speeds to apply varying degrees of shear. The degree of shear and time of shearing/blending has a major impact on the textural properties of the final cheese and is a useful means by which the product attributes (e.g. firmness, brittleness, plasticity) may be customised.

Sodium chloride is added at the level of ~0.5 to 1.0 g 100 g^{-1} to enhance flavour. While the addition of hydrocolloids is optional, they are generally added as texture modifiers at levels 0.2 to 0.5 g 100 g^{-1} in compliance with product specifications. Various types may be used including guar gum, traganth gum, xanthan gum, LBG, carboxymethyl cellulose, carrageenans and/or pectins (Hunt & Maynes, 1997; Anonymous, 2008a; CFR, 2008). In practice, proprietary blends affording customised consistency and texture are the most common; for example, a blend of xanthan and guar gums at a ratio of 1:1 gives a smooth, plastic texture when added at 0.3 g 100 g^{-1}; in contrast, guar on its own at 0.3 g 100 g^{-1} tends to give a firmer more gummy consistency. In addition to salt, various flavours (e.g. orange, lemon) and condiments and spices (e.g. preserved fish, vegetable, salt and pepper preparations) may also be added at this point.

Homogenisation, cooling and storage

Homogenisation is optional, its main function being to assist the manufacture of a product with uniform consistency, free from particles of undissolved hydrocolloid or defects such as a grainy or a sandy mouth-feel. It also tends to give a 'creaming effect', similar to that which occurs in processed cheese products on heating and shearing (Guinee *et al.*, 2004), and a firmer and more brittle end product. Typical first- and second-stage homogenisation pressures are in the range, 15 to 25, and 3 to 5 MPa, respectively. The hot curd (75–85°C) is mechanically filled into plastic containers or foil-wraps, which are stacked or packed, and placed at 4°C for storage.

8.4.3 *Recombination technology*

Various approaches appear in the patent literature for the production of cream cheeses or cream cheese style alternatives using one of the three recombination technologes:

- From reformed milks, prepared from dairy ingredients, that are converted to cream cheese using conventional technology involving fermentation, whey separation and curd treatments (Meurou & Andriot, 2007)

- From reformed milks with the same solids content as the final cheese (Guinee *et al.*, 2008; Sweley *et al.*, 2008)
- From formulated blends incorporating a fermented milk product base (such as cream cheese, cottage cheese, Quarg or yoghurt) and other ingredients (such as whey protein concentrates – WPCs) (Kent *et al.*, 2003; Brue *et al.*, 2005; Wolfschoon-Pombo *et al.*, 2006; Gutknecht & Ovitt, 2007; Kondo *et al.*, 2007; Gramza *et al.*, 2007; Brooks *et al.*, 2008; Sweley *et al.*, 2008; McPherson *et al.*, 2008)

In method 1, a 'reformed milk' is prepared from a mixture of butter or skimmed milk, skimmed milk powder, cream, butter and whey protein. It is pasteurised, homogenised and inoculated with starter culture. The resultant gel is subjected to separation to remove whey and yield a cheese base that may be treated as for conventionally made cream cheese. According to method 2, manufacture typically involves the preparation of 'reassembled milk' from a blend of milk protein concentrates, varying in casein-to-whey protein ratio and calcium-to-protein ratio, and cream or anhydrous butter oil (Guinee *et al.*, 2008). The reassembled milk, subjected to homogenisation or high shear mixing, is acidified to pH 4.6 to 4.8, using added starter culture, acidogen (such as glucono-δ-lactone) and/or food-grade acid. The resultant curd is then heated and treated with salt and hydrocolloid to yield cream cheese. Recombination by method 3 typically involves formulation of a cheese base from various ingredients (e.g. yoghurt, WPC, dry Cottage cheese, Bakers cheese curd, fermented skimmed milk concentrate at pH 4.8–5.0 and/or butter fat), acidification of the base, blending with emulsifying salts, bulking agents (e.g. buttermilk powder, corn syrup solids and/or starch) and various gums (e.g. carrageenan, gelatine and/or guar gum), followed by various heating, mixing and homogenisation steps (e.g. Schulz-Collins & Senge, 2004; Brue *et al.*, 2005; Wolfschoon-Pombo *et al.*, 2006).

While the volume of cream cheese made commercially using these approaches is relatively small, they may be used to afford certain functionalities (e.g. texture and mouth-feel, controlled meltability in heated dishes) and/or cost savings, especially in cream cheese products used as an ingredient in other foods such as desserts and savoury dishes. Other advantages of these methods is that cream cheese or cream cheese style products can be formulated readily from available ingredients, and precisely to meet legal requirements without excess solids or butterfat.

8.5 Basic characterisation of the structure and rheology of cream cheese

Electron microscopic evaluation of cream cheese shows that it is composed of a homogeneous network of compact casein/milk fat globule aggregates occluding large whey-containing areas, as well as partly coalesced milk fat globules. Protein was aggregated at the surface of the fat globule clusters or in their vicinity (Kaláb *et al.*, 1981). Protein aggregates are about 0.5 μm in diameter, with embedded fat globules typically ~1 μm in diameter, though some larger globules (up to 15 μm) are also present (Kaláb & Modler, 1985; Kimura *et al.*, 1986). A rigid protein matrix was not observed because of stirring and homogenisation of the curd during manufacture (Kaláb *et al.*, 1981; Kaláb & Modler, 1985; Sanchez *et al.*, 1996a). The corpuscular structure seems to be responsible for good spreadability with a high moisture content facilitating the mobility of the corpuscular constituents during spreading (Kaláb & Modler, 1985).

DCC exhibits viscoelastic–plastic rheological behaviour with an unusual flow curve; the shear stress in the ascending shear rate curve shows, depending on the manufacturing stages, two or three peaks. The first peak is commonly referred to as the static yield value (Sanchez *et al.*, 1994, 1996a, 1996b). The complex rheological behaviour of DCC indicates a 3-dimensional gel-like structure (Sanchez *et al.*, 1994). DCC shows time-dependent flow behaviour (partially thixotropic) and dynamic viscoelastic properties similar to those of nonchemically cross-linked polymers or pharmaceutical creams (Sanchez *et al.*, 1994, 1996a, 1996b). Dynamic low-amplitude osciallatory measurements indicate a predominantly solid-state behaviour of cream cheese and cream-based spreads even when the fat was completely melted at temperatures of 40 to 50°C (Brückner & Senge, 2008). However, reduction in temperature of cream cheese in the range 5 to 50°C leads to significant increases in the degree of elasticity (Brückner & Senge, 2008), fracture stress (Breidinger & Steffe, 2001) and viscosity (Sanchez *et al.*, 1996 a, 1996b). These trends are expected as crystallisation of the fat enclosed within the fat–protein composites comprising the network structure augments the rigidity and elastic response of the network *per se* to applied stress.

8.6 Factors affecting the properties of cream cheese

The microstructure and rheological properties of acid-induced milk gels are affected by many factors including *inter alia* milk protein concentration, ratio of casein-to-whey protein, fat content, homogenisation, pasteurisation temperature and time, pH of milk at heating, casein–whey protein interactions, addition of hydrocolloids, rennet treatment and gelation temperature (Heertje *et al.*, 1985; Gastaldi *et al.*, 1997; Cho *et al.*, 1999; Lucey *et al.*, 1998a, 1999; Sanchez *et al.*, 2000; O'Kennedy *et al.*, 2006; Pereira *et al.*, 2006; Oh *et al.*, 2007; Guyomarch *et al.*, 2007; Donato *et al.*, 2007). These have been reviewed by Guinee *et al.*, (1993), Lucey & Singh (1997), Lucey (2004a, 2004b) and Schulz-Collins & Senge (2004), and will not be discussed further here. In contrast, there have been only very few published studies on the properties of fresh acid curd cheeses and how these are affected by the structure of the acid gel after fermentation/acidification or by the treatments to which the gels subjected to – including heat treatment, centrifugal force, concentration (3–4 times), shearing and/or the addition of various materials, such as salt and hydrocolloids.

The following is a brief review of the major factors affecting the properties of cream cheese.

8.6.1 *Homogenisation of cheese milk*

As discussed earlier, homogenisation of cheese milk leads to a reformed milk fat globule membrane comprised of caseins and whey protein, and, thereby, a conversion of fat globules to pseudo-protein particles that become an integral part of the gel structure. Consequently, it is expected that a higher degree of homogenisation of the cheese milk would lead to smaller-sized emulsified fat globules that would be more stable to heat (Guinee *et al.*, 2000) and shear during heating and shearing of the gel post fermentation. Hence, in practice it is found that increasing first-stage homogenisation pressure from 5 to 25 MPa (while maintaining the first-stage pressure constant at 5 MPa) gives cream cheeses that are firmer, more brittle, less spreadable and flow less on heating.

8.6.2 *Holding of hot curd at high temperature while shearing*

At commercial level, alteration of the time of holding the hot curd at high temperature (while shearing and agitating) is a major lever by which the textural properties of cream cheese can be altered. The product becomes increasingly firmer, shorter and more brittle, with a tendency to crack as the holding time is increased. The response to holding is similar to that observed during the 'over creaming' of processed cheese, whereby the product becomes progressively thicker on prolonged processing (at temperatures of 75–85°C), and the finished processed cheese develops a firmer texture with a lower fracture strain and flows less when melted. This has been attributed to an increase in the degree of fat emulsification and an increase in protein aggregation, induced by hydrophobic interaction at the high temperature (Guinee *et al.*, 2004). These effects are consistent with the increases in firmness of cream cheese obtained on homogenisation and on slow cooling (compared to fast cooling), as discussed in the following sections (Sanchez *et al.*, 1994).

8.6.3 *Homogenisation of the heated cream cheese*

Increasing homogenisation pressures of the hot molten cream cheese curd in the range 0 to 20 MPa significantly increases firmness and viscosity and gives a lower perception of sandiness (Korolczuk & Mahaut, 1992; Sanchez *et al.*, 1996a, 1996b; Ortega-Fleitas *et al.*, 2000). Overall, it is concluded that homogenisation at 15 MPa gives the best textural quality. These trends are expected as a reduction in fat globule size, as affected by increasing homogenisation pressure, is expected to lead to lower displacement of the structure-forming protein-covered fat particles with a given solid-to-liquid fat ratio and, hence, of the cheese matrix as a whole, on the application of stress (assuming that the degree of displacement of a fat particle is proportional to its radius, $\approx \pi r$).

8.6.4 *Cooling rate*

An increase in cooling rate of cream cheese results in softer products, which tend to be stickier (Shulz-Collins & Senge, 2004; Sanchez *et al.*, 1994). This has been attributed to mechanical stress during dynamic cooling, resulting in a weaker structural organisation, as reflected by lower storage modulus, G'. Such a hypothesis is consistent with the decreasing strength of hydrophobic interactions between the protein and protein-covered fat particles as temperature is reduced, and the thixotrophic behaviour of cream cheese (Sanchez *et al.*, 1996a).

8.6.5 *Addition of whey protein*

Recent patents report the manufacture of cream cheeses with higher than normal firmness by replacing casein with polymerised whey proteins prepared by heating aqueous dispersions

(5–20 g 100 g^{-1}) at 70 to 95°C at pH values (7.0–9.0) conducive to the formation of intermolecular covalent disulphide bonds (Lindstrom *et al.*, 2005; Ma *et al.*, 2007).

In contrast, Sanchez *et al.* (1996c) reported that the addition of 1 g 100 g^{-1} of a commercial WPC (WPC, 65 g protein 100 g^{-1}) to hot curd during the heating stage (85°C) post fermentation significantly reduced the firmness of the resultant cream cheese (by ~15% reduction), as measured using cone pentrometry after 9 days storage at 5°C. Moreover, as the temperature to which the product was cooled before being placed in storage at 4°C was reduced from 40 to 15°C, cheese with the added WPC broke down to a greater degree (became more soft, shiny and sticky) than the control product that had a similar gross composition (apart from a lower pH). The effect of added WPC was attributed to the formation of whey protein aggregates, which interfered with network formation during cooling, leading to disruption of casein and casein-covered fat aggregates. The contrast between these studies demonstrates the importance of pH as a determinant of the type of interactions between heat-denatured whey proteins and their contribution to product structure.

8.6.6 *Hydrocolloids*

The most common gums used in the manufacture of hot-pack cream cheese include locust bean gum (LBG) (0.30–0.35 g 100 g^{-1}), xanthan gum, guar gum, tara gum and sodium alginate (Hunt & Maynes, 1997; Kosikowski & Mistry, 1997a). When considering the role of hydrocolloids in cream cheese, it is beneficial to define the product at the point of hydrocolloid addition: a low pH, a high-temperature system that has a high apparent protein concentration owing to the homogenisation of the milk before fermentation and gelation. Such a system is very prone to isoelectric aggregation of the caseins (and whey proteins if included), the formation of large aggregates of protein and/or protein-covered fat particles, and syneresis. The resultant product would be relatively heterogeneous, have a granular/grainy consistency and show some serum separation. Hence, the primary roles of hydrocolloid in cream cheese are as follows:

- Water binding to increase the viscosity of the aqueous phase and thereby restrict the movement of protein/protein-fat particles
- Product structuring to give desired degree of viscoelasticty and rheological behaviour, by way of firmness, brittleness, spreadability and/or shear thinning, for example, by (a) forming a weak gel or entanglement of polymers or (b) binding to the protein/ protein–fat particles and thereby creating a particle network
- Minimising the risk of granular or grainy structure by limiting the degree of protein/protein–fat particle interaction

Generally, a blend of hydrocolloids is best suited to these functions because of synergistic effects that ameliorate the blend functionality, for example, anionic hydrocolloids (e.g. κ-carageenan or xanthan gum) tend to limit the phase separation effects of some neutral hydrocolloids (e.g. guar gum, LBG) with casein, while the latter, in turn, by their limited phase separation, create the desired degree of 'body' (firmness) without giving excessive gumminess or rigidity. On their own, the anionic hydrocolloids bind with, and increase the negative charge on the casein, thereby favouring a more spreadable consistency with a reduction in rigidity/body. Moreover, variation in the ratio of galactose-to-mannose, which varies from 1:2 for guar gum to 1:4 for LBG (Hunt & Maynes, 1997), affords a convenient means of altering the ratio of elastic to viscous character in the final product. A lower content of galactose, which occurs in the form of branches off the mannose chain (backbone), as in LBG, permits an alignment of the chains that results in a more closely packed arrangement, a stronger intermolecular interaction and firmer, more elastic cream cheese than with guar or tara gum.

In cream cheese manufacture, a blend of xanthan with guar gum and/or LBG tends to give a smooth, creamy consistency and with good body and sliceability (Kovacs & Igoe, 1976). A similar blend is reported to optimise the baking performance of low protein cream cheese (Laye *et al.*, 2005). Hence, in practice, LBG, guar and xanthan gums are the most commonly used in cream cheese (Hunt & Maynes, 1997). The popularity of blends of these gums in practice is because a diverse range of texture and consistency (spreadable to short, gummy and firm) suitable for different applications may be easily obtained by variation in the relative levels in the blend.

8.6.7 *Composition*

An increase in the moisture content of cream cheese significantly softens the body and impairs sliceability (Ma *et al.*, 2005). Hence, in single-fat cream cheese that has a higher moisture level (e.g. ~65 compared to 55 g 100 g^{-1} in DCC), a different blend of hydrocolloids, among other factors, may be used to increase elasticity (Section 8.6.6). More recently, patent literature describes methods for the production of high moisture cheese with firmness comparable to or greater than control cheese with a lower moisture content; a principal feature of these processes is the use of high heat-treated whey protein polymerised at high pH (7–9) (Lindstrom *et al.*, 2005) or aggregated at low pH (3.5–4.5) (Laye *et al.*, 2002; Ma *et al.*, 2007) and subjected to high-pressure homogenisation, as a complete or partial substitute for casein.

The effect of increasing pH of cream cheese and North American-style Neufchatel cheese was investigated in model studies, in which thin slices of retail samples were exposed to ammonium hydroxide for 5 or 10 min to raise the pH from 4.6 to 5.4 (5-min exposure) or 5.8 (10-min exposure), before vacuum packing and storage at 4°C for 5 days (Almena-Aliste & Kindstedt, 2005; Almena-Aliste *et al.*, 2006). For both cheeses, there was a significant inverse relationship between pH and firmness. The softening effect at high pH was attributed to increased hydration of protein; hence, on re-acidification to pH 4.6 by exposure to hydrochloric acid, the firmness of the cheese was restored. The results suggest how the texture properties (e.g. firmness, spreadability, smoothness) may be modified by varying the pH away from the isoelectric point of the casein, where aggregation is strongest.

8.7 Related cheese varieties

8.7.1 *Mascarpone*

Mascarpone is a very smooth, creamy, spreadable, white-coloured cheese with a delicate, slightly cooked, nutty and fresh butter-like flavour; the product has a shelf-life of 1 to 3 weeks at <4°C. Compared to the other products in the group, it is characterised by a higher pH and relatively low contents of protein and calcium (Table 8.1); this is expected as the product is essentially a heavy cream, which is heated and acidified, and drained.

The manufacturing process described by Davis (1976) and Robindon and Wilbey (1998) involves the following steps:

- Heating of fresh cream (~35 g fat 100 g^{-1}) to 90 to 95°C.
- Acidification to pH ~5.5 to 5.7 with food grade organic acids such as acetic and tartaric, added to the cream while stirring.
- Pumping of the hot cream in cloth-lined moulds or muslin bags and allowing to drain for a period of hours.
- Optional salting of the cheese after removal from the mould, and then packaging.

Acidification at the high temperature does not lead to perceptible casein precipitation, 'curdiness' or granularity in the final cheese, because of the high volume fraction of the fat phase and low protein content of the cream.

8.7.2 *Neufchâtel and Petit-Suisse*

The manufacture of Neufchâtel and Petit-Suisse is generally assumed to be similar to that of cream cheese. However, little information is available and published methods differ somewhat in detail. According to Robinson and Wilbey (1998), starter culture is not normally added to the standardised cheese milk (~15–20°C) for Neufchâtel and rennet is added at a low level of ~1.5 mL 1000 L^{-1}. After a 20-h incubation, the resultant gel is strained through cheese cloth and the retained curd is pressed lightly, and afterwards passed through granite rollers to produce a smooth paste. Kosikowski and Mistry (1997b) described a process, which is quite similar to that for cream cheese up to the point of stirring the gel. The free whey is then drained and curd is then cooked by adding water at 50°C (similar to Cottage cheese), and stirring for ~15 min, after which the curd is placed in muslin cheese cloth bags as for cream cheese made by the traditional process. However, unlike cream cheese, the curd is not heated (to 80°C), sheared/homogenised or treated with hydrocolloid; instead the drained curd is packed and stored at 4°C.

The manufacture protocol for Petit-Suisse, as given by Robinson and Wilbey (1998), is similar to that for Neufchâtel, except that the set temperature is higher (~27°C) and the rennet dosage is much higher at 75 mL 1000 L^{-1}. While, the set time is not specified, it can be assumed to be much shorter than that of Neufchâtel but still relatively long compared to Cheddar cheese owing to the low temperature and low rennet dosage; for Cheddar, the rennet dosage is typically 180 mL 1000 L^{-1}, where the milk clotting activity of the rennet is ~60 chymosin units mL^{-1}.

8.7.3 *Kajmak*

Kajmak is an artisan farmhouse dairy product, indigenous to parts of the Balkans and Asia Minor and the Middle East; it is referred to as Kajmak in Serbia and Skorup in Montenegro (Dordevic, 1978; Pudja *et al.*, 2008). Though compositionally very variable, it is quite similar to cream cheese and Mascarpone, having a low protein level and a high FDM (Table 8.1). The latter difference in fat and protein increases with ripening time as a consequence of coalescence of fat and a gradual drainage of serum (containing protein and salts).

The traditional product, which has characteristics of both cheese and butter, is made by:

- Boiling raw milk and pouring the hot milk into open shallow vessels, a process that leads to the formation of dehydrated surface skin
- Allowing to cool and remain at 10 to 15°C for ~24 h during which time fat globules slowly rise and become embedded in the skin
- Scooping of the congealed mass (layer) of fat and protein from the surface of individual containers
- Adding salt for flavour and preservation
- Placing the layers of salted 'cheese' into storage vessels designed to allow excess residual serum to percolate off during storage
- Gradual layering and filling of the containers with small portions of Kajmak produced on consecutive days
- Storage of product

The product may be consumed fresh within a few days or may be allowed to ripen at 15 to 18°C for 15 to 20 days; the ripened product has a shelf-life of 3 to 6 months when stored at temperatures <8°C (Pudja *et al.*, 2008). The fresh product is an oil-in-water emulsion, is light yellow to ivory in colour, has a mild, cooked milk, butter-like flavour and has much similarity with cream cheese. On ageing, fat globules coalesce, resulting in the gradual transformation of the product to a water-in-oil emulsion, and the simultaneous percolation of serum. This results in an age-related drop in moisture content to 15 to 35 g 100 g^{-1} of that of the ripened product and an increase in the FDM (Table 8.1). On ageing, the product becomes darker yellow, an occurrence most likely due to the combined effects of the age-related phase transition, biochemical changes (e.g. in pH and redox potential), and the degradation of β-carotene as mediated by oxidation (oxygen- and light-induced) and biochemical changes (Mordi, 1993; Petersen *et al.*, 1999). With ageing, the product becomes more butter like in characteristics, with a progression to a buttery granular consistency (with some free serum), and then to a more uniform butter-like structure on advanced ageing (1 year) (Pudja *et al.*, 2008).

A recent patent (Pudja *et al.*, 2004) describes a procedure for the industrial production of traditional-style Kajmak with a more controlled microbiological status in a shorter time. The process involves standardisation of milk to the desired protein-to-fat ratio, pasteurisation, rapid skin formation on surface of hot milk by passing a stream of cool air over the surface, removal of bulk phase milk and replacing with pasteurised cooled cream, rapid cooling of skin and incorporation of fat, spraying of skin with a cream aerosol containing added culture, recovery of the congealed mass, salting and storage.

8.8 Conclusion

Cream cheese and related varieties are a relatively small group of soft/semi-soft cheeses. Compared to rennet–curd cheeses such as Cheddar, Gouda and Camembert, they have a relatively high fat content (generally >60 g 100 g^{-1} FDM), low levels of protein (generally <10 g 100 g^{-1}) and calcium (<100 mg 100 g^{-1}), soft smooth consistency and milder flavours. Yet, the group is quite diverse with respect to manufacturing procedure and sensory properties. Cream cheese, the largest variety, is traditionally made by slow quiescent starter culture-induced acidification, gelation, curd concentration by centrifuge, heating of the curd to ~85°C, addition of sodium chloride and hydrocolloid, and homogenisation. In that respect, the manufacturing protocol resembles natural cheese (culturing and gelation) and certain aspects (high heat treatment, shearing and homogenisation) of processed cheese but without emulsifying salts. Moreover, as for processed cheese foods and spreads, the texture and flavour characteristics can be greatly altered by varying the degree of shear and holding time at high temperature, type and level of hydrocolloid, extent of creaming effect and cooling rate. More recently, new methods are being developed for the manufacture of cream-style cheeses that are compliant with the Codex Alimentarius general standard for cheese (FAO/WHO, 2008b). The definition of cheese in this standard allows for unspecified technological changes that produce cheese of similar physical, chemical and organoleptic characteristics as those produced by more traditional production methods. Such developments are likely to be further stimulated by our increased understanding of the protein–protein and protein–hydrocolloid/starch interactions, the superimposed effects of process technology and environmental factors (such as heat, alkalinity/acidity, shear), and developments in analytical capability (such as microstructure) and formulation science. Undoubtedly, such approaches will provide more scope for engineering structure and functionality and for developing new variants of cream cheese (e.g. higher moisture, lower fat) and related varieties more cost-effectively.

It remains to be seen whether the compositional standards specified for cream cheese in the new Codex standard for cream cheese will be acceptable to and/or incorporated into the legislation of, individual countries, whose compositional standards differ, as outlined in this chapter. If this transpires, it will facilitate the further evolution of international brands with the same composition and labelling, rather than requiring products to be produced for specific markets to comply with the diversity of current national requirements.

References

Almena-Aliste, M. & Kindstedt, P. S. (2005) Effect of increasing pH on texture of full and reduced-fat cream cheese. *Australian Journal of Dairy Technology*, **60**, 225–230.

Almena-Aliste, M., Gigante, M. L. & Kindstedt, P. S. (2006) Impact of pH on the texture of cultured cream cheese: firmness and water phase characteristics. *Milchwissenschaft*, **61**, 400–404.

Anonymous (1952) International convention on the use of designations of origin and names for cheeses of 1 June 1951. *Journal Officiel de la République Française*, No. 5821–5824.

Anonymous (2001) Décret du 22 janvier 2001 relatif à l'appellation d'origine contrôlée ''Roquefort''. *Journal Officiel de la République Française*, 3(Texte n°35), 1283. (http://www.legifrance.gouv.fr/jopdf/common/jo_pdf.jsp?numJO=0&dateJO=20010125&numTexte=35&pageDebut=01283&pageFin=01285).

Anonymous (2006) Décret du 17 mai 2006 abrogeant le décret du 29 décembre 1986 relatif à l'appellation d'origine contrôlée ''Neufchâtel''. *Journal Officiel de la République Française*, 116(Texte 31 sur 117 NOR: AGRP0502365D), 7397–7400. (http://www.legifrance.gouv.fr/jopdf/common/jo_pdf.jsp?numJO=0&dateJO=20060519&numTexte=31&pageDebut=07397&pageFin=07399).

Anonymous (2007) Décret no 2007–628 du 27 avril 2007 relatif aux fromages et spécialités fromagères. *Journal Officiel de la République Française*, 0101 (Texte 14 sur 91 NOR: ECOC0750331D), 7628–7637. (http://www.legifrance.gouv.fr/jopdf/common/jo_pdf.jsp?numJO=0&dateJO=20070429&numTexte=14&pageDebut=07628&pageFin=07635); using the NOR and year at the above website accesses the text on the following link (http://www.legifrance.gouv.fr/rechTexte.do?reprise=true&page=1).

Anonymous (2008a) Décret du 22 janvier 2001 relatif à l'appellation d'origine contrôlée ''Roquefort'', Version consolidée au 18 Avril 2008. (http://www.legifrance.gouv.fr/affichTexte.do?cidTexte=JORFTEXT000000768632&dateTexte=20080826&fastPos=7&fastReqId=653197963&oldAction=rechTexte).

Anonymous (2008b) Paris Convention for the Protection of Industrial Property of March 20, 1883, as revised. (http://www.wipo.int/treaties/en/ip/paris/pdf/trtdocs_wo020.pdf).

Banon, S. & Hardy, J. (1992) A colloidal approach of milk acidification by glucon-delta-lactone. *Journal of Dairy Science*, **75**, 935–941.

Behr's Verlag (publisher) (2007) Käseverordnung 1986 – update to 13 December 2007. (http://www.behrs-kompakt.de/katalog/index.php?mode=index&id=125).

Bertozzi, L. & Panari, G. (1993) Cheeses with appellation d'origine contrôlée (AOC): factors that affect quality. *International Dairy Journal*, **3**, 297–312.

Breidinger, S. L. & Steffe, J. F. (2001) Texture map of cream cheese. *Journal of Food Science*, **63**, 453–456.

Brooks, T. V., Cha, A. S., Laye, I. M.-F., McPherson, A. E. & Zhao, E. (2008) Low fat, whey-based cream cheese product with carbohydrate-based texturizing system and methods of manufacture. **United States Patent Application**, 2008160133.

Brückner, M. & Senge, B. (2008) High-fat pasty milk products – investigation into mechanisms of structure and stability of cream cheese and cream-based spreads. *Milchwissenschaft*, **63**, 286–290.

Brue, N. L., Gutknecht, J. R. & Ovitt, J. B (2005) Yogurt cream cheeses. **United States Patent Application**, 7258886 B2.

Bundesministerium der Justiz (2008) Käseverordnung. (http://bundesrecht.juris.de/k_sev/index.html).

Cho, Y. H., Lucey, J. A. & Singh, H. (1999) Rheological properties of acid milk gels as affected by the nature of the fat globule surface material and heat treatment of milk. *International Dairy Journal*, **9**, 537–545.

CFR (2008) Part 133, Cheese and related products. *Food and Drugs 21 – Code of Federal Regulations, Parts 100–169*, US Government Printing Office, Washington DC. (http://www.foodsafety.gov/~lrd/fcf133.html).

Courtois, O. (2008). Norman cheeses: the Petit-suisse. (http://www.fromages.org/fdn/fdn_petit_suisse_en.html).

Creamer, L. K. (1985) Water absorption by renneted casein micelles. *Milchwissenschaft*, **40**, 589–559.

Dahlberg, A. C. (1927) A new method of manufacturing cream cheese of the Neufchatel type. *Journal of Dairy Science*, **10**, 106–116.

Dalgleish, D. G. & Law, A. J. R. (1988) pH-induced dissociation of bovine casein micelles. I. Analysis of liberated caseins. *Journal of Dairy Research*, **55**, 529–538.

Davies, F. L., Shankar, P. A., Brooker, B. E. & Hobbs, D. G. (1978) A heat-induced change in the ultrastructure of milk and its effect on gel formation in yoghurt. *Journal of Dairy Research*, **45**, 53–58.

Davis, J. G. (1966) The new cheese regulations. *Journal of the Society of Dairy Technology*, **19**, 119–120.

Davis, J. G. (1976) *Cheese*, Volume 3, pp. 586–592, 702–731, Churchill Livingstone, London.

Department of Justice Canada (2008) *Food and Drug Regulations CRC, c. 870*. (http://laws.justice.gc.ca/en/showtdm/cr/C.R.C.-c.870).

Donato, L., Guyomarch, F., Amiot, S. & Dalgleish, D. G. (2007) Formation of whey protein/kappa-casein complexes in heated milk: preferential reaction of whey protein with kappa-casein in the casein micelles. *International Dairy Journal*, **17**, 1161–1167.

Dordevic, J. (1978) Kajmak – problems of classifications and quality regulations. *Mljekarstvo*, **28**, 137–140.

ECJ (1987) Commission of the European Communities v Federal Republic of Germany. Failure of a State to fulfil its obligations – Purity requirement for beer in C 178/84. European Court of Justice.

Eden, K. J. (1993) History of German brewing. *Zymurgy*, **16** (No. 4 Special). (http://brewery.org/library/ReinHeit.html), (http://shop.beertown.org/brewers/product.asp?s_id=0&prod_name=Zymurgy+Magazine+Special+Issue+1993+Traditional+Brewing+Methods&pf_id=3220_763&dept_id=3220).

Ente Nazionale Italiano di Unificazione (1998) Fromaggio Mascarpone: Definizione di specificità, composizione, caratteristiche. **UNI 10710**.

EU (1976) Council Directive 76/118/EEC of 18 December 1975 on the approximation of the laws of the Member States relating to certain partly or wholly dehydrated preserved milk for human consumption. *Official Journal of the European Commission*, **L24**, 49–57.

EU (1985a) Completing the Internal Market: White Paper from the Commission to the European Council COM(85) 310, June 1985. (http://europa.eu/documents/comm/white_papers/pdf/com1985_0310_f_en.pdf).

EU (1985b) Council Directive 85/397/EEC of 5 August 1985 on health and animal-health problems affecting intra-Community trade in heat-treated milk. *Official Journal of the European Commission*, **L226**, 13–29.

EU (1988) Council Directive 88/388/EEC of 22 June 1988 on the approximation of the laws of the Member States relating to flavourings for use in foodstuffs and to source materials for their production. *Official Journal of the European Commission*, **L184**, 61–66.

EU (1989) Council Directive 89/107/EEC of 21 December 1988 on the approximation of the laws of the Member States concerning food additives authorized for use in foodstuffs intended for human consumption. *Official Journal of the European Commission*, **L40**, 27–33.

EU (1992a) Council Directive 92/46/EEC of 16 June 1992 laying down health rules for the production and placing on the market of raw milk, heat-treated milk and milk-based products. *Official Journal of the European Commission*, **L268**, 1–32.

EU (1992b) Council Regulation (EEC) No 2081/92 of 14 July 1992 on the protection of geographical indications and designations of origin for agricultural products and foodstuffs. *Official Journal of the European Commission*, **L208**, 1–8.

EU (1992c) Council Regulation (EEC) No 2082/92 of 14 July 1992 on certificates of specific character for agricultural products and foodstuffs. *Official Journal of the European Commission*, **L208**, 9–14.

EU (1994a) European Parliament and Council Directive 94/35/EC of 30 June 1994 on sweeteners for use in foodstuffs. *Official Journal of the European Commission*, **L237**, 3–12.

EU (1994b) European Parliament and Council Directive 94/36/EC of 30 June 1994 on colours for use in foodstuffs. *Official Journal of the European Commission*, **L237**, 13–29.

EU (1995a) Commission Directive 95/31/EC of 5 July 1995 laying down specific criteria of purity concerning sweeteners for use in foodstuffs. *Official Journal of the European Commission*, **L178**, 1–19.

EU (1995b) Commission Directive 95/45/EC of 26 July 1995 laying down specific purity criteria concerning colours for use in foodstuffs. *Official Journal of the European Commission*, **L226**, 1–45.

EU (1995c) European Parliament and Council Directive No 95/2/EC of 20 February 1995 on food additives other than colours and sweeteners. *Official Journal of the European Commission*, **L61**, 1–40.

EU (1996a) Commission Directive 96/77/EC of 2 December 1996 laying down specific purity criteria on food additives other than colours and sweeteners. *Official Journal of the European Commission*, **L339**, 1–69.

EU (1996b) Commission Regulation (EC) No 1107/96 of 12 June 1996 on the registration of geographical indications and designations of origin under the procedure laid down in Article 17 of Council Regulation (EEC) No 2081/92. *Official Journal of the European Commission*, **L148**, 1–10.

EU (1997) Commission Regulation (EC) No 2301/97 of 20 November 1997 on the entry of certain names in the 'Register of certificates of specific character' provided for in Council Regulation (EEC) No 2082/92 on certificates of specific character for agricultural products and foodstuffs. *Official Journal of the European Commission*, **L319**, 8–9.

EU (2000a) Directive 2000/13/EC of the European Parliament and of the Council of 20 March 2000 on the approximation of the laws of the Member States relating to the labelling, presentation and advertising of foodstuffs. *Official Journal of the European Commission*, **L109**, 29–42.

EU (2000b) White paper on food safety, (*COM (1999) 719 final*).

EU (2003) Corrigendum to Commission Directive 2002/72/EC of 6 August 2002 relating to plastic materials and articles intended to come into contact with foodstuffs. *Official Journal of the European Commission*, **L39**, 1–42.

EU (2004a). Commission Directive 2004/1/EC of 6 January 2004 amending Directive 2002/72/EC as regards the suspension of the use of azodicarbonamide as blowing agent. *Official Journal of the European Commission*, **L7**, 45–46.

EU (2004b) Commission Directive 2004/19/EC of 1 March 2004 amending Directive 2002/72/EC relating to plastic materials and articles intended to come into contact with foodstuffs. *Official Journal of the European Commission*, **L71**, 8–21.

EU (2004c) Regulation (EC) No 1935/2004 of the European Parliament and of the Council of 27 October 2004 on materials and articles intended to come into contact with food and repealing Directives 80/590/EEC and 89/109/EEC. *Official Journal of the European Commission*, **L338**, 4–17.

EU (2004d) Regulation (EC) No 852/2004 of the European Parliament and of the Council of 29 April 2004 on the hygiene of foodstuffs. *Official Journal of the European Commission*, **L226**, 3–21.

EU (2004e) Regulation (EC) No 853/2004 of the European Parliament and of the Council of 29 April 2004 laying down specific rules for food of animal origin. *Official Journal of the European Commission*, **L226**, 22–82.

EU (2004f) Regulation (EC) No 854/2004 of the European Parliament and of the Council of 29 April 2004 laying down specific rules for the organisation of official controls on products of animal origin intended for human consumption. *Official Journal of the European Commission*, **L139**, 83–127.

EU (2005a) Commission Regulation (EC) No 2073/2005 of 15 November 2005 on microbiological criteria for foodstuffs. *Official Journal of the European Commission*, **L338**, 1–26.

EU (2005b) Commission Regulation (EC) No 2074/2005 of 5 December 2005 laying down implementing measures for certain products under Regulation (EC) No 853/2004 of the European Parliament and of the Council and for the organisation of official controls under Regulation (EC) No 854/2004 of the European Parliament and of the Council and Regulation (EC) No 882/2004 of the European Parliament and of the Council, derogating from Regulation (EC) No 852/2004 of the European Parliament and of the Council and amending Regulations (EC) No 853/2004 and (EC) No 854/2004. *Official Journal of the European Commission*, **L338**, 27–58.

EU (2005c) Commission Regulation (EC) No 2076/2005 of 5 December 2005 laying down transitional arrangements for the implementation of Regulations (EC) No 853/2004, (EC) No 854/2004 and (EC) No 882/2004 of the European Parliament and of the Council and amending Regulations (EC) No 853/2004 and (EC) No 854/2004. *Official Journal of the European Commission*, **L338**, 83–88.

EU (2005d) *Guidance Document Implementation of Certain Provisions of Regulation (EC) No 853/2004 on the Hygiene of Food of Animal Origin*, European Commission: Health & Consumer Protection Directorate-General, Brussels.

EU (2005e) *Guidance Document on the Implementation of Certain Provisions of Regulation (EC) No 852/2004 on the Hygiene of Foodstuffs*, European Commission: Health & Consumer Protection Directorate-General, Brussels.

EU (2005f) *Guidance Document on the Implementation of Procedures Based on the HACCP Principles, and on the Facilitation of the Implementation of the HACCP Principles in Certain Food Businesses*, European Commission: Health & Consumer Protection Directorate-General, Brussels.

EU (2006a) Commission Regulation (EC) No 1662/2006 of 6 November 2006 amending Regulation (EC) No 853/2004 of the European Parliament and of the Council laying down specific hygiene rules for food of animal origin. *Official Journal of the European Commission*, **L320**, 1–10.

EU (2006b) Corrigendum to Regulation (EC) No 1924/2006 of the European Parliament and of the Council of 20 December 2006 on nutrition and health claims made on foods. *Official Journal of the European Union*, **L12**, 3–18.

EU (2006c) Guidelines for the development of Community guides to good practice for hygiene or for the application of the HACCP principles, in accordance with Article 9 of Regulation (EC) No 852/2004 on the hygiene of foodstuffs and Article 22 of Regulation (EC) No 183/2005 laying down requirements for feed hygiene. (http://ec.europa.eu/food/food/biosafety/hygienelegislation/guidelines_good_practice_en.pdf).

EU (2007a) Amended proposal for a Regulation of the European Parliament and of the Council establishing a common authorisation procedure for food additives, food enzymes and food flavourings (presented by the Commission pursuant to Article 250 (2) of the EC Treaty), (*COM (2007) 0672 final*). (http://eur-lex.europa.eu/LexUriServ/LexUriServ.do?uri=COM:2007:0672:FIN:EN:PDF).

EU (2007b) Amended proposal for a Regulation of the European Parliament and of the Council on flavourings and certain food ingredients with flavouring properties for use in and on foods and amending Council Regulation (EEC) No 1576/89, Council Regulation (EEC) No 1601/91, Regulation (EC) No 2232/96 and Directive 2000/13/EC (presented by the Commission pursuant to Article 250 (2) of the EC Treaty), (*COM (2007) 0671 final*). (http://eur-lex.europa.eu/LexUriServ/LexUriServ.do?uri=COM:2007:0671:FIN:EN:PDF).

EU (2007c) Amended proposal for a Regulation of the European Parliament and of the Council on food additives (presented by the Commission pursuant to Article 250 (2) of the EC Treaty), (*COM (2007) 0673 final*). (http://eur-lex.europa.eu/LexUriServ/LexUriServ.do?uri=COM:2007:0673:FIN:EN:PDF).

EU (2007d) Amended proposal for a Regulation of the European Parliament and of the Council on food enzymes and amending Council Directive 83/417/EEC, Council Regulation (EC) No 1493/1999, Directive 2000/13/EC of the European Parliament and of the Council, and Council Directive 2001/112/EC and Regulation (EC) No 258/97 of the European Parliament and of the Council (presented by the Commission pursuant to article 250 (2) of the EC Treaty), (*COM (2007) 0670 final*). (http://eur-lex.europa.eu/LexUriServ/LexUriServ.do?uri=COM:2007:0670:FIN:EN:PDF).

EU (2007e) Guidance on the implementation of Regulation No 1924/2006 on nutrition and health claims made on foods: conclusions of the Standing Committee on the food chain and animal health. (http://ec.europa.eu/food/food/labellingnutrition/claims/guidance_claim_14-12-07.pdf).

EU (2008a) Communication from the Commission to the European Parliament pursuant to the second subparagraph of Article 251 (2) of the EC Treaty concerning the common position of the Council on the adoption of a Regulation of the European Parliament and of the Council establishing a common authorisation procedure for food additives, food enzymes and food flavourings, (*COM/2008/0145 final*). (http://eur-lex.europa.eu/LexUriServ/LexUriServ.do?uri=COM:2008:0145:FIN:EN:PDF).

EU (2008b) Communication from the Commission to the European Parliament pursuant to the second subparagraph of Article 251 (2) of the EC Treaty concerning the common position of the Council on the adoption of a Regulation of the European Parliament and of the Council on food additives, (*COM (2008) 0143 final*). (http://eur-lex.europa.eu/LexUriServ/LexUriServ.do?uri=COM:2008:0143:FIN:EN:PDF).

EU (2008c) Communication from the Commission to the European Parliament pursuant to the second subparagraph of Article 251 (2) of the EC Treaty concerning the common position of the Council on the adoption of a Regulation of the European Parliament and of the Council on food enzymes and amending Council Directive 83/417/EEC, Council Regulation (EC) No 1493/1999, Directive 2000/13/EC, Council Directive 2001/112/EC and Regulation (EC) No 258/97, (*COM (2008) 0144 final*). (http://eur-lex.europa.eu/LexUriServ/LexUriServ.do?uri=COM:2008:0144:FIN:EN:PDF).

FAO/WHO (1995) *Codex General Standard for Food Additives CODEX STAN 192 – 1995.* (http://www.codexalimentarius.net/gsfaonline/CXS_192e.pdf).

FAO/WHO (1996) Report of the Second Session of the Codex Committee on Milk and Milk Products. (ftp://ftp.fao.org/docrep/fao/meeting/005/ W2198E/W2198e.pdf).

FAO/WHO (1998) Report of the Third Session of the Codex Committee on Milk and Milk Products, Alinorm 9/11, Annex IX, pp. 58–62 in Codex Alimentarius Commission, Montevideo. (http://www.fao.org/docrep/meeting/005/ w9503e/w9503e00.htm).

FAO/WHO (1999) Report of the Third Session of the Codex Committee on Milk and Milk Products. (http://www.codexalimentarius.net/web/archives.jsp?year=99).

FAO/WHO (2000) Codex international individual standard for cream cheese (Rahmfrischkase), Codex Stan C-31-1973. *Codex Alimentarius Milk and Milk Products*, Volume 12, pp. 92–93, Secretariat of the Joint FAO/WHO Food Standards Programme, Rome.

FAO/WHO (2006) *Understanding The Codex Alimentarius*, 3rd edn., pp. 1–39, Secretariat of the Joint FAO/WHO Food Standards Programme, Rome.

FAO/WHO (2007a) Codex Alimentarius Commission Thirtieth Session Report. (http://www.codexalimentarius.net/web/archives.jsp?year=07).

FAO/WHO (2007b) *Codex Alimentarius Milk and Milk Products*, 1st edn., Secretariat of the Joint FAO/WHO Food Standards Programme, Rome.

FAO/WHO (2007c) Codex general standard for cheese, Codex Stan A-6-1978, Rev. 1–1999, Amended 2006. *Codex Alimentarius Milk and Milk Products*, pp. 53–59, Communication Division FAO, Rome. (www.codexalimentarius.net/web/standard_list).

FAO/WHO (2007d) Codex group standard for unripened cheese including fresh cheese. *Codex Alimentarius Milk and Milk Products*, pp. 79–84, Communications Division of FAO, Rome. (www.codexalimentarius.net/web/standard_list).

FAO/WHO (2007e) Codex standard for cream cheese, CODEX STAN 275–1973. *Codex Alimentarius Milk and Milk Products*, pp. 158–164, Communication Division FAO, Rome. (www.codexalimentarius.net/web/standard_list).

FAO/WHO (2007f) Membership of the Codex Alimentarius Commission. *Codex Alimentarius Commission: Procedural Manual*, pp. 191–193, Codex Alimentarius Commission, Rome.

FAO/WHO (2007g) Norme Codex pour le Fromage à la Crème (ou ''Cream Cheese'') CODEX STAN 275–1973. (http://www.codexalimentarius.net/download/standards/216/CXS_275f.pdf).

FAO/WHO (2008a) *Codex Standard for Cream Cheese – CODEX STAN 275–1973.* FAO/WHO standards Codex Alimentarius, Current Official Standards, Rome. (www.codexalimentarius.net/web/standard_list).

FAO/WHO (2008b) *Codex General Standard for Cream Cheese – CODEX STAN 275–1973.* Codex General Standard for Cheese Codex Stan A-6-1978, Rev.1–1999, Amended 2006, Rome.

FDA (1906) *Federal Food and Drugs Act of 1906 (The ''Wiley Act'').* (http://www.fda.gov/opacom/laws/wileyact.htm).

FDA (1938) *Federal Food, Drug and Cosmetic Act 1938*, P.L. No. 717 of 1938. (http://www.fda.gov/opacom/laws/fdcact/fdcact1.htm#ftn2).

FDA (2006) Food Additive Status List. (http://www.cfsan.fda.gov/~dms/opa-appa.html).

Food Standards Australia New Zealand (2007) Australia New Zealand Food Standards Code. (http://www.foodstandards.gov.au/thecode/foodstandardscode.cfm).

Fox, P. F., Guinee, T. P., Cogan, T. M. & McSweeney, P. L. H. (2000) Fresh acid-curd cheese varieties. *Fundamentals of Cheese Science*, pp. 363–387, Aspen Publishers Inc., Gaithersburg.

Fox, P. F. & Brodkorb, A. (2008) The casein micelle: historical aspects, current concepts and significance. *International Dairy Journal*, **18**, 677–684.

French, M. & Phillips, J. (2000) The evolution and operation of the Food and Drugs Acts 1875–1907. *Cheated Not Poisoned?: Food Regulation in the United Kingdom, 1875–1938*, pp. 31–65, Manchester University Press, Manchester.

Gastaldi, E., Lagaude, A., Marchesseau, S. & Tarodo-de-la-Fuente, B. (1997) Acid milk gel formation as affected by total solids content. *Journal of Food Science*, **62**, 671–675.

Gastaldi, E., Pellegrini, O., Lagaude, A. & Tarodo-de-la-Fuente, B. (1994) Functions of added calcium in acid milk coagulation. *Journal of Food Science*, **59**, 310–312, 320.

Gillis, J.-C. (2000) Definitions of cheese and standardisation. *Cheesemaking – from Science to Quality Assurance*, (eds. A. Eck & J.-C. Gillis), pp. 788–790, Lavoiser Publishing, Paris.

Gramza, E. M., Loza, O., Bahrani, R. A., Reinhart, B. M., Galloway, G. I., Meibach, R. L. & Metzger, V. (2007) Intermediate dairy mixture and a method of manufacture thereof. **European Patent Application**, EP1825758 A2.

Guinee, T. P., Auty, M. A. E., Mullins, C., Corcoran, M. O. & Mulholland, E. O. (2000) Preliminary observations on effects of fat content and degree of fat emulsification on the structure–functional relationship of Cheddar-type cheese. *Journal of Texture Studies*, **31**, 645–663.

Guinee, T. P., Caríc, M. & Kaláb, M. (2004) Pasteurized processed cheese and substitute/imitation cheese products. *Cheese, Chemistry, Physics and Microbiology*, (eds. P. F. Fox, P. L. H. McSweeney, T. M. Cogan, & T. P. Guinee), 3rd edn., Volume 2 *Major Cheese Groups*, pp. 349–394, Elsevier Academic Press, Amsterdam.

Guinee, T. P., Kelly, P., O'Kennedy, B. T. & Mounsey, J. (2008) Cheeses from micellar casein powders with different calcium levels. Filed for patent.

Guinee, T. P., Pudja, P. D. & Farkye, N. Y. (1993) Fresh acid-curd cheese varieties. *Cheese Chemistry, Physics and Microbiology/Major Cheese Groups*, 3rd edn., Volume 2, pp. 363–419, Elsevier Academic Press, Amsterdam.

Guinee, T. P., Pudja, P. D., Reville, W. J., Harrington, D., Mulholland, E. O., Cotter, M. & Cogan, T. M. (1995) Composition, microstructure and maturation of semi-hard cheeses from high protein ultrafiltered milk retentates with different levels of denatured whey protein. *International Dairy Journal*, **5**, 543–568.

Gutknecht, J. R. & Ovitt, J. B. (2007) Process for making yoghurt cream cheese, and the resulting products. **European Patent Application**, EP 1827120.

Guyomarch, F., Mahieux, O., Renan, M., Chatriot, M., Gamerre, V. & Famelart, M. H. (2007) Changes in the acid gelation of skim milk as affected by heat-treatment and alkaline pH conditions. *Lait*, **87**, 119–137.

Harbutt, J. (1998) *The World Encyclopedia of Cheese*, pp. 25, Lorenz Books, New York.

Harwalkar, V. R. & Kaláb, M. (1980) Milk gel structure – XI. Electron microscopy of glucono-δ-lactone-induced skim milk gels. *Journal of Texture Studies*, **11**, 35–49.

Harwalkar, V. R. & Kaláb, M. (1985). Microstructure of isoelectric precipitates from β-lactoglobulin solutions heated at various pH values. *Milchwissenschaft*, **40**, 665–668.

Heertje, I., Visser, J. & Smits, P. (1985) Structure formation in acid milk gels. *Food Microstructure*, **4**, 267–277.

Hicks, C. L., Marks, F., O'Leary, J. & Langlois, B. E. (1985) Effect of culture media on cheese yield. *Cultured Dairy Products Journal*, **20**(3), 9–11.

HMSO (1860) Adulteration of Food and Drink Act 1860, 23 & 24 Vict. c.84, Her Majesty Stationary Office, London.

HMSO (1875) Sale of Food and Drugs Act 1875, 38 & 39 Vict. c. 63, Her Majesty Stationary Office, London.

HMSO (1907) The Butter and Margarine Act 1907, Her Majesty Stationary Office, London.
HMSO (1915) Milk and Dairies (Consolidation) Act 1915, 5 & 6 Geo. 5 c.66, Her Majesty Stationary Office, London.
HMSO (1922a) Milk and Dairies (Amendment) Act 1922, 13 & 13 Geo. V c.54, Her Majesty Stationary Office, London.
HMSO (1922b) The Milk (Special Designations) Order, 1922, SRO 1922 No. 1332, Her Majesty Stationary Office, London.
HMSO (1923a) The Public Health (Condensed Milk) Regulations 1923, SRO 1923 No. 509, Her Majesty Stationary Office, London.
HMSO (1923b) The Public Health (Dried Milk) Regulations, 1923, SRO 1923 No. 1323, Her Majesty Stationary Office, London.
HMSO (1938) The Sale of Food & Drugs Act 1938, 1 & 2 Geo VI c.56, Her Majesty Stationary Office, London.
HMSO (1955) The Sale of Food & Drugs Act 1955, 4 & 5 Eliz II c.16, Her Majesty Stationary Office, London.
HMSO (1965) The Cheese Regulations 1965, SI No. 2199 of 1965, Her Majesty Stationary Office, London.
HMSO (1970) The Cheese Regulations 1970, SI 1970 No. 94, Her Majesty Stationary Office, London.
HMSO (1974) The Cheese (Amendment) Regulations 1974, SI 1974 No. 1122, Her Majesty Stationary Office, London.
HMSO (1984) The Cheese (Amendment) Regulations 1984, SI 1984 No. 649, Her Majesty Stationary Office, London.
HMSO (1990) Food Safety Act 1990, c. 16, Her Majesty Stationary Office, London.
HMSO (1995) The Cheese and Cream Regulations 1995, SI 1995 No. 3240, Her Majesty Stationary Office, London.
HMSO (1996) The Food Labelling Regulations 1996, SI 1996 No. 1499, Her Majesty Stationary Office, London.
van Hooydonk, A. C. M., Hagedoorn, H. G. & Boerrigter, I. J. (1986) pH-induced physico-chemical changes of casein micelles in milk and their effect on renneting. I. Effects of acidification on physico-chemical properties. *Netherlands Milk and Dairy Journal*, **40**, 281–296.
Hunt, C. C. & Maynes, J. R. (1997) Current issues in the stabilization of cultured dairy products. *Journal of Dairy Science*, **80**, 2639–2643.
International Food Information Council (2005) *Food Ingredients and Colors*, International Food Information Council, Washington, DC.
Jana, A. H. & Upadhyay, K. G. (1993) Homogenisation of milk for cheesemaking – a review. *Australian Journal of Dairy Technology*, **47**, 72–79.
Kaláb, M. & Modler, H. W. (1985) Milk gel structure – XV. Electron microscopy of whey protein-based cream cheese spread. *Milchwissenschaft*, **40**, 193–196.
Kaláb, M., Sargant, A. G. & Froehlich, D. A. (1981) Electron microscopy and sensory evaluation of commercial cream cheese. *Scanning Electron Microscopy*, **3**, 473–482.
Kent, C., Kijowski, M., Campbell, B., Pfeifer, J. K., Smith, C. B. M., Bahrani, R., Lee, J. E., Nellenback, T., Byrd, R. & Zaikos, W. (2003) Process for incorporating whey protein into cheese. **United States Patent Application**, 6 558 716 B1.
Kimura, T., Sagara, Y. & Tanimoto, M. (1986) Microstructure of cream cheese observed by cryo-SEM. *Effects of Melting Salts and Shear Rate on Cheese Structure during Processing*, pp. 43–54, Reports of Research Laboratory, Snow Brand Milk Products Co., Tokyo.
Kondo, H., Ueda, H. & Kawsaki, I. (2007) Cream cheese-like food product and method for producing same. **Japanese Patent Application**, JP 2007151418 A.
Korolczuk, J. & Mahaut, M.(1992) Effect of homogenization of milk on the consistency of UF fresh cheeses. *Milchwissenschaft*, **47**, 225–227.

Kosikowski, F. V. (1974) Cheesemaking by ultrafiltration. *Journal of Dairy Science*, **57**, 488–491.

Kosikowski, F. V. (1977) Bakers', Neufchatel and Cream cheese. *Cheese and Fermented Milk Foods*, pp. 144–167, Edwards Brothers, Inc., Ann Arbor, MI.

Kosikowski, F. V. & Mistry, V. V. (1997a) Bakers', Neufchatel, Cream, Quark, and Ymer. *Cheese and Fermented Milk Foods – Origins and Principles*, Volume 2, 3rd edn., pp. 147–161, F.V. Kosikowski, L. L. C., Westport.

Kosikowski, F. V. & Mistry, V. V. (1997b) Bakers', Neufchatel, Cream, Quark, and Ymer. *Cheese and Fermented Milk Foods – Procedures and Analysis*, Volume 2:, 3rd edn., pp. 42–54, F.V. Kosikowski, L. L. C., Westport.

Kovacs, P. & Igoe, I. S. (1976) Xanthan gum galactomannan system improves functionality of cheese spreads. *Food Product Development*, **10**(8), 32–38.

Kraft Foods (2008) *Philadelphia Brand History*. (http://www.philadelphia.co.uk/philadelphia2/page?siteid=philadelphia2-prd&locale=uken1&PagecRef=583).

LACOTS (1997) Labelling of Food – Cheese – Query LAC 17 97 14. (http://www.lacors.gov.uk/lacors/search.aspx?Nso=1&prev=41+32+53+4003+4000&Ne=4000+0+2000+3000+5000+6000+7000+8000+9000+10000+11000&N=41+32+53+4003&No=1240&Ns=DOC_PUBLISHED&tl=10000&id).

Lakemond, C. M. M. & van Vliet, T. (2008) Rheological properties of acid skim milk gels as affected by the spatial distribution of the structural elements of and the interaction forces between them. *International Dairy Journal*, **18**, 585–593.

Laye, I. M.-F., Loh, J. P., Pechak, D. G., Cha, A. S., Campbell, B. E., Lindstrom, T. R. & Zwolfer, M. (2002) High moisture cream cheese texture control. **European Patent Application**, EP 1214884A2.

Laye, I. M-F., Cha, A. S., Loh, J. P., Lindstrom, T. R. & Rodriguez, A. P. (2005) Low protein cream cheese. **European Patent Application**, EP 1579769A1.

Lindstrom, T. R., Mehring, A. D. & Hudson, H. M. (2005) Cream cheese made from whey protein polymers. **European Patent Application**, EP 1561383 A1.

Link, O. J. (1942) Cream cheese manufacture. **United States Patent Application**, 2 387 296.

Les Autorités Fédérales de la Confédération Suisse (2008) Convention internationale sur l'emploi des appellations d'origine et dénominations de fromages (Etat le 24 mai 2005) 0.817.142.1. (http://www.admin.ch/ch/f/rs/0_817_142_1/index.html).

Lucey, J. A. (2004a) Formation, structural properties and rheology of acid-coagulated milk gels. *Cheese: Chemistry, Physics and Microbiology – General Aspects*, (eds. P. F. Fox, P. L. H. McSweeney, T. M. Cogan & T. P. Guinee), 3rd edn., Volume 1, pp. 105–122, Elsevier Academic Press, Amsterdam.

Lucey, J. A. (2004b) Cultured dairy products: an overview of their gelation and texture properties. *International Journal of Dairy Technology*, **57**, 77–84.

Lucey, J. A. & Singh, H. (1997) Formation and physical properties of acid milk gels. *Food Research International*, **30**, 529–542.

Lucey, J. A., Munro, P. A. & Singh, H. (1998a) Rheological properties and microstructure of acid milk gels as affected by fat content and heat treatment. *Journal of Food Science*, **63**, 660–664.

Lucey, J. A., Tamehana, M., Harjinder, S. & Munro, P. A. (1998b) Effect of interactions between denatured whey proteins and casein micelles on the formation and rheological properties of acid skim milk gels. *Journal of Dairy Research*, **65**, 555–567.

Lundstedt, E. (1954) Manufacture of quality cream cheese: a means of utilizing some of our excess milk fat. *Journal of Dairy Science*, **37**, 714–716.

Ma, Y., Lindstrom, T. R., Laye, I., Rodriguez, A. P., Schmidt, G. M. & Doyle, M. C. (2007) High moisture, low fat cream cheese with maintained product quality and method for making same. **European Patent Application**, EP 1769683A1.

McMahon, D. J. & Oommen, B. S. (2008) Supramolecular structure of the casein micelle. *Journal of Dairy Science*, **91**, 1709–1721.

McPherson, A. E., Laye, I. M.-F., Brooks, T. V. & Cha, A. S. (2008) Low fat, whey-based cream cheese product with carbohydrate-based texturizing system and methods of manufacture. **Canadian Patent Application**, CA2613422.

McSweeney, P. L. H., Ottogalli, G. & Fox, P. F. (2004) Diversity of cheese varieties: an overview. *Cheese Chemistry, Physics and Microbiology – Major Cheese Groups*, (eds. P. F. Fox, P. L. H. McSweeney, T. M. Cogan & T. P. Guinee), 3rd edn., Volume 2, pp. 1–22, Elsevier Academic Press, Amsterdam.

Meurou, G. & Andriot, P. (2007) Procede de fabrication d'un fromage de type a pate fraiche et fromage obtenu par le procede. **French Patent Application**, FR 2889794 A1.

Ministère de la Justice Canada (2008) Loi sur les aliments et drogues (L.R., 1985, ch. F-27). (http://laws.justice.gc.ca/fr/showtdm/cs/F-27).

Ministerialtidende Danmark (2004) Bekendtgørelse om mælkeprodukter. BEK nr 335 af 10/05/2004. Copenhagen

Mordi, R. C. (1999) Mechanism of β-carotene degradation. *Biochemical Journal*, **292**, 310–312.

Monier-Williams, G. W. (1951) Historical aspects of the pure food laws. *British Journal of Nutrition*, **5**, 363–367.

National Health and Medical Research Council (1999) Standard H9 Cheese and Cheese Products. *Food Standards Code 1987 Incorporating Amendments up to and Including Amendment 44*, pp. 10–900-10–910, ANZFA, Canberra.

O'Kennedy, B. T. & Mounsey, J. S. (2008) The dominating effect of ionic strength on the heat-induced denaturation and aggregation of β-lactoglobulin in simulated milk ultrafiltrate. *International Dairy Journal*. (http://www.sciencedirect.com/science).

O'Kennedy, B. T., Mounsey, J. S., Murphy, F., Duggan, E. & Kelly, P. M. (2006) Factors affecting the acid gelation of sodium caseinate. *International Dairy Journal*, **16**, 1132–1141.

Oh, H. E., Wong, M., Pinder, D. N., Hemar, Y. & Anema, S. G. (2007) Effect of pH adjustment at heating on the rheological properties of acid skim milk gels with added potato starch. *International Dairy Journal*, **17**, 1384–1392.

Ortega-Fleitas, F. O., Reinery, P. L., Rocamora, Y., Real-del-Sol, E., Cabrera, M. C. & Casals, C. (2000) Effect of homogenization pressure on texture properties of soy cream cheese. *Alimentaria*, **313**, 125–127.

Patocka, J., Renz-Schauen, A., & Jelen, P. (1986) Protein coagulation in sweet and acid wheys upon heating in highly acidic conditions. *Milchwissenschaft*, **41**, 490–494.

Petersen, M., Wiking, L. & Stapelfeldt, H. (1999) Light sensitivity of two colorants for Cheddar cheese. Quantum yields for photodegradation in an aqueous model system in relation to light stability of cheese in illuminated display. *Journal of Dairy Research*, **66**, 599–607.

Pereira, R., Matia-Merino, L., Jones, V. & Singh, H. (2006) Influence of fat on the perceived texture of set acid milk gels: a sensory perspective. *Food Hydrocolloids*, **20**, 305–313.

Porter, D. V. & Earl, R. O. (1992) Contextual factors affecting the regulation of misbranded food. *Food Labelling: Toward National Uniformity*, (eds. D. V. Porter & R. O. Earl), pp. 35–62, National Academy Press, Washington, DC.

Pudja, P., Starcevic, V. & Radovanovic, M. (2004) A procedure for industrial production of traditional quality Kajmak and Kajmak produced thereof. **World Intellectual Property Organisation Patent Application**, PCT WO2004/023882A3.

Pudja, P., Starcevic, V. & Radovanovic, M. (2008) An autochthonous Serbian product – Kajmak. Characteristics and production procedures. *Dairy Science and Technology*, **88**, 163–172.

Randell, A. & Race, J. (1996) Regulatory and legal aspects of functional foods: an international perspective. *Nutrition Reviews*, **54**, 152–155.

Rieck, N. (2008) Reinheitsgebot: German Beer Purity Law, 1516. (http://www3.sympatico.ca/n.rieck/docs/Reinheitsgebot.html).

Rieke, J. W. (2004) Geographic indicators – a fight to keep traditional product names. (http://www.idfa.org/meetings/presentations/reike_df2004.ppt).

Robinson, R. K. & Wilbey, R. A. (1998) *Cheesemaking Practice – R. Scott*, 3rd edn., pp. 327–437, Aspen Publishers, Inc., Gaithersburg.

Roefs, S. P. F. M., Walstra, P., Dalgleish, D. G. & Horne, D. S. (1985) Preliminary note on the change in casein micelles caused by acidification. *Netherlands Milk and Dairy Journal*, **39**, 119–122.

Rynne, N. M., Beresford, T. P., Kelly, A. L. & Guinee, T. P. (2004) Effect of milk pasteurization temperature and *in situ*whey protein denaturation on the composition, texture and heat-induced functionality of half-fat Cheddar cheese. *International Dairy Journal*, **14**, 989–1001.

Sanchez, C., Zuniga-Lopez, R., Schmitt, C., Despond, S. & Hardy, J. (2000) Microstructure of acid-induced skim milk-locust bean gum-xanthan gels. *International Dairy Journal*, **10**, 199–212.

Sanchez, C., Beauregard, J. L., Bimbenet, J. J. & Hardy, J. (1996c) Flow properties, firmness and stability of double cream cheese containing whey protein concentrate. *Journal of Food Science*, **61**, 840–843, 846.

Sanchez, C., Beauregard, J. L., Bride, M., Buchheim, W. & Hardy, J. (1996b) Rheological and microstructural characterization of double cream cheese. *Nahrung*, **40**, 108–116.

Sanchez, C., Beauregard, J. L., Chassagne, M. H., Bimbenet, J. J. & Hardy, J. (1996a). Effects of processing on rheology and structure of double cream cheese. *Food Research International*, **28**, 547–552.

Sanchez, C., Beauregard, J. L., Chassagne, M. H., Duquenoy, A. & Hardy, J. (1994) Rheological and textural behaviour of double cream cheese – II. Effect of curd cooling rate. *Journal of Food Engineering*, **23**, 595–608.

Schmidt, D. G. & Poll, J. K. (1986) Electrokinetic measurements on unheated and heated casein micelle systems. *Netherlands Milk and Dairy Journal*, **40**, 269–280.

Schulz-Collins, D. & Senge, B. (2004) Acid- and acid/rennet-curd cheeses, part A.: Quark, cream cheese and related varieties. *Cheese Chemistry, Physics and Microbiology – Major Cheese Groups*, (eds. P. F. Fox, P. L. H. McSweeney, T. M. Cogan & T. P. Guinee), 3rd edn., Volume 2, pp. 301–328, Elsevier Academic Press, Amsterdam.

Scott, R. (1981) *Cheesemaking Practice*, pp. 335–461, Applied Science Publishers Ltd., London.

Singh, H., Roberts, M. S. Munro, P. A. & Cheng, T. T. (1996) Acid-induced dissociation of casein micelles in milk: effects of heat treatment. *Journal of Dairy Science*, **79**, 1340–1346.

Swann, J. P. (2008) History of the FDA. (http://www.fda.gov/oc/history/historyoffda/fulltext.html).

Sweley, J. C., Smith, G. F., Apel, L. J., Kincaid, C. M. & Kalamas, T. M. (2008). Cultured dairy products obtained without a whey separation step. **New Zealand Patent,** NZ547883.

Syndicat de Défense et de Qualité du Fromage Neufchâtel (2008) Historique; Le plus vieux des fromages normands. (http://www.neufchatel-aoc.org/module4/index.html).

The Stationery Office Dublin (1924) The Dairy Produce Act 1924, No. 58/1924, Dublin.

The Stationery Office Dublin (1935) Sale of Food and Drugs (Milk) Act, 1935, No. 3/1935, Dublin.

The Stationery Office (1970a) The Cheese (Scotland) Regulations 1970, SI 1970 No. 108, Her Majesty Stationary Office, London.

The Stationery Office (1970b) The Cheese Regulations (Northern Ireland) 1970, SR & O (NI) 1970 No. 14, Belfast.

The Stationery Office (1984) The Cheese (Scotland) Amendment Regulations 1984, SI 1984 No. 847, Her Majesty Stationary Office, London.

The Stationery Office (1995) The Cheese (Scotland) Amendment Regulations 1974, SI 1995 No. 1337, Her Majesty Stationary Office, London.

The Stationery Office (1996) The Cheese and Cream Regulations (Northern Ireland) 1996, SR 1996 No. 52, Belfast.

Tuckey, S. L. (1967) Unripened soft cheeses. *Cottage Cheese and other Cultured Milk Products*, (eds. D. B. Emmons & S. Tuckey), pp. 99–110, Chas. Pfizer & Co, Inc., New York.

USA National Archives and Records Administration (2008) US Code of Federal Regulations 2008. The Office of the Federal Register National Archives and Records Administration, Washington, DC. (http://www.access.gpo.gov/cgi-bin/cfrassemble.cgi?title=200821).

USDA (1976) Composition of foods – dairy and egg products: raw, processed and prepared. *Agriculture HandBook* No. 8-1, Agricultural Research Service, United States Department of Agriculture, Washington, DC.

USDA (2008) *Yoghurt Production: Dairy Products: Per Capita Consumption*. (http://www.nass.usda.gov/Statistics_by_State/Wisconsin/Publications/Annual_Statistical_Bulletin/page62.pdf).

Visser, J., Minihan, A., Smits, P., Tjan, S. B. & Heertje, I. (1986) Effects of pH and temperature on the milk salt system. *Netherlands Milk and Dairy Journal*, **40**, 351–368.

van Vliet, T. & Dentener-Kikkert, A. (1982) Influence of the composition of the milk fat globule membrane on the rheological properties of acid-milk gels. *Netherlands Milk Dairy Journal*, **36**, 261–265.

Wade, T., Beattie, J. K., Rowlands, W. N. & Augustin, M. A. (1996) Electroacoustic determination of size and zeta potential of casein micelles in skim milk. *Journal of Dairy Research*, **63**, 387–404.

Walstra, P. & Jenness, R. (1984) *Dairy Chemistry and Physics*, John Wiley & Sons, Inc., New York.

Winwood, J. (1983) Quarg production methods – past, present and future. *Journal of the Society of Dairy Technology*, **36**, 107–109.

Wolfschoon-Pombo, A, Rose, M., Muxfeldt, D. & Eibel, H. (2006) Cream cheese product and its method of preparation. **European Patent Application**, EP 1698231 A1.

9 Microbial Production of Bioactive Metabolites

S. Mills, R.P. Ross, G. Fitzgerald and C. Stanton

9.1 Introduction

The once mooted theory that diet could influence the health status of an individual has recently become a widely accepted reality, leading to an explosion in the functional foods market (Sharma, 2005). This is primarily fuelled by a rapid increase in lifestyle illnesses and the desire to prolong health and evade disease. 'Functional foods' is a term describing foods or food components that have beneficial effects on human health above that expected on the basis of nutritive value (Milner, 1999). In recent years, the health promoting properties of dairy fats have been highlighted by several scientific investigations (Stanton *et al.*, 2003; Lock & Bauman, 2004; Bhattacharya *et al.*, 2006). In particular, conjugated linoleic acids (CLAs) of dairy origin have received considerable attention on account of their biological properties. CLA is produced through the actions of the anaerobic rumenic bacteria when dietary linoleic and linolenic acids are converted to *trans*-vaccenic acid by the enzyme linoleic acid isomerase, a process called biohydrogenation (Stanton *et al.*, 2003). *Trans*-vaccenic acid is then converted to CLA (mainly *cis*-9, *trans*-11) in the mammary gland via the enzyme stearoyl-CoA desaturase and is thus excreted in milk (Griinari & Bauman, 1999). CLA has been shown to reduce the risk of heart disease and several types of cancer and to enhance bone formation and immune function in both animal models and human cell culture studies. Indeed, investigations using animal models have demonstrated that CLA is anticarcinogenic for many types of cancer (Whigham *et al.*, 2000; Belury, 2002; Parodi, 2002; Ip *et al.*, 2002). Furthermore, experimental animal studies have indicated that CLA may have beneficial effects on the atherosclerotic process, including reduced LDL (low-density lipoproteins) cholesterol level, reduced oxidation of LDL cholesterol, slowed development of atherosclerotic lesions and reduced severity of pre-existing lesions (Kritchevsky *et al.*, 2004). Other fatty acids, besides CLA, have also been shown to exert a range of beneficial biological effects. The short-chain fatty acid (SCFA) butyric acid or butyrate, which is found in the milk fat of ruminant animals (Stanton *et al.*, 2003), has been shown to exert a range of physiological effects ranging from anti-proliferative to the regulation of fat absorption. Indeed, butyrate is not unique, for other SCFAs have also demonstrated therapeutic potential. In addition to being associated with dairy fats, butyrate is one of the major end products of bacterial metabolism in the human large intestine along with other SCFAs such as acetate, propionate, valerate, caproate, iso-butyrate and iso-valerate. SCFAs are produced following fermentation of non-digestible foods, aptly termed prebiotics, by certain anaerobic bacteria (Brusseze *et al.*, 2006). Indeed, the concept of prebiotics has been partially fuelled by the desire to increase fatty acid production by the intestinal microflora. It is of no consequence therefore that these metabolites and the strains producing them have become a major target for the development of functional food products. Interestingly, human milk has also been shown to harbour a range of non-digestible oligosaccharides that promote the growth of beneficial bacteria, which in turn convey many benefits on the host, ranging from immunomodulatory to the production of SCFAs (Coppa *et al.*, 2004). Specific SCFAs may reduce the risk of developing gastrointestinal disorders, cancer and cardiovascular disease. In addition, SCFAs are readily absorbed into the bloodstream where they have been shown to exert a positive effect on the cardiovascular system; however, they may also play a role in cholesterol synthesis. Therefore, methods to increase SCFA concentrations in the body by gut microflora through the use of prebiotics or through the consumption of certain probiotic bacteria would be expected to exert a positive effect on host health. This concept comes at a time when consumer awareness of colonic health to overall wellbeing has been heightened.

Finally, this review could not be complete without mentioning the microbial metabolite gamma amino butyric acid (GABA). Interestingly, GABA is ubiquitous among single cell organisms, plants and higher animals. In microorganisms, GABA is predicted to exert a protective role against hostile pH environments. In vertebrates, GABA is the major inhibitory neurotransmitter of the central nervous system (CNS). While the blood–brain barrier is impermeable to GABA (Kuriyama & Sze, 1971), orally administered GABA exerts beneficial effects on the body through alleviation of stress and stimulation of the immune response among others.

The purpose of this review is to examine the biological properties of these metabolites in terms of health and disease of which an overview is provided in Table 9.1. Examples are provided of both *in vivo* and *in vitro* studies documenting the therapeutic properties of these agents. In addition, the impact of diet will be assessed and methods to increase production or to embellish food products with such molecules will be considered.

9.2 Short-chain fatty acids

9.2.1 *Background*

SCFAs are organic fatty acids with 1 to 6 carbon atoms in the acyl chain. They occur naturally in certain foods including fruits, vegetables and, in particular, milk fat (Newmark & Young, 1995). SCFAs are also the major anions arising from the anaerobic bacterial fermentation of polysaccharide, oligopolysaccharide, and protein in both the rumens and large bowels of plant-eating animals, where they can reach millimolar concentrations (Cummings *et al.*, 1987). However, carbohydrates play the most important role in the formation of SCFAs (Miller & Wolin, 1979) where the primary metabolite is pyruvate. These SCFAs occur mainly as acetate (C2), propionate (C3) and butyrate (C4), but valerate (C5), caproate (C6), iso-butyrate (iC4) and iso-valerate (iC5) are also present (Cummings *et al.*, 1987). SCFAs are rapidly absorbed from the human colon and stimulate salt and water absorption. They also have a major effect on epithelial cell growth. Indeed, butyrate is the major energy source for colonocytes, whereas propionate is mainly taken up by the liver, whilst acetate enters the peripheral circulation to be metabolised by peripheral tissues (Wong *et al.*, 2006). In recent years, it has been demonstrated that certain SCFAs, in particular butyrate, acetate and

Table 9.1 Overview of associated health effects of SCFAs.

Metabolite	Proposed effect on host health

SCFAs	• Reduce colonic pH:
	  • *Prevents outgrowth of pH-sensitive bacteria.*
	  • *Decreases solubility of secondary bile acids.*
	  • *Inhibits 7α-dehydroxylase which degrades primary bile acids to secondary bile acids.*
	• Increase colonic blood flow:
	  • *Enhances tissue oxygenation and transport of absorbed nutrients.*
	• Increase absorption of colonic water:
	  • *Treatment for antibiotic-induced and infectious diarrhoea.*
	• Increase calcium absorption in the gut:
	  • *Enhances bone health.*
	• Increase secretion of GLP-1,
	• Regulation of the expression of the proglucagon gene:
	  • *Aids in the management of diabetes and obesity.*
	• Modulate intestinal motility:
	  • *Aids in the management of GERD.*

Acetate	• Prevent DNA and cell damage induced by H_2O_2.
	• Substrate for cholesterol synthesis.
	• Activation of GPR43
	  • *Induction of immune and inflammatory responses*
	  • *Stimulation of lipid accumulation and anti-lipolysis activity.*

Orally-administered acetate	• Inhibition of lipogenesis in the liver, Increase in faecal bile acid excretion
	  • *Reduction of serum total cholesterol and triacylglycerol*
	  • Improvement of diabetic parameters.

Propionate	• Inhibition of cholesterol synthesis
	  • *Competitive exclusion of acetate transport into hepatocytes*
	  • *Inhibits 3-hydroxy-3-methylglutaryl-CoA reductase & 3-hydroxy-3-methylglutaryl-CoA synthase.*
	• Induction of cell proliferation.
	• Anti-proliferative
	  • *Inhibition of histone deacetylase, induction of p21.*
	• Activation of GPR43 and GPR41
	  • *Induction of immune and inflammatory responses*
	  • *-Stimulation of lipid accumulation and anti-lipolysis activity.*
	• Anti-inflammatory
	  • *Inhibition of chemokine IL-8 production.*

Butyrate	• Induction of cell proliferation. • Anti-proliferative • *Inhibition of histone deacetylase, induction of p21, down-regulation of cell cyclins, Induction of GADD45α & GADD45β, upregulation of VDR, Inhibition of Akt/protein kinase B.*

(continued)

Table 9.1 *(continued).*

Metabolite	Proposed effect on host health

	• Induction of GSTs.
	• Anti-inflammatory
	  • *Induction of cathelicidin peptide LL-37*
	  • *Inhibition of chemokines IL-8, IL-6 and cyclooxygenase-2 production*
	  • *Stimulation of ICAM-1prouction*
	  • *Inhibition of interferon γ/STAT1 signalling*
	  • *Inhibition of TNF-α-induced inflammatory responses, inhibition of NF-κB*
	  • *Induction of PPARα & PPARγ*
	• Regulation of fat absorption and circulating lipoproteins
	  • *Increase in synthesis and secretion of ApoA-IV*
	  • *Decrease the incorporation of acetate into exported cholesterol ester, diminishing delivery of chylomicrons & very low-density lipoproteins.*

propionate, reduce the risk of developing diseases such as colon cancer and cardiovascular disease.

9.2.2 *Production of short-chain fatty acids in the colon*

With a microbial population in the region of 10^{10} to 10^{11} colony forming units (cfu) g^{-1} wet weight (Hill, 1995), the colonic microflora can be considered an organ in itself. The dominant organisms of the colon are anaerobes, including bacteroides (30% of total), bifidobacteria, eubacteria, fusobacteria, streptococci and lactobacilli, as well as enterobacteria in fewer numbers. Non-digestible food components such as carbohydrates (plant cell wall polysaccharides: cellulose, pectins and hemicelluloses) and proteins are a source of substrate for fermentation by such anaerobic microflora, because they are resistant to hydrolysis and digestion in the stomach and small intestine, and eventually enter the colon for fermentation. In addition, other fermentation products such as lactate, ethanol and succinate, which are intermediates in the global fermentation process in the microbiota, are to varying extents metabolised to SCFA by cross-feeding species in the ecosystem (Bernalier *et al.*, 1999; Belenguer *et al.*, 2006). Interestingly, bacterial numbers, fermentation, and proliferation are highest in the proximal colon where substrate availability is greatest (Macfarlane *et al.*, 1992; Macfarlane & Gibson, 1995). Indeed, the total amount of SCFA in the proximal colon is estimated to range from 70 to 140 mM (Cook & Sellin, 1998; Topping & Clifton, 2001; Roberfroid, 2005) and falls to 20 to 70 mM in the distal colon (Topping & Clifton, 2001). Carbohydrates are fermented by saccharolytic bacteria primarily producing linear SCFAs, H_2 and CO_2 (Macfarlane & Macfarlane, 2003; Roberfroid, 2005), whereas proteins and amino acids are fermented by proteolytic bacteria yielding branched SCFAs, H_2, CO_2, CH_4, phenols and amines (Roberfroid, 2005). In contrast to carbohydrates, protein fermentation contributes very little (±5%) of the SCFAs produced in the colon. However, Macfarlane *et al.* (1992) and Macfarlane & Macfarlane (1995) demonstrated both *in vitro* and *in vivo* that long transit times in the large intestine can have profound effects on bacterial physiology and metabolism, leading to protein breakdown and amino acid fermentation, thus increasing the SCFA supply in the colon. Indeed, iso-butyrate, iso-valerate, and 2-methyl butyrate are branched-chain SCFAs arising, respectively, from the fermentation of the amino acids valine, leucine and isoleucine (Topping & Clifton, 2001).

Overall, the rate and amount of SCFA production depends on the numbers and types of microflora present in the colon (Roberfroid, 2005), the substrate source (Cook & Sellin, 1998) and gut transit time (Argenzio & Southworth, 1974; Owens & Isaacson, 1977; Cook & Sellin, 1998). Many host-related factors other than diet also affect bacterial metabolism and SCFA formation, such as ageing, neuroendocrine system activity, stress, pancreatic and other secretions in the digestive tract, mucus production, disease, drugs, antibiotics and epithelial cell turn-over times (Macfarlane & Macfarlane, 2003). Interestingly, chemostat studies using pure cultures of saccharolytic gut microorganisms demonstrated that carbon availability and growth rate strongly affected the outcome of fermentation with acetate and formate being the major bifidobacterial fermentation products formed during growth under carbon limitation, and acetate and lactate being produced when carbon was in excess (Macfarlane & Macfarlane, 2003). In the case of *Bacteroides fragilis*, acetate and succinate were the major fermentation products when carbon was abundant, but succinate was decarboxylated to propionate when carbon and energy sources were limiting (Macfarlane & Macfarlane, 2003). In addition, by comparing the fermentation patterns of different fibre sources by intestinal microflora, Pylkas *et al.* (2005) demonstrated that SCFA production varied among the different fibres with hydrolysed guar gum and galactomannan producing the greatest amounts of total SCFAs. Lactose has also been shown to have a profound effect on the microbial community in a colon simulator, leading to significant increases in SCFA production (Makivuokko *et al.*, 2006). Moreover, oligofructose was shown to selectively stimulate the bacterial conversion of acetate and lactate to butyrate (Morrison *et al.*, 2006). Indeed, such carbohydrates have been termed butyrogenic prebiotics, because of their additional functionality. On the other hand, L-rhamnose was shown to have a profound effect on the levels of serum propionate in humans (Vogt *et al.*, 2004), whereas ascorbose was shown to enhance colonic butyrate production (Weaver *et al.*, 1997). Interestingly, all of these substances are catabolised in a relatively small number of biochemical pathways (Macfarlane & Macfarlane, 2003). Studies of this nature are vital in understanding and developing methods to improve SCFA production in the gut, especially when one considers the vast health effects exuded by SCFAs.

9.2.3 *Role of short-chain fatty acids in health and disease*

SCFAs are rapidly absorbed from the human colon with similar absorption rates in all regions and both active (apical carrier-mediated) and passive (non-ionic) transport have been proposed (Roy *et al.*, 2006). They are then metabolised at three major sites in the body: the colonocytes which utilise butyrate (via β-oxidation) as a major source of energy; the liver cells which metabolise residual butyrate and propionate for glucogenesis, in addition, 50 to 70% of acetate is also taken up by the liver; the muscle cells that generate energy from the oxidation of residual acetate (Roberfroid, 2005). Indeed, the oxidation of SCFA, especially butyrate, supplies the colon with some 60 to 70% of the energy needs of isolated colonocytes (Roediger, 1995) while supplying the body with 7 to 10% of its energy needs (Roy *et al.*, 2006). Moreover, SCFAs have been proposed to play a key role in the maintenance of colonic homeostasis. While SCFAs are relatively weak acids with p*Ka* values of ~4.8, raising their concentrations through fermentation lowers digesta pH, believed to prevent the outgrowth of pH-sensitive pathogenic bacteria, although this is based largely on *in vitro* studies (Topping & Clifton, 2001). Indeed, propionate and formate have been shown to kill *Escherichia coli* and *Salmonella* at high (pH 5) acidity (Cherrington *et al.*, 1991), which has also been reported in other animal studies (Prohaszka *et al.*, 1990). Furthermore, the drop in colonic pH is also known to decrease the solubility of free bile acids, which may decrease the potential tumour promoter activity of secondary bile acids (Grubben *et al.*, 2001) and may also inhibit the colonic bacterial enzyme 7α-dehydroxylase, which degrades primary bile acids to secondary bile acids (Thornton, 1981).

SCFAs have also been shown to exert an effect on blood flow. Rectal infusion of SCFAs into human surgical patients led to a 1.5- to 5.0-fold greater splanchnic blood flow (Mortensen *et al.*, 1991). Similarly, colonic blood flow was increased following infusion of acetate, propionate or butyrate in the denervated canine large bowel (Kvietys & Granger, 1981). This greater rate of blood flow possibly enhances tissue oxygenation and transport of absorbed nutrients (Topping & Clifton, 2001).

It also appears that SCFAs can assist in the management of antibiotic-induced and infectious diarrhoea (Topping & Clifton, 2001; Raghupathy *et al.*, 2006; Roy *et al.*, 2006). By increasing the expression of the apical Na^+/H^- exchanger (Rioux *et al.*, 2005) SCFAs increase the absorption of colonic water and have thus found many applications in the treatment of various diarrhoeal conditions.

SCFAs may also be responsible for enhancing bone health through promotion of calcium absorption in the gut (Griffin *et al.*, 2002; Coxam, 2005; Weaver, 2005). One hypothesis suggests that the low pH induced by SCFAs in the gut leads to increased mineral solubility and hence enhanced calcium absorption, or it may simply be that SCFAs directly increase transcellular calcium absorption (Cashman, 2003) due to enlargement of the gut's absorptive surface as a result of increased proliferation of enterocytes mediated by SCFAs (Scholz-Ahrens *et al.*, 2007). Other studies have concluded that SCFAs contribute to enhanced absorption of calcium via a cation exchange mechanism (Lutz & Scharrer, 1991; Trinidad *et al.*, 1996).

Obesity and diabetes represent two of the most progressive illnesses in Western countries. However, SCFAs may also have a role in the management of both diseases. The addition of the dietary fibre inulin-type fructans to the diet of rats has been shown to promote the secretion of the endogenous gastrointestinal peptide, glucagon-like peptide-1 (GLP-1), which is involved in the regulation of pancreatic secretion of insulin and in the differentiation and maturation of β-cells (Brubaker & Drucker, 2004). While the molecular mechanisms involved remain to be elucidated, other studies have suggested a role of SCFA in the regulation of the expression of the intestinal proglucagon gene in both *in vitro* and *in vivo* studies in animals (Tappenden *et al.*, 1998; Drozdowski *et al.*, 2002).

Intestinal motility is also modulated by SCFAs (Dass *et al.*, 2007) and has been shown to have a beneficial effect on patients with gastroesophageal reflux disease (GERD). Indeed, colonic fermentation of indigestible carbohydrates increased the rate of transient lower esophageal sphincter relaxations, the number of acid reflux episodes and the symptoms of GERD (Piche *et al.*, 2003). The authors speculate that excess release of GLP-1 may account, at least in part, for some of the effects.

Moreover, SCFAs have exhibited protective properties with regard to irritable bowel disease (IBD) and colon carcinogenesis (Whitehead *et al.*, 1987; Mcintire *et al.*, 1993;

Hinnebusch *et al.*, 2002; Galvez *et al.*, 2005). Indeed, several studies have reported that irritable bowel syndrome is associated with impairment in SCFA production. However, the anti-inflammatory efficacy of the individual SCFAs differs substantially, thus each will be discussed in more detail in the next section where the distinct physiological effects of the SCFAs, acetate, propionate and butyrate, are discussed in depth.

Acetate

Acetate is the principal SCFA produced in the colon, but exits the colon to be metabolised by the liver, muscle and other peripheral tissues (Kien *et al.*, 1992; Puchowicz *et al.*, 1999). Indeed, colonic fermentation is generally considered to be the major source of blood acetate (Ballard, 1972; Scheppach *et al.*, 1991) and uptake and use of acetate by many tissues constitute the main route whereby the body obtains energy from undigested and unabsorbed carbohydrates in the small bowel (Roy *et al.*, 2006). Current evidence suggests that colonic acetate provides the body with about 1.5 to 2 kcal g^{-1} (Topping & Clifton, 2001). It has also been demonstrated recently that acetate and butyrate are the preferred carbon sources for *de novo* lipogenesis in isolated rat colonic epithelial cells (Zambell *et al.*, 2003). Moreover, it has been shown that acetate has the ability to prevent DNA and cell damage of freshly isolated colon cells induced by hydrogen peroxide and exhibited the same level of protection as butyrate (Abrahamse *et al.*, 1999). Yet, acetate is also the primary building block for cholesterol synthesis (Bloch, 1965), one of the main factors contributing to the aetiology of cardiovascular disease. When acetate enters the hepatocyte, it is mainly activated by the cytosolic acetyl-CoA synthetase 2, and then enters the cholesterogenesis and lipogenesis pathways (Delzenne & Williams, 2002). Thus, there is considerable evidence that increased availability of colonic acetate raises blood lipids. Indeed, infusion of acetate into the rectum acutely raised serum cholesterol and triacylglycerol concentrations (Wolever *et al.*, 1989, 1991) and significantly more carbon atoms from labelled acetate appeared in serum cholesterol and triacylglycerol after rectal rather than intravenous infusion of [^{13}C]acetate (Wolever *et al.*, 1995). In addition, an increase in serum cholesterol levels in a test group following digestion of the rapidly fermented dietary fibre, lactulose, was also attributed to the increased SCFA concentrations and thus an increase in the hepatic uptake of acetate (Jenkins *et al.*, 1991). In a more recent study, Wolever *et al.* (2002) demonstrated that changes in serum acetate in subjects with type-2 diabetes consuming a high fibre diet over a 6-month period were significantly related to changes in serum concentrations of cholesterol and triacylglycerol, observed between 3 and 6 months of the diet period. On the other hand, less fermentable fibre sources such as psyllium may be very effective in reducing serum lipids (Thacker *et al.*, 1981; Anderson *et al.*, 1988). This is presumably related to their effects in increasing faecal losses of bile acids and may also be due to the production of propionate, which is known to inhibit the incorporation of acetate into cholesterol and fatty acids (Nishina & Freedland, 1990). Indeed, rectal infusions of 180 mmol of propionate had no effect on serum lipids or triglycerides in healthy young men and women (Wolever *et al.*, 1991). However, when 60 mmol of propionate was infused with 180 mmol of acetate, free fatty acids decreased by an additional 10% and negated the increase in total and LDL cholesterol observed when acetate was given alone (Wolever *et al.*, 1991). It is possible that one of the actions of propionate on serum lipids is the ratio of propionate to acetate (Wolever *et al.*, 1991, 1996). This action of propionate could involve a competition between propionate and acetate for the transporter of acetate into hepatocytes (Delzenne & Williams, 2002).

In contrast to the studies cited earlier, recent research has demonstrated that oral administration of acetic acid reduced serum total cholesterol and triacylglycerol in rats fed with a diet containing cholesterol, due to the inhibition of lipogenesis in the liver and the increase in faecal bile acid excretion (Fushimi *et al.*, 2006). In addition, oral administration of acetic acid improved various diabetic parameters in KK-A(y) mice (Sakakibara *et al.*, 2006). The authors postulated that the hypoglycemic effect of the orally administered acetic acid was possibly due to the activation of 5'-AMP-activated protein kinase (AMPK) in the liver, which mediates glucose uptake and free fatty acid oxidation in skeletal muscle and inhibits gluconeogenesis, glycolysis, lipogenesis and cholesterol formation in the liver (Hardie *et al.*, 2003; Hardie, 2004). This phenomenon can be explained by the fact that AMPK is activated by a high AMP/ATP ratio in the cytosol (Kawaguchi *et al.*, 2002). Thus, when acetate is metabolised by acetyl-CoA synthetase, AMP is simultaneously produced: acetate + CoA + ATP ↔ acetyl-CoA + AMP + pyrophosphate (Fushimi *et al.*, 2006). The increase in AMP and the concomitant decrease in ATP activates AMPK, which subsequently inhibits activation of biosynthetic enzymes to conserve ATP (Hardie *et al.*, 1998). It may be possible that absorption and transport of orally administered acetic acid follows a different time course to that of the acid produced by the gut microflora as suggested by Illman *et al.* (1988) with regard to dietary propionate. Likewise, dietary acetic acid may exert its effect following absorption in the upper gut and subsequent clearing by the liver.

Recent evidence has demonstrated that SCFAs are also involved in complex signalling pathways where acetate, propionate and butyrate have been shown to act as specific agonists of the orphan G-protein-coupled receptors (GPCRs), GPR41 and GPR43 (Brown *et al.*, 2003). GPCRs constitute one of the largest gene families identified to date (Bockaert & Pin, 1999). These so far uncharacterised receptors are referred to as orphan GPCRs. Although little information is available concerning the receptors, GPR41, which is more widely distributed in tissue than GPR43, has been shown to induce apoptosis via a p53/Bax pathway (Kimura *et al.*, 2001). In contrast, GPR43 is mainly expressed in leukocyte populations, particularly progenitor neutrophils, suggesting that this receptor could have an important function in the differentiation and/or activation of leukocytes (Senga *et al.*, 2003). Interestingly, acetate was 100-fold less potent on GPR41 as compared with propionate and butyrate, whereas the three ligands activated GPR43 with a similar potency (Le Poul *et al.*, 2003). However, the average concentrations of propionate and butyrate in blood are considered too low to activate GPR41 or GPR43, but the blood concentrations reached by acetate are well within the range for GPR43 (Le Poul *et al.*, 2003). The specific expression of GPR43 in neutrophils suggests an early role in the induction of immune and inflammatory responses, with a possible involvement in inflammatory bowel disease. Neutrophils play a key role in the early steps of a number of inflammatory processes of the gastrointestinal tract. Following migration from the systemic circulation into the mucosal interstitial space, neutrophils undergo activation leading to the perpetuation of the inflammatory response, as well as contributing to the ultimate mucosal injury (Le Poul *et al.*, 2003). Le Poul *et al.* (2003) postulated that GPR43 could potentially be a target for modulation of immune responses in certain pathological situations such as IBD or alcoholism-associated immune depression. In subsequent studies it was demonstrated that SCFAs also activate GPR41 and stimulate leptin production in both cultured adipocytes and whole animals (Xiong *et al.*, 2004). Leptin, an adipose-derived hormone, regulates a wide variety of physiological processes, including feeding behaviour, metabolic rate, sympathetic nerve activity, reproduction and immune response. In contrast, however, Hong *et al.* (2005) demonstrated that acetate and propionate act on lipid accumulation and inhibition of lipolysis in adipose tissue mainly through GPR43 and not GPR41. The study reported that GPR43 plays an important role in adipocyte differentiation and development and, further, that acetate and propionate stimulate lipid accumulation and have anti-lipolysis activity. However, as suggested by the authors, further studies are required to clarify the mechanisms of action of SCFAs in adipogenesis.

Propionate

Unlike acetate, metabolism of propionate in humans is less well understood. However, it is the main glucose precursor in ruminants. Propionate is also a substrate for hepatic gluconeogenesis in humans but appears to have two competing and opposite effects, being both a substrate and an inhibitor (Wong *et al.*, 2006). It has been suggested that the inhibitory effect of propionate may be related to its metabolic intermediaries (Baird *et al.*, 1984). Some human and animal studies have shown that propionate may have a favourable impact on cholesterol levels as discussed previously. This hypothesis has been the subject of several *in vivo* and *in vitro* investigations (Chen *et al.*, 1984; Illman *et al.*, 1988; Nishina & Freedland, 1990; Venter *et al.*, 1990; Wright *et al.*, 1990; Wolever *et al.*, 1991, 1995; Beaulieu & McBurney, 1992; Cameron-Smith *et al.*, 1994; Levrat *et al.*, 1994; Demigne *et al.*, 1995; Lin *et al.*, 1995b; Hara *et al.*, 1998, 1999; Cheng & Lai, 2000; Delzenne & Kok, 2001; Lopez *et al.*, 2001). Observations in animals suggest that propionate inhibits cholesterol synthesis by inhibiting both 3-hydroxy-3-methylglutaryl-CoA synthase and 3-hydroxy-3-methylglutaryl-CoA reductase (Bush & Milligan, 1971; Rodwell *et al.*, 1976; Levrat *et al.*, 1994). However, while data obtained from animal studies have shown convincing lipid-lowering properties for propionate, studies conducted in humans yield more conflicting results (Beylot, 2005); thus further studies are required in the future to resolve this variation.

Interestingly, propionate has also demonstrated proliferative properties when incubated with biopsy specimens of normal human caecal mucosa (Scheppach *et al.*, 1992). Indeed, caecal crypt proliferation was raised in all experiments, with both butyrate and propionate being the most effective, leading to increases in cell proliferation of 89 and 70%, respectively, whereas acetate exerted only a minor effect (31%). It is important at this point to understand the architecture of the normal colonic mucosa. The spatial organisation of the colonic mucosa implies that *in vivo* cells normally undergo a sequence of events that includes proliferation, differentiation, apoptosis and extrusion or cellular shedding (Heerdt *et al.*, 1994). During transformation of colonic mucosa, terminal differentiation is abnormal or blocked and the region of proliferation expands (Lipkin, 1989), almost certainly culminating in chronic inflammation and/or colonic carcinomas. Interestingly, propionate has been shown to cause significant growth suppression on human colonic carcinoma cell lines (Siavoshian *et al.*, 1997; Beyer-Sehlmeyer *et al.*, 2003). However, it is important to express that the effect of propionate is not as marked as that of butyrate. Several studies have indicated that propionate, like butyrate, can inhibit histone deacetylase and, hence, induce histone acetylation (Cousens *et al.*, 1979; Samuels *et al.*, 1980; Kiefer *et al.*, 2006). Such alterations in histones have considerable influence on the organisation of chromatin and on the control of gene expression and cell growth (Grunstein, 1997; Kuo & Allis, 1998). Indeed, inhibition of histone deacetylase and consequent acetylation of histones result in neutralisation of their positive charge, disrupting their ionic interaction with DNA and thereby allowing transcription factors to access and activate specific genes (Grunstein, 1997). Histone hyperacetylation as a result of treatment with propionate and valerate subsequently resulted in changes in the expression of the cell cycle regulators $p21^{WAF1/Cip1}$ and cyclin B1 (CB1) (Hinnebusch *et al.*, 2002). Protein cyclins such as CB1 are coupled with cyclin-dependent kinase (Cdk) and accumulate throughout the cell cycle, whereas $p21^{WAF1/Cip1}$ is a Cdk inhibitor (Morgan, 1995) that brings about a block in the cell cycle at G1, allowing DNA checkpoint-mediated repair of genomic instability or mutations (Young *et al.*, 2005). Interestingly, human colon cancers have been shown to express abnormally high levels of CB1 (Wang *et al.*, 1997). In another study, the therapeutic effects of propionic acid were demonstrated following the accelerated healing of chemically induced colitis in rats that were administered whey culture containing *Propionibacterium freudenreuchii* (Uchida & Mogami, 2005) where propionic acid was considered the active substance. Indeed, the low pH generated by *P. freudenreuchii* as a result of the production of propionate and acetate has been shown to have a major impact on the mode of propionibacterial-SCFA-induced cell death (Lan *et al.*, 2006). Propionibacterial-SCFA-triggered apoptosis was in the pH range 5.5 to 7.5. At the latter pH, SCFA induced cell cycle arrest in the G2/M phase, followed by a sequence of cellular events characteristic of apoptosis. At the lower pH of 5.5, the effect was even more rapid and drastic.

As discussed previously, propionate like acetate has been shown to stimulate the orphan GPCRs, GPR41 and GPR43 (Brown *et al.*, 2003; Le Poul *et al.*, 2003; Xiong *et al.*, 2004; Hong *et al.*, 2005). However, further studies are required to determine the exact physiological effects mediated through the SCFA activation of these receptors.

Butyrate: introduction

Of all the SCFAs produced in the colon, butyrate is the most interesting for its array of effects on the colonic mucosa. Butyrate serves as the principal energy source for colonic epithelial cells, but is also known to regulate a number of genes associated with cell differentiation, proliferation and apoptosis, and thus has a profound influence on the aetiology of various colonic diseases (Scheppach & Weiler, 2004). On one hand butyrate induces cell growth and proliferation and on the other hand, induces growth arrest and apoptosis, a phenomenon known as the 'butyrate paradox'. This discrepancy may be explained by differences between *in vitro* and *in vivo* environments, the timing of butyrate administration, the source of butyrate and interaction with dietary fibre (Lupton, 2004). Interestingly, rats that were administered increasing amounts of wheat bran from 0 to 20% demonstrated an inverse correlation between luminal butyrate levels and colonic cell proliferation and a positive linear correlation between luminal butyric acid levels and colon epithelial cell histone acetylation (Boffa *et al.*, 1992).

Because butyrate is known to nourish the colonic mucosa and prevent cancer of the colon by promoting cell differentiation, cell cycle arrest and apoptosis, much effort has focused on identifying butyrate-producing bacteria in the colon. Butyrate production in the human colon is widely distributed among bacteria belonging to Clostridial clusters IV and V including many undescribed species related to *Eubacterium, Roseburia, Faecalibacterium* and *Coprococcus*, which are highly oxygen-sensitive anaerobes (Pryde *et al.*, 2002). Interestingly, cluster-specific probes used to identify butyrate-producing bacteria in faecal samples from ten individuals detected bacteria related to *Roseburia intestinalis*, *Faecalibacterium prausnitzii* and *Eubacterium hallii* at mean populations of 2.3, 3.8 and 0.6%, respectively, of the total human faecal microbiota (Hold *et al.*, 2003), where *E. hallii* has been shown to utilise lactate as a substrate for butyrate production (Duncan *et al.*, 2004). Recently, a CoA-transferase gene, butyryl-CoA transferase, was identified from *Roseburia* species, which was shown to play a key role in butyrate formation in the human colon (Charrier *et al.*, 2006). Moreover, bacteria that are able to produce a higher amount of butyrate are less prone to metabolising primary bile acids to secondary bile acids, the latter of which are considered tumour promoters (Zampa *et al.*, 2004).

The cellular effect of butyrate and cancer

Several studies have yielded results demonstrating the anti-proliferative and apoptotic properties of butyrate (Whitehead *et al.*, 1987; Barnard & Warwick, 1993; Hague *et al.*, 1993, 1995; Heerdt *et al.*, 1994; Kiefer *et al.*, 2006). Butyrate is transported across the colonic luminal membrane via MCT1 (monocarboxylate transporter 1), which is significantly down-regulated during the transition from normality to malignancy in the human colon (Lambert *et al.*, 2002) and exerts its anti-proliferative effect upstream of acetyl-CoA synthesis (Leschelle *et al.*, 2000). Butyrate inhibits histone deacetylase resulting in hyperacetylation of histones H3 and H4 (Sealy & Chalkley, 1978). Like propionate, butyrate induces the cell cycle regulator $p21^{WAF1/Cip1}$ (Archer *et al.*, 1998; Chai *et al.*, 2000; Siavoshian *et al.*, 2000), which has been shown to involve a large complex of factors including a Krüppel-type zinc finger transcription factor ZBP-89, Sp family members and coactivators (Bai & Merchant, 2000). In addition, it down-regulates CB1 expression (Hinnebusch *et al.*, 2002; Archer *et al.*, 2005). Interestingly, it was demonstrated that butyrate and the histone deacetylase inhibitor, trichostatin (TSA), both modulated the expression of 23 genes in a colorectal cancer cell line (Della Ragione *et al.*, 2001) where butyrate was shown to block cells mainly at the G0/G1 phase of the cell cycle, and arrested cells at the G2 phase. Even so, both down-regulated the cyclins D1 and D3 on the K562 cell line (Li *et al.*, 2003). However, as research into the area has progressed, the extent of the molecular effects of butyrate has been further elucidated. Both butyrate and TSA were shown to induce GADD45α and β gene expression in human colon carcinoma cells, genes known to be involved in cell growth arrest and apoptosis (Chen *et al.*, 2002). In addition, butyrate-induced growth arrest in Caco-2 cells, a human colon cancer cell line, was mediated by up-regulation of the vitamin D receptor (VDR), which was followed by induction of p21 and down-regulation of cdk 6 and cyclin A, both of which are involved in cell cycle progression (Gaschott & Stein, 2003). More recently, butyrate was shown to induce apoptosis in human cancer cells through inhibition of Akt/protein kinase B (Chen *et al.*, 2006). Akt is a serine/threonine protein kinase, which, when constitutively active, can block apoptosis (Kennedy *et al.*, 1997) through phosphorylating transcription factors that control the expression of pro-survival and anti-apoptotic genes.

However, global gene expression profiles have probably provided the greatest amount of data indicating the extent of the effect of butyrate on cellular gene expression. Indeed, gene expression profiles of bovine kidney epithelial cells demonstrated that 450 genes, relating to multiple signalling pathways involved in cell cycle and apoptosis, were significantly regulated by sodium butyrate (Li & Li, 2006). Moreover, Daly and Shirazi-Beechey (2006) examined the effect of butyrate on 19 400 genes in a human colonic epithelial cell line and found 221 butyrate-responsive genes. Of the 59 up-regulated and the 162 down-regulated genes, all are involved in the key processes of proliferation, differentiation and apoptosis, and many of the genes were found to be oppositely expressed to that reported in colon cancer tissue.

Although a small proportion of colorectal tumours are caused by inherited genetic alterations (Lynch & de la Chapelle, 2003), the greatest number of tumours are sporadic and probably the result of life-long accumulation of genetic alterations in somatic tissues (Fearson & Vogelstein, 1990; Fodde *et al.*, 2001). The ability to prevent DNA damage by inactivating dietary carcinogens contributes to an individual's risk of developing cancer (Pool-Zobel *et al.*, 2005). A family of enzymes that plays an important role in detoxification is glutathione S-transferases (GSTs; EC 2.5.1.18), which catalyse the conjugation of many electrophilic compounds with reduced glutathione (GSH). It is thought that most GST substrates are xenobiotics or products of oxidative stress, including some environmental carcinogens (Hayes & Pulford, 1995). Pool-Zobel (2005) demonstrated that butyrate is an efficient inducer of several GSTs in cells from all three stages of malignancy: primary human colon tissue, premalignant LT97 adenoma and HT29 tumour cells. However, this situation may give rise to the 'double-edged sword' effect of butyrate. While induction of GSTs in normal cells is straightforward and favourable, GST induction in tumour and adenoma cells could counteract cancer chemotherapy by causing resistance to therapeutic agents, thus enhancing the survival of transformed cells (Ebert *et al.*, 2001). However, according to authors Pool-Zobel *et al.* (2005) this adverse situation may not be possible *in vivo*, since the luminal millimolar concentrations of butyrate could be much too high to result in GST induction and would instead impair tumour cell or adenoma cell growth, decreasing the availability of such cells for GST induction.

While the effects of butyrate *in vitro* are extremely promising, there are no human studies of the effect *in vivo* on colorectal neoplasia and animal studies are few in number. A review of the evidence to date by Sengupta *et al.* (2006) indicated that while an adequate amount of butyrate to the appropriate site protects against tumourigenic events, new directions for investigation and research are needed, focusing on risk stratification and the development of strategies for butyrate delivery in humans.

Butyrate and inflammatory bowel disease

The term IBD comprises two closely related pathologies, ulcerative colitis (UC) and Crohn's disease (CD), which are characterised by chronic and spontaneously relapsing inflammation of the gut (Galvez *et al.*, 2005). This exacerbated inflammatory response results from an inappropriate reaction towards a luminal agent, most probably driven by the intestinal microflora (Farrell & Peppercorn, 2002). Interestingly, Schauber *et al.* (2003) found that induction of the antimicrobial peptide cathelicidin LL-37, produced by the single layer of the colonic epithelium, in four colon carcinoma cell lines was more pronounced in response to *n*-butyrate than to the other SCFAs. Moreover, impaired butyrate metabolism has been identified in patients with UC (Roediger, 1980; Chapman *et al.*, 1994, 1995) but this result has not been observed ubiquitously (Clausen & Mortensen, 1995; Allan *et al.*, 1996). However, Den Hond *et al.* (1998) found that UC patients had decreased colonic butyrate oxidation compared to patients in remission. The anti-inflammatory effects of butyrate have been studied in various *in vitro* systems. Studies performed in HT-29 cells have indicated that both butyrate and propionate are able to inhibit the production of the chemokine, IL-8, when stimulated by lipopolysaccharides (LPSs) (Rodriguez-Cabezas *et al.*, 2002). Similarly, Ogawa *et al.* (2003) demonstrated that butyrate inhibited IL-6 and cyclooxygenase-2 expression in response to LPS stimulation and augmented the expression of the intracellular adhesion molecule (ICAM-1) in human intestinal microvascular endothelial cells, the latter of which has been described to play an important role in the pathogenesis of different inflammatory conditions, including IBD (Rutgeerts *et al.*, 2003). In HCT116 and Hke colon cancer cells, butyrate was shown to inhibit interferon γ/STAT1 signalling (Klampfer *et al.*, 2003). In addition, butyrate has also been shown to block the TNF-α-induced inflammatory responses (Andoh *et al.*, 2003). It is believed that the mechanisms by which butyrate inhibits TNF-α may be closely associated with inhibition of the transcription factor NF-κB, which is known to activate a variety of genes involved in immune, inflammatory and anti-apoptotic responses (Andoh *et al.*, 2003). Indeed, treatment of UC patients with butyrate enemas for 4 to 8 weeks resulted in a significant inhibition of NF-κB activation and correlated with minor mucosal inflammation (Luhrs *et al.*, 2002). Butyrate has also been shown to enhance expression of the nuclear lipid-responsive receptors, peroxisome proliferator-activated receptors (PPAR), PPARα (Zapolska-Downar *et al.*, 2004) and PPARγ (Wachtershauser *et al.*, 2000). PPARs control a variety of genes in several pathways of lipid metabolism (Desvergne & Wahli, 1999) and have gained considerable interest recently as a crucial target for the control of inflammation associated with IBD (Katayama *et al.*, 2003; Bassaganya-Riera *et al.*, 2004).

While evidence of the beneficial effects of butyrate in human trials is limited, Vernia *et al.* (2003) demonstrated that the combined treatment of butyrate with conventional therapy, 5-ASA/steroid, was significantly more effective in the management of refractory distal colitis in UC patients than 5-ASA alone. In conclusion, butyrate provides great promise as a therapeutic for the treatment of IBD patients through the modulation of the immune response and control of the intestinal microbial balance.

Butyrate and lipid metabolism

It has been demonstrated that butyrate has a role in regulating fat absorption and circulating lipoproteins. Nazih *et al.* (2001) demonstrated that butyrate significantly increased the synthesis and secretion of the major glycoprotein component apolipoprotein A-IV (ApoA-IV) in Caco-2 cells, resulting in a 20 to 30% increase in cholesterol efflux. Indeed, several functions have been assigned to ApoA-IV including promotion of cholesterol efflux (Stein *et al.*, 1986; von Eckardstein *et al.*, 1995), activation of lecithin:cholesterol acyltransferase (Emmanuel *et al.*, 1994), modulation of the activation of lipoprotein lipase by ApoCII (Goldberg *et al.*, 1990) and regulation of food intake (Fujimoto *et al.*, 1993), among others. Furthermore, overexpression of either human or mouse *apoA-IV* in transgenic mice confers significant protection against diet-induced atherosclerosis in cholesterol fed animals and *apoE*-deficient mice (Duverger *et al.*, 1996; Cohen *et al.*, 1997). Additional research indicated that butyrate regulated intestinal fat absorption and circulation of lipoproteins in Caco-2 cells by reducing secreted triglycerides and phospholipids, decreasing the incorporation of acetate into exported cholesteryl ester while diminishing the delivery of chylomicrons and very low density lipo-proteins LDL (Marcil *et al.*, 2002, 2003).

Conclusion

While *in vitro* studies and indeed some animal studies have provided promising evidence for the therapeutic effects of acetate, propionate and in particular butyrate in the various pathological disorders including inflammatory and cancer conditions of the colon and the control of lipid metabolism, future research should now focus on carefully designed human intervention trials.

9.3 Gamma amino butyric acid

9.3.1 *Introduction*

While it may not fall directly into the category of short-chain fatty acid, GABA, also a product of microbial metabolism, warrants discussion due to its importance to various organisms, from single cells to mammalians. Indeed, GABA is the major inhibitory neurotransmitter of the vertebrate CNS (Roberts & Frankel, 1950). It is produced primarily from the irreversible α-decarboxylation of L-glutamate by the enzyme glutamate decarboxylase (GAD), resulting in the ω-amino acid, which is unable to form peptide links (figure 9.1). Interestingly, GAD, a pyridoxal 5'-phosphate (PLP)-dependent enzyme, has been found in bacteria, animals and higher plants (Ueno, 2000). In both *E. coli* and *Shigella*, GAD, along with arginine decarboxylase, has been shown to play a role in acid resistance (Lin *et al.*, 1995a; Castanie-Cornet *et al.*, 1999). In the case of both enzymes, amino acid decarboxylation consumes a proton at low pH_i, after which a membrane antiporter exchanges the resulting product for the amino acid substrate in the medium (Bearson *et al.*, 1997). Several studies have also reported the presence of a *gad* gene in lactic acid bacteria (LAB) (Ueno *et al.*, 1997; Nomura *et al.*, 1998,1999a, 1999b; Park & Oh, 2004). Indeed, *Lactococcus lactis* has been shown to encode a glutamate-dependent acid resistance mechanism similar to that of *E. coli* and *Shigella* (Sanders *et al.*, 1998). In this instance, both the decarboxylase, *gadB* and the antiporter, *gadC* are transcribed from a chloride-inducible promoter. Furthermore, GABA is also produced by cheese starters during cheese ripening (Nomura *et al.*, 1998). Interestingly, Higuchi *et al.* (1997) demonstrated that decarboxylation of glutamate by *Lactobacillus* strain E1 was also associated with the synthesis of ATP, suggesting that a proton motive force arises from the cytoplasmic proton consumption that accompanies glutamate decarboxylation. More recently, a novel *gad* gene was isolated from *Lactobacillus brevis* OPK-3, which consisted of an open reading frame of 1401 bp encoding a protein of 467 amino acid residues (Park & Oh, 2007).

However, as previously mentioned, GABA is also the most abundant and well-studied inhibitory neurotransmitter in the vertebrate CNS; in addition, both GABA and GABA receptors have been found in several peripheral tissues (Erdo, 1985; Erdo & Wolff, 1990;

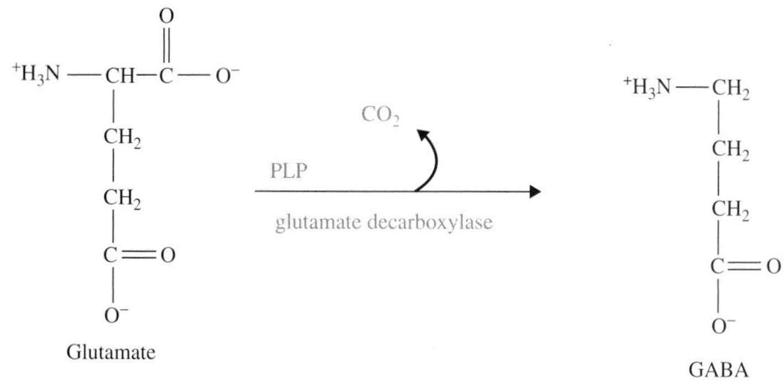

$^{+}H_3N—CH—C(=O)—O^{-}$, CH_2, CH_2, $C=O$, O^{-}

Glutamate

CO_2

PLP

glutamate decarboxylase

$^{+}H_3N—CH_2$, CH_2, CH_2, $C=O$, O^{-}

GABA

Fig. 9.1 GABA production from glutamate by the enzyme glutamate decarboxylase (pyridoxal 5′ -phosphate - PLP).

Ong & Kerr, 1990). Nerve cells that synthesise, store, specifically accumulate and release GABA are defined as GABAergic neurons (Erdo & Wolff, 1990). Interestingly, it has been estimated that 17% of the synapses in the mammalian brain are GABAergic (Halasy & Somogyi, 1993). Since many clinical conditions including psychiatric disorders appear to involve an imbalance in excitation and inhibition (Olsen, 2002), the GABA system is the target of a wide range of drugs active on the CNS, including anxiolytics, sedative-hypnotics, general anaesthetics and anticonvulsants (Macdonald & Olsen, 1994). In addition, GABA is involved in the regulation of cardiovascular conditions, such as blood pressure and heart rate, and plays a role in the sensations of pain and anxiety (Mody *et al.*, 1994). Unique among neurotransmitters, GABA is an electroneutral zwitterion at physiological pH (Mody *et al.*, 1994). It exerts its inhibitory effect on three types of receptors, the ionotropic $GABA_A$ (Bormann, 1988; Gage, 1992) and $GABA_C$ receptors (Zhang *et al.*, 1995; Chebib, 2004; Schlicker *et al.*, 2004), which are directly coupled to chloride channels and mediate fast synaptic inhibition, and the metabotropic $GABA_B$ receptor (Bowery, 1993; Couve *et al.*, 2000). Indeed, GAD is considered unique among enzymes involved in neurotransmitter synthesis because both its substrate and product exhibit opposite actions: L-glutamate is the major neurotransmitter for fast excitatory synaptic transmission, whereas GABA is the major inhibitory neurotransmitter (Ueno, 2000).

9.3.2 *Gamma amino butyric acid effects*

GABA exerts numerous positive effects on the body. Several studies have examined the role of GABA in the control of growth hormone secretion (Cavagnini *et al.*, 1980a, 1980b; Racagni *et al.*, 1982; McCann *et al.*, 1984; McCann & Rettori, 1986; Volpi *et al.*, 1997; Aquilar *et al.*, 2005) where GABA has been shown to play a stimulatory role. Indeed, oral administration of GABA in healthy volunteers has shown rapid increases in circulating human growth hormone (Cavagnini *et al.*, 1980b).

GABA has also been shown to have an effect on the modulation of cardiovascular function (Stanton, 1963; Gillis *et al.*, 1980; Giuliani *et al.*, 1986). Indeed, GABA has been shown to reduce blood pressure in both animals and humans (Takahashi *et al.*, 1955; Elliott & Hobbiger, 1959; Billingsley & Suria, 1982). Interestingly, the blood–brain barrier is impermeable to GABA (Kuriyama & Sze, 1971); thus the antihypertensive effects seen following i.p. or i.v. administration of GABA are due to its actions within the peripheral tissues. Hayakawa *et al.* (2002) demonstrated that the antihypertensive effect of GABA in hypertensive rats was due to its inhibition of noradrenaline release from sympathetic nerves in the mesenteric arterial bed via presynaptic $GABA_B$ receptors.

In addition, GABA has been shown to exert a protective effect against glycerol-induced acute renal failure in rats (Kim *et al.*, 2004). Oh *et al.* (2003) also demonstrated that brown rice extracts containing a high level of GABA may have a nutraceutical role in the recovery from and prevention of chronic alcohol-related diseases.

Anti-proliferative properties have also been established for GABA. Indeed, GABA reduced the induced migratory activity of SW480 colon carcinoma cells (Joseph *et al.*, 2002). Interestingly, this inhibitory effect was mediated by the $GABA_B$ receptor and coincided with a decrease in the intracellular cyclic AMP concentration. Other studies have also yielded similar promising results. Indeed, Oh & Oh (2004) demonstrated that brown rice extracts with enhanced levels of GABA had an inhibitory action on leukaemia cell proliferation and a stimulatory action on cancer cell apoptosis in both human and mouse cancer cell lines.

GABA has also been shown to exert a stimulatory effect on the immune response (Oh *et al.*, 2003). Brown rice extracts with enhanced levels of GABA were shown to increase the viability of mesenteric lymph node cells up to 55 to 110% of the control. The authors postulated that germinated brown rice was effective for the improvement of immunoregulatory action, possibly due to the enhanced levels of GABA and/or the combined effects of several components including GABA. Moreover, Abdou *et al.* (2006) investigated the effect of orally administered GABA on relaxation and immunity during stressful conditions in humans. The authors demonstrated that GABA, which was produced by natural fermentation by a specific LAB strain, could work as a natural relaxant, exerting its effect within 1 hour of administration to diminish stress, worry and anxiety. Furthermore, GABA could elevate the immune response through its relaxant and anxiolytic effects.

Interestingly, GABA is also found in the gastrointestinal tract (Glassmeier *et al.*, 1998). Studies by Thwaites *et al.* (2000) demonstrated that GABA can undergo rapid transepithelial transport across the intestinal wall via a pH-dependent and Na^+-independent H^+/GABA symporter, providing a potential route for oral absorption of GABA. Knowledge of this nature, in addition to the ability of LAB to produce GABA from glutamic acid, has spurred research into the development of fermented dairy products containing GABA. Inoue *et al.* (2003) developed a fermented milk product containing GABA using two kinds of starters, *Lb. casei* strain Shirota, which hydrolysed milk protein into glutamic acid, and *L. lactis* YIT 2027, which converted glutamic acid into GABA. Indeed, the fermented milk product was shown to have a blood pressure-lowering effect in spontaneously hypertensive rats (Hayakawa *et al.*, 2004) and in mildly hypertensive people (Inoue *et al.*, 2003). Several studies have thus investigated means of enhancing microbial production of GABA. Indeed, GABA production increased significantly in *Bifidobacterium longum* following expression of a rice glutamate decarboxylase gene in the strain (Park *et al.*, 2005). Likewise, a *gad* gene from *Lb. brevis* was successfully expressed in a *Bacillus subtilis* strain used in the production of Chungkukjang, a traditional Korean fermented soybean product, resulting in a product with a significantly higher amount of GABA than the control (Park & Oh, 2006). More recently, Park & Oh (2007) developed a yoghurt with enhanced levels of GABA through the use of the lactic acid bacterium, *Lb. brevis* OPY-1 and germinated soybean.

In conclusion, both the functional and nutritive value of foods can be vastly enhanced through the addition of GABA-producing LAB. In addition, orally administered GABA is a highly attractive therapeutic considering that it exerts a positive effect on the most prominent illnesses of modern society, that is, hypertension, stress and cancer.

9.4 Overall conclusion

Research to date has certainly demonstrated that the microbial metabolites discussed in this review have the potential to enhance host health and even alleviate disease. However, there are issues remaining that must be tackled before one can assume the medicinal capabilities of these agents. While both animal studies and human cell culture studies have provided promising results, the outcome of human intervention trials on a global scale remains to be seen. A healthy gut flora, derived from an adequate dietary supply of appropriate prebiotic substances, should ensure that the host has an adequate supply of SCFAs delivered to the target site, that is, the colon, whereas oral administration of such bioactives does not guarantee targeted delivery. In this nutrigenomics era, there is no doubt that bioactive microbial metabolites will hold an important role for individualised health and nutritional benefit.

9.5 Acknowledgements

This work was funded by the Science Foundation Ireland (SFI) funds of the Irish Government under the National Development Plan 2000–2006.

References

Abdou, A. M., Higashiguchi, S., Horie, K., Kim, M., Hatta, H. & Yokogoshi, H. (2006) Relaxation and immunity enhancement of γ-aminobutyric acid (GABA) administration in humans. *BioFactors*, **26**, 201–208.

Abrahamse, S. L., Pool-Zobel, B. L. & Rechkemmer, G. (1999) Potential of short chain fatty acids to modulate the induction of DNA damage and changes in the intracellular calcium concentration by oxidative stress in isolated rat distal colon cells. *Carcinogenesis*, **20**, 629–634.

Allan, E. S., Winter, S., Light, A. M. & Allan, A. (1996) Mucosal enzyme activity for butyrate oxidation: no defect in patients with ulcerative colitis. *Gut*, **38**, 886–893.

Anderson, J. W., Zettwoch, N., Feldman, T., Tietyen-Clark, J., Oeltgen, P. & Bishop, C. W. (1988) Cholesterol-lowering effects of psyllium hydrophillic mucilloid. Adjunct therapy to a prudent diet for hypercholesterolemic men. *Archives of Internal Medicine*, **148**, 292–296.

Andoh, A., Tsujikawa, T. & Fujiyama, Y. (2003) Role of dietary fibre and short chain fatty acids in the colon. *Current Pharmaceutical Design*, **9**, 347–358.

Aquilar, E., Tena-Sempere, M. & Pinilla, L. (2005) Role of excitatory amino acids in the control of growth hormone secretion. *Endocrine Reviews*, **28**, 295–302.

Archer, S. Y., Johnson, J., Kim, H.-J., Ma, Q., Mou, H., Daesety, V., Meng, S. & Hodin, R. A. (2005) The histone deacetylase inhibitor downregulates cyclin B1 gene expression via a p21/WAF-1-dependent mechanism in human colon cancer cells. *American Journal of Physiology. Gastrointestinal and Liver Physiology*, **289**, 696–703.

Archer, S. Y., Meng, S., Shei, A. & Hodin, R. A. (1998) p21^{WAF1} is required for butyrate-mediated growth inhibition of human colon cancer cells. *Proceedings of the National Academy of Sciences of the United States of America*, **95**, 6791–6796.

Argenzio, R. A. & Southworth, M. (1974) Sites of organic acid production and absorption in gastrointestinal tract of the pig. *American Journal of Physiology*, **226**, 454–460.

Bai, L. & Merchant, J. L. (2000) Transcription factor ZBP-89 cooperates with histone acetyltransferase p300 during butyrate activation of p21^{waf1} transcription in human cells. *The Journal of Biological Chemistry*, **275**, 30725–30733.

Baird, G. D., Lomax, M. A., Symonds, H. W. & Shaw, S. R. (1980) Net hepatic and splanchnic metabolism of lactate, pyruvate and propionate in dairy cow *in vivo* in relation to lactation and nutrient supply. *The Biochemical Journal*, **186**, 47–57.

Ballard, F. J. (1972) Supply and utilisation of acetate in mammals. *The American Journal of Clinical Nutrition*, **25**, 773–779.

Barnard, J. A. & Warwick, G. (1993) Butyrate rapidly induces growth inhibition and differentiation in HT-29 cells. *Cell Growth and Differentiation*, **4**, 495–501.

Bassaganya-Riera, J., Reynolds, K., Martino-Catt, S., Cui, Y., Hennighausen, L., Gonzalez, F., Rohrer, J., Benninghoff, A. U. & Hontecillas, R. (2004) Activation of PPAR-γ and -δ by conjugated linoleic acid mediates protection from experimental inflammatory bowel disease. *Gastroenterology*, **127**, 777–791.

Bearson, S., Bearson, B. & Foster, J. W. (1997) Acid stress responses in enterobacteria. *FEMS Microbiology Letters*, **147**, 173–180.

Beaulieu, K. E. & McBurney, M. I. (1992) Changes in pig serum lipids, nutrient digestibility and sterol excretion during cecal infusion of propionate. *The Journal of Nutrition*, **122**, 214–245.

Belenguer, A., Duncan, S. H., Calder, A. G., Holtrop, G., Louis, P., Lobley, G. & Flint, H. J. (2006) Two routes of metabolic cross-feeding between *Bifidobacterium adolescentis* and butyrate-producing anaerobes from the human gut. *Applied and Environmental Microbiology*, **72**, 3593–3599.

Belury, M. A. (2002) Inhibition of carcinoegenesis by conjugated linoleic acid: potential mechanisms of action. *The Journal of Nutrition*, **132**, 2995–2998.

Bernalier, A., Dore, J. & Durund, M. (1999) Biochemistry of fermentation. *Colonic Microbiota, Nutrition and Health*, (eds. G. R. Gibson & M. B. Roberfroid), pp. 37–53, Kluwer Academic Publishers, Dordrecht.

Beyer-Sehlmeyer, G., Glei, M., Hartmann, E., Hughes, R., Persin, C., Bohm, V., Rowland, I., Schubert, R., Jahreis, G. & Pool-Zobel, B. L. (2003) Butyrate is only one of several growth inhibitors produced during gut flora-mediated fermentation of dietary fibre sources. *The British Journal of Nutrition*, **90**, 1057–1070.

Beylot, M. (2005) Effects of inulin-type fructans on lipid metabolism in man and in animal models. *The British Journal of Nutrition*, **93**, S163–S168.

Bhattacharya, A., Banu, J., Rahman, M., Causey, J. & Fernandes, G. (2006) Biological effects of conjugated linoleic acids in health and disease. *The Journal of Nutrition*, **17**, 789–810.

Billingsley, M. L. & Suria, A. (1982) Effects of peripherally administered GABA and other amino acids on cardiopulmonary responses in anaesthetized rats and dogs. *Archives Internationales de Pharmacodynamie et de Therapie*, **255**, 131–140.

Bloch, K. (1965) The biological synthesis of cholesterol. *Science*, **150**, 19–28.

Bockaert, J. & Pin, J. P. (1999) Molecular tinkering of G-coupled protein receptors: an evolutionary success. *The EMBO Journal*, **18**, 1723–1729.

Boffa, L. C., Lupton, J. R., Mariani, M. R., Ceppi, M., Newmark, H. L., Scalmati, A. & Lipkin, M. (1992) Modulation of colonic epithelial cell proliferation, histone acetylation, and luminal short chain fatty acids by variation of dietary fibre (wheat bran) in rats. *Cancer Research*, **52**, 5906–5912.

Bormann, J. (1988) Electrophysiology of $GABA_A$ and $GABA_B$ receptor subtypes. *Trends in Neurosciences*, **11**, 112–116.

Bowery, N. G. (1993) $GABA_B$ receptor pharmacology. *Annual Review of Pharmacology and Toxicology*, **33**, 109–147.

Brown, A. J., Goldsworthy, S. M., Barnes, A., Eilert, M. E., Tcheang, L., Daniels, D., Miuir, A. I., Wiggelsworth, M. J., Kinghorn, I., Fraser, N. J., Pike, N. B., Strum, J. C., Steplewski, K. M., Murdock, P. R., Holder, J. C., Marshall, F. H., Szekeres, P. G., Wilson, S., Ignar, D. M., Foord, S. M., Wise, A. & Dowell, S. J. (2003) The orphan G protein coupled receptors GPR41 and GPR43 are activated by propionate and other short chain carboxylic acids. *The Journal of Biological Chemistry*, **278**, 11312–11319.

Brubaker, P. L. & Drucker, D. J. (2004) Glucagon-like peptides regulate cell proliferation and apoptosis in the pancreas, gut and central nervous system. *Endocrinology*, **145**, 2653–2659.

Bruzzese, E., Volpicelli, M., Squaglia, M., Tartaglione, A. & Guarino, A. (2006) Impact of prebiotics on human health. *Digestive Liver Disease*, **38**, S283–S287.

Bush, R. S. & Milligan, L. P. (1971) Study of the mechanism of inhibition of ketogenesis by propionate in bovine liver. *Canadian Journal of Animal Science*, **51**, 121–127.

Cameron-Smith, D., Collier, G. R. & O' Dea, K. (1994) Effect of propionate on *in vivo* carbohydrate metabolism in streptozotocin-induced diabetic rats. *Metabolism*, **43**, 728–734.

Cashman, K. (2003) Prebiotics and Ca bioavailability. *Current Issues in Intestinal Microbiology*, **4**, 21–32.

Castanie-Cornet, M.-P., Penfound, T. A., Smith, D., Elliott, J. F. & Foster, J. W. (1999) Control of acid resistance in *Escherichia coli*. *Journal of Bacteriology*, **181**, 3525–3535.

Cavagnini, F., Invitti, C., Pinto, M., Maraschini, C., Di Landro, A., Dubini, A. & Marelli, A. (1980b) Effect of acute and repeated administration of gamma aminobutyric acid (GABA) on growth hormone and prolactin secretion in man. *Acta Endocrinologica*, **93**, 149–154.

Cavagnini, F., Benetti, G., Invitti, C., Ramella, G., Pinto, M., Lazza, M., Dubini, A., Marelli, A. & Muller, E. E. (1980a) Effect of gamma-aminobutyric acid on growth hormone and prolactin secretion in man: influence of pimozide and domperidone. *The Journal of Clinical Endocrinology and Metabolism*, **51**, 789–792.

Chai, F., Evdokiou, A., Young, G. P. & Zalewski, P. D. (2000) Involvement of $p21^{WAF1/Cip1}$ and its cleavage by DEVD-caspase in apoptosis of colorectal cancer cells by butyrate. *Carcinogenesis*, **21**, 7–14.

Chapman, M. A., Grahn, M. F., Boyle, M. A., Hutton, M., Rogers, J. & Williams, N. S. (1994) Butyrate oxidation is impaired in the colonic mucosa of sufferers of quiescent ulcerative colitis. *Gut*, **35**, 73–76.

Chapman, M. A., Grahn, M. F., Hutton, M. & Williams, N. S. (1995) Butyrate metabolism in the terminal ileal mucosa of patients with ulcerative colitis. *The British Journal of Surgery*, **82**, 36–38.

Charrier, C., Duncan, G. J., Reid, M. D., Rucklidge, G. J., Henderson, D., Young, P., Russell, V. J., Aminov, R. I., Flint, H. J. & Louis, P. (2006) A novel class of CoA-transferases involved in short-chain fatty acid metabolism in butyrate-producing colonic bacteria. *Microbiology*, **152**, 179–185.

Chebib, M. (2004) $GABA_C$ receptor ion channels. *Clinical and Experimental Pharmacology and Physiology*, **31**, 800–804.

Chen, J., Ghazawi, F. M., Bakkar, W. & Li, Q. (2006) Valproic acid and butyrate induce apoptosis in human cancer cells through inhibition of gene expression of Akt/protein kinase B. *Molecular Cancer*, **5**, 20.

Chen, W. J., Anderson, J. W. & Jennings, D. (1984) Propionate may mediate the hypocholesterolemic effects of certain soluble plant fibres in cholesterol-fed rats. *Proceedings of the Society for Experimental Biology and Medicine*, **175**, 215–218.

Chen, Z., Clark, S., Birkeland, M., Sung, C.-M., Lago, A., Liu, R., Kirkpatrick, R., Johanson, K., Winkler, J. D. & Hu, E. (2002) Induction and superinduction of growth arrest and DNA damage gene 45 (GADD45) α and β messanger RNAs by histone deacetylase inhibitors trichostatin A (TSA) and butyrate in SW260 human colon carcinoma cells. *Cancer Letters*, **188**, 127–140.

Cheng, H. H. & Lai, M. H. (2000) Fermentation of resistant starch produces propionate reducing serum and hepatic cholesterol in rats. *The Journal of Nutrition*, **130**, 1991–1995.

Cherrington, C. A., Hinton, M., Pearson, G. R. & Chopra, I. (1991) Short-chain organic acids at pH 5.0 kill *Escherichia coli* and *Salmonella* spp. without causing membrane perturbation. *The Journal of Applied Bacteriology*, **70**, 161–165.

Clausen, M. R. & Mortensen, P. B. (1995) Kinetic studies on colonocyte metabolism of short chain fatty acids and glucose in ulcerative colitis. *Gut*, **34**, 684–689.

Cohen, R. D., Castellani, L. W., Qiao, J. H., van Lenten, B. J., Lusis, A. J. & Reue, K. (1997) Reduced aortic lesions and elevated high density lipoprotein levels in transgenic mice over-expressing mouse apolipoprotein AIV. *The Journal of Clinical Investigation*, **99**, 1906–1916.

Cook, S. I. & Sellin, J. H. (1998) Review article: short chain fatty acids in health and disease. *Alimentary Pharmacology and Therapeutics*, **12**, 499–507.

Coppa, G. V., Bruni, S., Morelli, L., Soldi, S. & Gabrielli, O. (2004) The first prebiotics in humans: human milk oligosaccharides. *Journal of Clinical Gastroenterology*, **38**, S80–S83.

Cousens, L. S., Gallwitz, D. & Alberts, B. M. (1979) Different accessibilities in chromatin to histone acetylase. *The Journal of Biological Chemistry*, **254**, 1716–1723.

Couve, A., Moss, S. J. & Pangalos, M. N. (2000) $GABA_B$ receptors: a new paradigm in G protein signaling. *Molecular and Cellular Neurosciences*, **16**, 296–312.

Coxam, V. (2005) Inulin-type fructans and bone health: state of the art and perspectives in the management of osteoporosis. *The British Journal of Nutrition*, **93**, S111–S123.

Cummings, J. H., Pomare, E. W., Branch, W. J., Naylor, C. P. E. & Macfarlane, G. T. (1987) Short chain fatty acids in human large intestine, portal, hepatic and venous blood. *Gut*, **38**, 870–877.

Daly, K. & Shirazi-Beechey, S. P. (2006) Microarray analysis of butyrate regulated genes in colonic epithelial cells. *DNA and Cell Biology*, **25**, 49–62.

Dass, N. B., John, A. K., Bassil, A. K., Crumbley, C. W., Shehee, W. R., Maurio, F. P., Moore, G. B. T., Taylor, C. M. & Sanger, G. J. (2007) The relationship between the effects of short-chain fatty acids on intestinal motility *in vitro* and GPR43 receptor activation. *Neurogastroenterology and Motility*, **19**, 66–74.

Della Ragione, F., Criniti, V., Della Pietra, V., Borriello, A., Oliva, A., Indaco, S., Yamamoto, T. & Zappia, V. (2001) Genes modulated by histone acetylation as new effectors of butyrate activity. *FEBS Letters*, **499**, 199–204.

Delzenne, N. & Williams, C. (2002) Prebiotics and lipid metabolism. *Current Opinion in Lipidology*, **13**, 61–67.

Delzenne, N. M. & Kok, N. (2001) Effects of fructans-type prebiotics on lipid metabolism. *The American Journal of Clinical Nutrition*, **73**, 456S–458S.

Demigne, C., Morand, C., Levrat, M., Besson, C., Moundras, C. & Remesy, C. (1995) Effect of propionate on fatty acid and cholesterol synthesis and on acetate metabolism in isolated rat hepatocytes. *The British Journal of Nutrition*, **74**, 209–219.

Den Hond, E., Hiele, M., Evenepoel, P., Peeters, M. Ghoos, Y. & Rutgeerts, P. (1998) *In vivo* butyrate metabolism and colonic permeability in extensive ulcerative colitis. *Gastroenterology*, **115**, 584–590.

Desvergne, B. & Wahli, W. (1999) Peroxisome proliferator-activated receptors: nuclear control of metabolism. *Endocrine Reviews*, **20**, 649–688.

Drozdowski, L. A., Dixon, W. T., McBurney, M. I. & Thomson, A. B. (2002) Short-chain fatty acids and total parenteral nutrition affect intestinal gene expression. *Journal of Parenteral and Enteral Nutrition*, **26**, 145–150.

Duncan, S. H., Louis, P. & Flint, H. J. (2004) Lactate-utilising bacteria, isolated from human feces, that produce butyrate as a major fermentation product. *Applied and Environmental Microbiology*, **70**, 5810–5817.

Duverger, N., Tremp, G., Caillaud, J. M., Emmanuel, F., Castro, G., Fruchart, J. C., Steinmetz, A. & Denefle, P. (1996) Protection against atherogenesis in mice mediated by human apolipoprotein A-IV. *Science*, **273**, 966–968.

Ebert, M. N., Beyer-Sehlmeyer, G., Liegibel, U. M., Kautenburger, T., Becker, T. W. & Pool-Zobel, B. L. (2001) Butyrate induces glutathione S-transferase in human colon cells and protects from genetic damage by 4-hydroxynonenal. *Nutrition and Cancer*, **41**, 156–164.

Elliott, C. A. K. & Hobbiger, F. (1959) Gamma aminobutyric acid: circulatory and respiratory effects in different species; re-investigation of the anti-strychnine action in mice *The Journal of Physiology (London)*, **146**, 70–84.

Emmanuel, F., Steinmetz, A., Rosseneu, M., Brasseur, R., Gosselet, N., Attenot, F., Cuine, S., Seguret, S., Latta, M., Fruchart, J. C. & Denefle, P. (1994) Identification of specific amphipathic alpha-helical sequence of human apolipoprotein A-IV involved in lecithin:cholesterol acyltransferase activation. *The Journal of Biological Chemistry*, **269**, 29883–29890.

Erdo, S. L. (1985) Peripheral GABAergic mechanisms. *Trends in Pharmacological Sciences*, **6**, 205–208.

Erdo, S. L. & Wolff, J. R. (1990) γ-Aminobutyric acid outside the mammalian brain. *Journal of Neurochemistry*, **54**, 363–372.

Farrell, R. J. & Peppercorn, M. A. (2002) Ulcerative colitis. *Lancet*, **359**, 331–340.

Fearson, E. R. & Vogelstein, B. (1990) A genetic model for colorectal tumorigenesis. *Cell*, **61**, 759–767.

Fodde, R., Smits, R. & Clevers, H. (2001) APC signal transduction and genetic instability in colorectal cancer. *Nature Reviews Cancer*, **1**, 55–67.

Fujimoto, K., Fukagawa, K., Sakata, T. & Tso, P. (1993) Suppression of food intake by apolipoprotein A-IV is mediated through the central nervous system in rats. *The Journal of Clinical Investigation*, **91**, 1830–1833.

Fushimi, T., Suruga, K., Oshima, Y., Fukihara, M., Tsukamoto, Y. & Goda, T. (2006) Dietary acetic acid reduces serum cholesterol and triacylglycerols in rats fed a cholesterol-rich diet. *The British Journal of Nutrition*, **95**, 916–924.

Gage, P. W. (1992) Activation and modulation of neuronal K^+ channels by GABA. *Trends in Neurosciences*, **15**, 46–51.

Galvez, J., Rodriguez-Cabezas, M. E. & Zarzuelo, A. (2005) Effects of dietary fibre on inflammatory bowel disease. *Molecular Nutrition and Food Research*, **49**, 601–608.

Gaschott, T. & Stein, J. (2003) Short-chain fatty acids and colon cancer cells: the vitamin D receptor-butyrate connection. *Recent Results in Cancer Research*, **164**, 247–257.

Gillis, R. A., DiMicco, J. A., Williford, D. T., Hamilton, B. & Gale, K. N. (1980) Importance of CNS GABAergic mechanisms in the regulation of cardiovascular function. *Brain Research Bulletin*, **5**, 303S–315S.

Giuliani, S., Maggi, C. A. & Meli, A. D. (1986) Differences in cardiovascular responses to peripherally administered GABA as influenced by basal conditions and type of anaesthesia. *British Journal of Clinical Pharmacology*, **88**, 659–670.

Glassmeier, G., Herzig, K. H., Hoefner, M., Lemmer, K., Jansen, A. & Schruebl, H. (1998) Expression of functional GABAA receptors cholecystokinin-secreting gut neuroendocrine murine STC-1 cells. *Journal Physiology (London)*, **510**, 805–814.

Goldberg, I. J., Scheraldi, C. A., Yacoub, L. K., Saxena, U. & Bisgaier, C. L. (1990) Lipoprotein apoCII activation of lipoprotein lipase. Modulation of apolipoprotein A-IV. *The Journal of Biological Chemistry*, **265**, 4266–4272.

Griffin, I. J., Davila, P. M. & Abrams, S. A. (2002) The effect of adaptation to moderate dietary intakes of oligosaccharides on calcium absorption in children. *The British Journal of Nutrition*, **87**, S187–S191.

Griinari, J. M. & Bauman, D. E. (1999) Biosynthesis of conjugated linoleic acid an its incorporation into meat and milk in ruminants. *Advances in Conjugated Linoleic Acid*, (eds. M. P. Yurawecz, M. M. Mossoba, J. K. G. Kramer, M. W. Pariza & G. J. Nelson), pp. 180–200, AOCS Press, Champaign.

Grubben, M. J., van den Braak, C. C., Essenberg, M., Olthof, M., Tangerman, A., Katan, M. B. & Nagangast, F. M. (2001) Effect of resistant starch on potential biomarkers of colonic cancer risk in patients with colonic adenomas: a controlled trial. *Digestive Diseases and Sciences*, **46**, 750–756.

Grunstein, M. (1997) Histone acetylation in chromatin structure and transcription. *Nature*, **389**, 349–352.

Hague, A., Elder, D. J., Hicks, D. J. & Paraskeva, C. (1995) Apoptosis in colorectal tumor cells: induction by short chain fatty acids butyrate, propionate acetate and by the bile salt deoxycholate. *International Journal of Cancer*, **60**, 400–406.

Hague, A., Manning, A. M., Hanlon, K. A., Huschtscha, L. I., Hart, D. & Paraskeva, C. (1993) Sodium butyrate induces apoptosis in human colonic tumor cell lines in a p53-independant pathway: implications for the possible role of dietary fibre in the prevention of large bowel cancer. *International Journal of Cancer*, **55**, 498–505.

Halasy, K. & Somogyi, P. (1993) Distribution of GABAergic synapses and their targets in the dentate gyrus of rat: a quantitative immunoelectron microscopic analysis *Journal für Hirnforschung*, **34**, 299–308.

Hara, H., Haga, S., Aoyama, Y. & Kiriyama, S. (1999) Short-chain fatty acids suppress cholesterol synthesis in rat liver and intestine. *The Journal of Nutrition*, **129**, 942–948.

Hara, H., Haga, S., Kasai, T. & Kiriyama, S. (1998) Fermentation products of sugar-beet fibre by cecal bacteria lower plasma cholesterol concentration in rats. *The Journal of Nutrition*, **128**, 688–693.

Hardie, D. G. (2004) AMP-activated protein kinase: a master switch in glucose and lipid metabolism. *Reviews in Endocrine and Metabolic Disorders*, **5**, 119–125.

Hardie, D. G., Carling, D. & Carlson, M. (1998) The AMP-activated/SNF1 protein kinase subfamily: metabolic sensors of the eukaryotic cell? *Annual Review of Biochemistry*, **67**, 821–855.

Hardie, D. G., Scott, J. W., Pan, D. A. & Hudson, E. R. (2003) Management of cellular energy by the AMP-activated protein kinase system. *FEBS Letters*, **546**, 113–120.

Hayakawa, K., Kimura, M. & Kamata, K. (2002) Mechanism underlying γ-aminobutyric acid-induced antihypertensive effect in spontaneously hypertensive rats. *European Journal of Pharmacology*, **438**, 107–113.

Hayakawa, K., Kimura, M., Kasaha, K., Matsumoto, K., Sansawa, H. & Yamori, Y. (2004) Effect of γ-aminobutyric acid-enriched dairy product on the blood pressure of spontaneously hypertensive and normotensive Wistar-Kyoto rats. *The British Journal of Nutrition*, **92**, 411–417.

Hayes, J. D. & Pulford, D. J. (1995) The glutathione S-transferase supergene family: regulation of GST and the contribution of the isoenzymes to cancer chemoprotection and drug resistance. *Critical Reviews in Biochemistry and Molecular Biology*, **30**, 445–600.

Heerdt, B. G., Houston, M. A. & Augenlicht, L. H. (1994) Potentiation of specific short chain fatty acids of differentiation and apoptosis in human colonic carcinoma cell lines. *Cancer Research*, **54**, 3288–3294.

Higuchi, T., Hayashi, H. & Abe, K. (1997) Exchange of glutamate and γ-aminobutyrate in a *Lactobacillus* strain. *Journal of Bacteriology*, **179**, 3362–3364.

Hill, M. J. (1995) Bacterial fermentation of complex carbohydrate in the human colon. *European Journal of Cancer Prevention*, **4**, 353–358.

Hinnebusch, B. F., Meng, S., Wu, J. T., Archer, S. Y. & Hodin, R. A. (2002) The effects of short-chain fatty acids on human colon cancer cell phenotype are associated with histone hyperacetylation. *Journal of Nutrition*, **132**, 1012–1017.

Hold, G. L., Schwiertz, A., Aminov, R. I., Blaut, M. & Flint, H. J. (2003) Oligonucleotide probes that detect quantitatively significant groups of butyrate-producing bacteria in human feces. *Applied and Environmental Microbiology*, **69**, 4320–4324.

Hong, Y., Nishimura, Y., Hishikawa, D., Tsuzuki, H., Miyahara, H., Gotoh, C., Choi, K., Feng, D., Chen, C., Lee, H., Katoh, K., Roh, S. & Sasaki, S. (2005) Acetate and propionate short chain fatty acids stimulate adipogenesis via GPCR43. *Endocrinology*, **146**, 5092–5099.

Illman, R. J., Topping, D. L., McIntosh, G. H., Trimble, R. P., Storer, G. B., Taylor, M. N. & Cheng, B.-Q. (1988) Hypocholesterolaemic effects of dietary propionate studies in whole animals and in perfused rat liver. *Annals of Nutrition and Metabolism*, **32**, 97–107.

Inoue, K., Shirai, T., Ochiai, H., Kasao, M., Hayakawa, K., Kimura, M. & Sansawa, H. (2003) Blood-pressure-lowering effect of a novel fermented milk containing γ-aminobutyric acid (GABA) in milk hypertensives. *European Journal of Clinical Nutrition*, **57**, 490–495.

Ip, C., Dong, Y., Ip, M. M., Banni, S., Carta, G., Angioni, G., Murru, E., Spada, S., Melis, M. P. & Saebo, A. (2002) Conjugated linoleic acid isomers and mammary cancer prevention. *Nutrition and Cancer*, **43**, 52–58.

Jenkins, D. J., Wolever, T. M., Jenkins, A., Brighenti, F., Vuksan, V., Rao, A. V., Cunnane, S. C., Ocana, A., Corey, P. & Vezina, C. (1991) Specific types of colonic fermentation may raise low-density-lipoprotein-cholesterol concentrations. *The American Journal of Clinical Nutrition*, **54**, 141–147.

Joseph, J., Niggemann, B., Zaenker, K. S. & Entschladen, F. (2002) The neurotransmitter γ-aminobutyric acid is an inhibitory regulator for the migration of SW480 colon carcinoma cells. *Cancer Research*, **62**, 6467–6469.

Katayama, K., Wada, K., Nakajima, K., Mizuguchi, H., Hayakawa, T., Nakagawa, S., Kadowaki, T., Nagai, R., Kamisaki, Y., Blumberg, R. S. & Mayumi, T. (2003) A novel PPAR-γ gene therapy to control inflammation associated with inflammatory bowel disease in a murine model. *Gastroenterology*, **124**, 1315–1324.

Kawaguchi, T., Osatomi, K., Yamashita, H., Kabashima, T. & Uyeda, K. (2002) Mechanism for fatty acid ''sparing'' effect on glucose-induced transcription. *The Journal of Biological Chemistry*, **277**, 3829–3835.

Kennedy, S. G., Wagner, A. J., Conzen, S. D., Jordan, J., Bellacosa, A., Tsichlis, P. N. & Hay, N. (1997) The P1 3-kinase/Akt signaling pathway delivers an anti-apoptotic signal. *Genes and Development*, **11**, 701–713.

Kiefer, J., Beyer-Sehlmeyer, G. & Pool-Zobel, B. L. (2006) Mixtures of SCFA, composed according to physiologically available concentrations in the gut lumen, modulate histone acetylation in human HT29 colon cancer cells. *The British Journal of Nutrition*, **96**, 803–810.

Kien, C. L., Kepner, J., Grotjohn, J., Ault, K. & McClead, R. E. (1992) Stable isotope model for estimating colonic acetate production in premature infants. *Gastroenterology*, **102**, 1458–1466.

Kim, H. Y., Yokozawa, T., Nakagawa, T. & Sasaki, S. (2004) Protective effect of gamma-aminobutyric acid against glycerol-induced acute renal failure in rats. *Food and Chemical Toxicology*, **42**, 2009–2014.

Kimura, M., Mizukami, Y., Miura, T., Fujimoto, K., Kobayashi, S. & Matsuzaki, M. (2001) Orphan G protein-coupled receptor, GPR41, induces apoptosis via a p53/Bax pathway during ischemic hypoxia and neoxygenation. *The Journal of Biological Chemistry*, **276**, 26453–26460.

Klampfer, L., Huang, J., Sasazuki, T., Shirasawa, S. & Augenlicht, L. (2003) Inhibition of interferon gamma signalling by the short chain fatty acid butyrate. *Molecular Cancer Research*, **1**, 855–862.

Kritchevsky, D., Tepper, S. A., Wright, S., Czarnecki, S. K., Wilson, T. A. & Nicolosi, R. J. (2004) Conjugated linoleic acid isomer effects in atherosclerosis: growth and regression of lesions. *Lipids*, **39**, 611–616.

Kuo, M. H. & Allis, C. D. (1998) Roles of histone acetyltransferases and deacetylases in gene regulation. *BioEssays*, **20**, 615–626.

Kuriyama, K. & Sze, P. Y. (1971) Blood-brain barrier to 3H-γ-aminobutyric acid in normal and amino oxyacetic acid-treated animals. *Neuropharmacology*, **10**, 103–108.

Kvietys, P. R. & Granger, D. N. (1981) Effect of volatile fatty acids on blood flow and oxygen uptake by the dog colon. *Gastroenterology*, **80**, 962–969.

Lambert, D. W., Wood, I. S., Ellis, A. & Shirazi-Beechey, S. P. (2002) Molecular changes in the expression of human colonic nutrient transporters during the transition from normality to malignancy. *British Journal of Cancer*, **86**, 1262–1269.

Lan, A., Lagadic-Gossmann, D., Lemaire, C., Brenner, C. & Jan, G. (2006) Acidic extracellular pH shifts colorectal cancer cell death from apoptosis to necrosis upon exposure to propionate and acetate, major end-products of the human probiotic propionibacteria. *Apoptosis*, **12**, 573–591.

Le Poul, E., Loison, C., Struyf, S., Springael, J., Lannoy, V., Decobecq, M., Brezillon, S., Dupriez, V., Vassart, G., Van Damme, J., Parmentier, M. & Detheux, M. (2003) Functional characterisation of human receptors for short chain fatty acids and their role in polymorphonuclear cell activation. *The Journal of Biological Chemistry*, **278**, 25481–25489.

Leschelle, A., Delpal, S., Goubern, M., Blottiere, H. M. & Blachier, F. (2000) Butyrate metabolism upstream and downstream acetyl-CoA synthesis and growth control of human colon carcinoma cells. *European Journal of Biochemistry*, **267**, 6435–6442.

Levrat, M.-A., Favier, M.-L., Moundras, C., Remesy, C., Demigne, C. & Morand, C. (1994) Role of dietary propionic acid and bile acid excretion in the hypocholesterolemic effects of oligosaccharides in rats. *The Journal of Nutrition*, **124**, 531–538.

Li, C., Liu, W., F., M., Huang, W., Zhou, J., Sun, H. & Feng, Y. (2003) Effect and comparison of sodium butyrate and trichostatin A on the proliferation/differentiation of K562. *Journal of Huazhong University of Science and Technology. Medical Sciences*, **23**, 249–253.

Li, R. W. & Li, C. (2006) Butyrate induces profound changes in gene expression related to multiple pathways in bovine kidney epithelial cells. *BMC Genomics*, **7**, 234–247.

Lin, J., Lee, I. S., Frey, J., Slonczewski, J. L. & Foster, J. W. (1995a) Comparative analysis of extreme acid survival in *Salmonella typhimurium, Shigella flexineri* and *Escherichia coli*. *Journal of Bacteriology*, **177**, 4097–4104.

Lin, Y., Vonk, R. J., Slooff, M. J. H., Kuipers, F. & Smit, M. J. (1995b) Differences in propionate-induced inhibition of cholesterol and triacylglycerol synthesis between human and rat hepatocytes in primary culture. *The British Journal of Nutrition*, **74**, 197–207.

Lipkin, M. (1989) Intermediate biomarkers of increased susceptibility to cancer of the large intestine. *Cell and Molecular Biology of Colon Cancer*, (ed. L. H. Augenlicht), pp. 97–109, CRC Press Inc, Boca Raton, Fl..

Lock, A. L. & Bauman, D. E. (2004) Modifying milk fat composition of dairy cows to enhance fatty acids beneficial to human health. *Lipids*, **39**, 1197–1206.

Lopez, H. W., Levrat-Verny, M. A., Coudray, C., Besson, C., Krespine, V., Messager, A., Demigne, C. & Remesy, C. (2001) Class 2 resistant starches lower plasma and liver lipids and improve mineral retention in rats. *The Journal of Nutrition*, **131**, 1283–1289.

Luhrs, H., Gerke, T., Muller, J. G., Melcher, R., Schauber, J., Boxberge, F., Scheppach, W. & Menzel, T. (2002) Butyrate inhibits NF-κB activation in lamina propria macrophages of patients with ulcerative colitis *Scandinavian Journal of Gastroenterology*, **37**, 458–466.

Lupton, J. R. (2004) Microbial degradation products influence colon cancer risks: the butyrate controversy. *The Journal of Nutrition*, **134**, 479–482.

Lutz, T. & Scharrer, F. (1991) The effect of SCFA on Ca absorption by the rat colon. *Experimental Physiology*, **76**, 615–618.

Lynch, H. T. & de la Chapelle, A. (2003) Hereditary colorectal cancer. *The New England Journal of Medicine*, **348**, 919.

Macdonald, R. L. & Olsen, R. W. (1994) $GABA_A$ receptor channels. *Annual Review of Neuroscience*, **17**, 569–602.

Macfarlane, G. T. & Gibson, G. R. (1995) Microbiological aspects of production of short chain fatty acids in the large bowel. *Physiological and Clinical Aspects of Short Chain Fatty Acids*, (eds. J. H. Cummings, J. L. Rombeau & T. Sakata, pp. 87–105, Cambridge University Press, Cambridge, New York.

Macfarlane, G. T., Gibson, G. R. & Cummings, J. H. (1992) Comparison of fermentation reactions in different regions of the human colon. *The Journal of Applied Bacteriology*, **72**, 57–64.

Macfarlane, S. & Macfarlane, G. T. (1995) Proteolysis and amino acid fermentation. *Human Colonic Bacteria: Role in Nutrition, Physiology and Pathology*, (eds. G. R. Gibson & G. T. Macfarlane), pp. 75–100,CRC Press, Boca Raton, FL.

Macfarlane, S. & Macfarlane, G. T. (2003) Regulation of short chain fatty acid production. *The Proceedings of the Nutrition Society*, **62**, 67–72.

Makivuokko, H., Saarinen, M.T., Ouwehand, A.C. & Rautonen, N.E. (2006) Effects of lactose on colon microbial community structure and function in a four-stage semi-continuous culture system. *Bioscience, Biotechnology, and Biochemistry*, **70**, 2056–2063.

Marcil, V., Delvin, E., Garofalo, C. & Levy, E. (2003) Butyrate impairs lipid transport by inhibiting microsomal triglyceride transfer protein in Caco-2 cells. *The Journal of Nutrition*, **133**, 2180–2183.

Marcil, V., Delvin, E., Seidman, E., Poitras, L., Zoltowska, M., Garofalo, C. & Levy, E. (2002) Modulation of lipid synthesis, apolipoprotein biogenesis, and lipoprotein assembly by butyrate. *American Journal of Physiology. Gastrointestinal and Liver*, **283**, G340–G346.

McCann, S.M. & Rettori, V. (1986) Gamma aminobutyric acid (GABA) controls anterior pituitary hormone secretion. *Advances in Biochemical Psychopharmacology*, **42**, 173–189.

McCann, S.M., Vijayan, E., Negro-Vilar, A., Mizunuma, H. & Mangat, H. (1984) Gamma aminobutyric acid (GABA), a modulator of anterior pituitary hormone secretion by hypothalamic and pituitary action. *Psychoneuroendocrinology*, **9**, 97–106.

Mcintire, A., Gibson, P.R. & Young, G.P. (1993) Butyrate production from dietary fibre and protection against large bowel cancer in a rat model. *Gut*, **34**, 386–391.

Miller, T.L. & Wolin, M.J. (1979) Fermentations by saccharolytic intestinal bacteria. *The American Journal of Clinical Nutrition*, **32**, 164–172.

Milner, J.A. (1999) Functional foods and health promotion. *The Journal of Nutrition,* **129**, 1395S–1397S.

Mody, I., De Koninck, Y., Otis, T.S. & Soltesz, I. (1994) Bridging the cleft at GABA synapses in the brain. *Trends in Neurosciences,* **17**, 517–525.

Morgan, D.O. (1995) Principles of CDK regulation. *Nature (London)*, **374**, 131–134.

Morrison, D.J., Mackay, W.J., Edwards, C.A., Preston, T., Dodson, B. & Weaver, L.T. (2006) Butyrate production from oligofructose fermentation by the human fecal flora: what is the contribution of extracellular acetate and lactate? *The British Journal of Nutrition*, **96**, 570–577.

Mortensen, F.V., Hessov, I., Birke, H., Korsgaad, N. & Nielsen, H. (1991) Microcirculatory and trophic effects of short chain fatty acids in the human rectum after Hartmann's procedure. *The British Journal of Surgery,* **78**, 1208–1211.

Nazih, H., Nazih-Sanderson, F., Krempf, M., Huvelin, J.M., Mercier, S. & Bard, J.M. (2001) Butyrate stimulates apoA-IV-containing lipoprotein secretion in differentiated Caco-2 cells: role in cholesterol efflux. *Journal of Cellular Biochemistry,* **83**, 230–238.

Newmark, H.L. & Young, C.W. (1995) Butyrate and phenylacetate as differentiating agents: practical problems and opportunities. *Journal of Cellular Biochemistry,* **22**, S247–S253.

Nishina, P.M. & Freedland, R.A. (1990) Effects of propionate on lipid biosynthesis in isolated rat hepatocytes. *The British Journal of Nutrition,* **120**, 668–673.

Nomura, M., Kimoto, H., Someya, Y. & Suzuki, I. (1999a) Novel characteristic for distinguishing *Lactococcus lactis* subsp. *lactis* from subsp. *cremoris*. *International Journal of Systematic Bacteriology,* **49**(Pt 1), 163–166.

Nomura, M., Kimoto, H., Someya, Y., Furukawa, S. & Suzuki, I. (1998) Production of gamma-aminobutyric acid by cheese starters during cheese ripening. *Journal of Dairy Science,* **81**, 1486–1491.

Nomura, M., Nakajima, I., Fujita, Y., Kobayashi, M., Kimoto, H., Suzuki, I. & Aso, H. (1999b) *Lactococcus lactis* contains only one glutamate decarboxylase gene. *Microbiology,* **145**(Pt 6), 1375–1380.

Ogawa, H., Rafiee, P., Fisher, P.J., Johnson, N.A., Otterson, M.F. & Bioion, D.G. (2003) Butyrate modulates gene and protein expression in human intestinal endothelial cells. *Biochemical and Biophysical Research Communications,* **309**, 512–519.

Oh, C.-H. & Oh, S.-H. (2004) Effects of germinated brown rice extracts with enhanced levels of GABA on cancer cell proliferation and apoptosis. *Journal of Medicinal Food,* **7**, 19–23.

Oh, S.-H., Soh, J.R. & Cha, Y.S. (2003) Germinated brown rice extract shows a neutraceutical effect in the recovery of chronic alcohol-related symptoms. *Journal of Medicinal Food,* **6**, 115–121.

Olsen, R. W. (2002) GABA. *Neuropsychopharmacology: The Fifth Generation of Progress.*, (eds. K. L. Davis, D. Charney & J. T. Coyle), pp. 159–168, Lippincott, Williams & Wilkins, Philadelphia.

Ong, J. & Kerr, D. B. (1990) GABA-receptors in peripheral tissues. *Life Sciences,* **46**, 1489–1501.

Owens, F. N. & Isaacson, H. R. (1977) Ruminal microbial yields. *Federation Proceedings,* **36**, 198–202.

Park, K. B. & Oh, S. H. (2004) Cloning and expression of a full length glutamate decarboxylase gene from *Lactobacillus plantarum. International Journal of Food Sciences and Nutrition*, **9**, 324–329.

Park, K. B. & Oh, S. H. (2006) Enhancement of γ-aminobutyric acid production in Chungkukjang by applying a *Bacillus subtilis* strain expressing glutamate decarboxylase from *Lactobacillus brevis. Biotechnology Letters,* **28** 1459–1463.

Park, K. B. & Oh, S. H. (2007) Cloning, sequencing and expression of a novel glutamate decarboxylase gene from a newly isolated lactic acid bacterium, *Lactobacillus brevis* OPK-3 *Bioresource Technology*, **98**, 312–319.

Park, K. B., Ji, G.-E., Park, M.-S. & Oh, S.-H. (2005) Expression of rice glutamate decarboxylase in *Bifidobacterium longum* enhances γ-aminobutyric acid production. *Biotechnology Letters*, **27**, 1681–1684.

Parodi, P. W. (2002) Health benefits of conjugated linoleic acid. *Food Industries Journal*, **5**, 222–259.

Piche, T., des Varannes, S. B., Sacher-Huvelin, S., Holst, J. J., Cuber, J. C. & Galmiche, J.-P. (2003) Colonic fermentation influences lower esophageal sphincter function in gastroesophageal reflux disease. *Gastroenterology*, **124**, 894–902.

Pool-Zobel, B. L., Selvaraju, V., Sauer, J., Kautenburger, T., Kiefer, J., Richter, K. K., Soom, M. & Wolfl, S. (2005) Butyrate may enhance toxicological defence in primary, adenoma and tumor human colon cells by favourable modulating expression of glutathione S-transferase genes, an approach in nutrigenomics. *Carcinogenesis*, **26**, 1064–1076.

Prohaszka, L., Jayarao, B. M., Fabian, A. & Kovacs, S. (1990) The role of intestinal volatile fatty acids in the *Salmonella* shedding of pigs. *Zentralblatt für Veterinärmedizin. Reihe B*, **37**, 570–574.

Pryde, S. E., Duncan, S. H., Hold, G. L., Stewart, C. S. & Flint, H. J. (2002) The microbiology of butyrate formation in the human colon. *FEMS Microbiology Letters,* **217**, 133–139.

Puchowicz, M. A., Bederman, I. R., Comte, B. Yang, D., David, F., Stone, E., Jabbour, K., Wasserman, D. H. & Brunengraber, H. (1999) Zonation of acetate labelling across the liver: implications for studies of lipogenesis by MIDA. *The American Journal of Physiology,* **277**, E1022–E1027.

Pylkas, A. M., Raj Juneja, L. & Slavin, J. L. (2005) Comparison of different fibers for *in vitro* production of short chain fatty acids by intestinal microflora. *Journal of Medicinal Food,* **8**, 113–116.

Racagni, G., Apud, J. A., Cocchi, D., Locatelli, V. & Muller, E. E. (1982) GABAergic control of anterior pituitary hormone secretion. *Life Sciences,* **31**, 823–838.

Raghupathy, P., Ramakrishna, B. S., Oommen, S. P., Ahmed, M. S., Priyaa, G., Dziura, J., Young, G. P. & Binder, H. J. (2006) Amylase-resistant starch as adjunct to oral rehydration therapy in children with diarrhoea. *Journal of Pediatric Gastroenterology and Nutrition*, **42**, 362–368.

Rioux, K. P., Madsen, K. L. & Fedorak, R. N. (2005) The role of enteric microflora in inflammatory bowel disease: human and animal studies with probiotics and prebiotics. *Gastroenterology Clinics of North America,* **34**, 465–482.

Roberfroid, M. B. (2005) *Inulin-Type Fructans: Functional Food Ingredients*, CRC Press, Boca Raton, FL.

Roberts, E. & Frankel, S. (1950) Gamma-aminobutyric acid in brain: its formation from glutamic acid. *The Journal of Biological Chemistry*, **187**, 55–63.

Rodriguez-Cabezas, M. E., Galvez, J., Lorente, M. D., Concha, A., Camuesco, D., Azzouz, A., Osuna, A., Redondo, L. & Zarvuelo, A. (2002) Dietary fibre down-regulates colonic tumor necrosis factor alpha and nitric oxide production in trinitrobenzenesulfonic acid-induced colitis in rats. *The Journal of Nutrition*, **132**, 3263–3271.

Rodwell, V. W., Nordstrom, J. L. & Mitschelen, J. J. (1976) Regulation of HMG CoA reductase. *Advances in Lipid Research,* **14**, 1–74.

Roediger, W. E. W. (1980) The colonic epithelium in ulcerative colitis: an energy deficiency disease? *Lancet*, **2**, 712–715.

Roediger, W. E. W. (1995) The place of short-chain fatty acids in colonocyte metabolism in health and ulcerative colitis: the impaired colonocyte barrier. *Physiological and Clinical Aspects of Short-Chain Fatty Acids.*, (eds. J. H. Cummings, J. L. Rombeau & T. Sakata), pp. 337–351, Cambridge University Press, Cambridge.

Roy, C. C., Kien, C. L., Bouthillier, L. & Levy, E. (2006) Short-chain fatty acids: ready for prime time. *Nutrition in Clinical Practice,* **21**, 351–366.

Rutgeerts, P., Van Deventer, S. & Schreiber, S. (2003) Review article: the expanding role of biological agents in the treatment of inflammatory bowel disease – focus on selective adhesion molecule inhibition. *Alimentary Pharmacology and Therapeutics,* **17**, 1435–1450.

Sakakibara, S., Yamauchi, T., Oshima, Y., Tsukamoto, Y. & Kadowaki, T. (2006) Acetic acid activates hepatic AMPK and reduces hyperglycemia in diabetic KK-A(y) mice. *Biochemical and Biophysical Research Communications*, **344**, 597–604.

Samuels, H. H., Stanley, F., Casanova, J. & Shao, T. C. (1980) Thyroid hormone nuclear receptor levels are influenced by the acetylation of chromatin-associated proteins. *The Journal of Biological Chemistry,* **255**, 2499–2508.

Sanders, J. W., Leenhouts, K., Burghoorn, J., Brands, J. R., Venema, G. & Kok, J. (1998) A chloride-inducible acid resistance mechanism in *Lactococcus lactis* and its regulation. *Molecular Microbiology,* **27**, 299–310.

Schauber, J., Svanholm, C., Termen, S., Iffland, K., Menzel, T., Scheppach, W., Melcher, R., Agerberth, B., Luhrs, H. & Gudmundsson, G. H. (2003) Expression of the cathelicidin LL-37 is modulated by short chain fatty acids in colonocytes: relevance of signalling pathways. *Gut*, **52**, 735–741.

Scheppach, W. & Weiler, F. (2004) The butyrate story: old wine in new bottles? *Current Opinion in Clinical Nutrition and Metabolic Care,* **7**, 563–567.

Scheppach, W., Pomare, W. E., Elia, M. & Cummings, J. H. (1991) The contribution of the large intestine to blood acetate in man. *Clinical Science*, **80**, 177–182.

Scheppach, W., Bartram, P., Richter, A., Richter, F., Liepold, H., Dusel, G., Hofstetter, G., Ruthlein, J. & Kasper, H. (1992) Effect of short chain fatty acids on human colonic mucosa *in vitro*. *Journal of Parenteral and Enteral Nutrition,* **16**, 43–48.

Schlicker, K., Boller, M. & Schmidt, M. (2004) $GABA_C$ receptor mediated inhibition in acutely isolated neurons of the rat dorsal lateral geniculate nucleus. *Brain Research Bulletin,* **63**, 91–97.

Scholz-Ahrens, K. E., Ade, P., Marten, B., Weber, P., Timm, W., Gluer, A. C.-C. & Schrezenmeir, J. (2007) Prebiotics, probiotics, and synbiotics affect mineral absorption, bone mineral content, and bone structure. *The Journal of Nutrition*, **137**, 838S–846S.

Sealy, L. & Chalkley, R. (1978) The effect of sodium butyrate on histone modification. *Cell*, **14**, 115–121.

Senga, T., Iwamoto, S., Yoshida, T., Yokota, T., Adachi, K., Azuma, E., Hamaguchi, M. & Iwamoto, T. (2003) LSSIG is a novel murine leukocyte-specific GPCR that is induced by the activation of STAT3. *Blood*, **101**, 1185–1187.

Sengupta, S., Muir, J. G. & Gibson, G. R. (2006) Does butyrate protect from colorectal cancer? *Journal of Gastroenterology and Hepatology*, **21**, 209–218.

Sharma, R. (2005) Market trends and opportunities for functional dairy beverages. *Australian Journal of Dairy Technology*, **60**, 196–199.

Siavoshian, S., Blottiere, H. M., le Foll, E., Kaeffer, B., Cherbut, C. & Galmiche, J.-P. (1997) Comparison of the effect of different short chain fatty acids on the growth and differentiation of human colonic carcinoma cell lines *in vitro*. *Cell Biology International*, **21**, 281–287.

Siavoshian, S., Segain, J. P., Kornprobst, M., Bonnet, C., Cherbut, C., Galmiche, J.-P. & Blottiere, H. M. (2000) Butyrate and trichostatin A effects on the proliferation/differentiation of human intestinal epithelial cells: induction of cyclin D3 and p21 expression. *Gut*, **46**, 507–514.

Stanton, C., Murphy, J., McGrath, E. & Devery, R. (2003) Animal feeding strategies for conjugated linoleic acid enrichment of milk. *Advances in Conjugated Linoleic Acid Research*, (eds. J. L. Sebedio, W. W. Christie & R. O. Adlof), pp. 123–145, AOCS Press, Champaign, IL.

Stanton, H. C. (1963). Mode of action of gamma amino butyric acid on the cardiovascular system. *Archives Internationales de Pharmacodynamie et de Thérapie*, **143**, 195–204.

Stein, O., Stein, Y., Lefevre, M. & Roheim, P. S. (1986) The role of apolipoprotein A-IV in reverse cholesterol transport with cultured cells and liposomes derived from an ether analog of phosphatidylcholine. *Biochimica et Biophysica Acta*, **878**, 7–13.

Takahashi, H., Tiba, M., Lino, M. & Takayasu, T. (1955) The effect of γ-aminobutyric acid on blood pressure. *The Japanese Journal of Physiology*, **5**, 334–341.

Tappenden, K. A., Drozdowski, L. A., Thomson, A. B. & McBurney, M. I. (1998) Short-chain fatty acid-supplemented total parenteral nutrition alters intestinal structure, glucose transporter 2 (GLUT2) mRNA and protein, and proglucagon mRNA abundance in normal rats. *The American Journal of Clinical Nutrition*, **68**, 118–125.

Thacker, P. A., Salomons, M. O., Aherne, F. X., Milligan, L. P. & Bowland, J. P. (1981) Influence of propioinc acid on the cholesterol metabolism of pigs fed hypercholesterolemic diets. *Canadian Journal of Animal Science*, **61**, 969–975.

Thornton, J. R. (1981) High colonic pH promotes colorectal cancer. *Lancet*, **1**, 1081–1083.

Thwaites, D. T., Basterfield, L., McCleave, P. M. J., Carter, S. M. & Simmons, N. (2000) Gamma-aminobutyric acid (GABA) transport across human intestinal epithelial (Caco-2) cell monolayers. *British Journal of Clinical Pharmacology*, **129**, 457–464.

Topping, D. L. & Clifton, P. M. (2001) Short chain fatty acids and human colonic function: roles of resistant starch and non-starch polysaccharides. *Physiological Reviews*, **81**, 1031–1064.

Trinidad, T. P., Wolever, T. M. S. & Thompson, L. U. (1996) Effect of acetate and propionate on Ca absorption from the rectum and distal colon of humans. *The American Journal of Clinical Nutrition*, **63**, 574–578.

Uchida, M. & Mogami, O. (2005) Milk whey culture with *Propionibacterium freudenreichii* ET-3 is effective on the colitis induced by 2,4,6-trinitrobenzene sulfonic acid in rats. *Journal of Pharmacological Sciences*, **99**, 329–334.

Ueno, H. (2000) Enzymatic and structural aspects on glutamate decarboxylase. *Journal of Molecular Catalysis. B, Enzymatic*, **10**, 67–79.

Ueno, Y., Hayakawa, K., Takahashi, S. & Oda, K. (1997) Purification and characterization of glutamate decarboxylase from *Lactobacillus brevis* IFO 12005. *Bioscience, Biotechnology, and Biochemistry*, **61**, 1168–1171.

Venter, C. S., Vorster, H. H. & Cummings, J. H. (1990) Effects of dietary propionate on carbohydrate and lipid metabolism in healthy volunteers. *The American Journal of Gastroenterology*, **85**, 549–553.

Vernia, P., Annese, V., Bresci, G., d'Albasio, G., D'Inca, R., Giaccari, S., Ingrosso, M., Mansi, C., Riegler, C., Valpani, D. & Caprilli, R. (2003) Topical butyrate improves efficacy of 5-ASA in refractory ulcerative colitis: results of a multicentre trial. *European Journal of Clinical Investigation*, **33**, 244–248.

Vogt, J. A., Pencharz, P. B. & Wolever, T. M. (2004) L-Rhamnose increases serum propionate in humans. *The American Journal of Clinical Nutrition*, **80**, 89–94.

Volpi, R., Chiodera, P., Caffarra, P., Scaglioni, A., Saccani, A. & Coiro, V. (1997) Different control mechanisms of growth hormone (GH) secretion between gamma-amino- and gamma-hydroxy-butyric acid: neuroendocrine evidence in Parkinson's disease. *Psychoneuroendocrinology*, **22**, 531–538.

von Eckardstein, A., Huang, Y., Wu, S., Sarmadi, A. S., Schwarz, T., Steinmetz, A. & Assmann, G. (1995) Lipoproteins containing apolipoprotein A-IV but not apolipoprotein A-I take up and esterify cell-derived cholesterol in plasma. *Arteriosclerosis, Thrombosis, and Vascular Biology*, **15**, 1755–1763.

Wachtershauser, A., Loitsch, S. M. & Stein, J. (2000) PPAR-γ is selectively upregulated in Caco-2 cells by butyrate. *Biochemical and Biophysical Research Communications*, **272**, 380–385.

Wang, A., Yoshimi, N., Ino, N., Tanaka, T. & Mori, H. (1997) Overexpression of cyclin B1 in human colorectal cancers. *Journal of Cancer Research and Clinical Oncology*, **123**, 124–127.

Weaver, C. M. (2005) Inulin, oligofructose and bone health: experimental approaches and mechanisms. *The British Journal of Nutrition*, **93**, S99–S103.

Weaver, G. A., Tangel, C., Krause, J. A., Parfitt, M. M., Jenkins, P. L., Rader, J. M., Lewis, B. A., Miller, T. L. & Wolin, M. J. (1997) Ascorbose enhances human colonic butyrate production. *The Journal of Nutrition*, **127**, 717–723.

Whigham, L. D., Cook, M. E. & Atkinson, R. L. (2000) Conjugated linoleic acid: implications for human health. *Pharmacological Research*, **42**, 503–510.

Whitehead, R. H., Young, G. P. & Bhathal, P. S. (1987) Effects of short chain fatty acids on a new human colon carcinoma cell line (LIM 1215). *Gut*, **27**, 1457–1463.

Wolever, T. M., Fernandes, J. & Rao, A. V. (1996) Serum acetate: propionate ratio is related to serum cholesterol in men but not women. *The Journal of Nutrition*, **126**, 2790–2797.

Wolever, T. M., Schrade, K. B., Vogt, J. A., Tsihlias, E. B. & McBurney, M. I. (2002) Do colonic short-chain fatty acids contribute to the long-term adaptation of blood lipids in subjects with type 2 diabetes consuming a high-fibre diet. *The American Journal of Clinical Nutrition*, **75**, 1023–1030.

Wolever, T. M. S., Spadafora, P. & Eshuis, H. (1991) Interaction between colonic acetate and propionate in humans. *The American Journal of Clinical Nutrition*, **53**, 681–687.

Wolever, T. M. S., Spadafora, P., Cunnane, S. C. & Pencharz, P. B. (1995) Propionate inhibits incorporation of colonic [1,2–^{13}C]acetate into plasma lipids in humans. *The American Journal of Clinical Nutrition*, **61**, 1241–1247.

Wolever, T. M. S., Brighenti, F., Royall, D., Jenkins, A. L. & Jenkins, D. J. A. (1989) Effect of rectal infusion of short chain fatty acids in human subjects. *The American Journal of Gastroenterology*, **84**, 1027–1033.

Wong, W. M. J., de Souza, R., Kendall, C. W. C., Emam, A. & Jenkins, D. J. A. (2006) Colonic health: fermentation and short chain fatty acids. *Clinics in Gastroenterology*, **40**, 235–243.

Wright, R. S., Anderson, J. W. & Bridges, S. R. (1990) Propionate inhibits hepaocyte lipid synthesis. *Proceedings of the Society for Experimental Biology and Medicine*, **195**, 26–29.

Xiong, Y., Miyamoto, N., Shibata, K., Valasek, M. A., Motoike, T., Kedzierski, R. M. & Yanagisawa, M. (2004) Short-chain fatty acids stimulate leptin production in adipocytes through the G-protein coupled receptor GPR41. *Proceedings of the National Academy of Sciences of the United States of America*, **101**, 1045–1050.

Young, G. P., Hu, Y., Le Leu, R. K. & Nyskohus, L. (2005) Dietary fibre and colorectal cancer: a model for environment–gene interactions. *Molecular Nutrition and Food Research*, **49**, 571–584.

Zambell, K. L., Fitch, M. D. & Fleming, S. E. (2003) Acetate and butyrate are the major substrates for *de novo* lipogenesis in rat colonic epithelial cells. *The Journal of Nutrition*, **133**, 3509–3515.

Zampa, A., Silvi, S., Fabiani, R., Morozzi, G., Orpianesi, C. & Cresci, A. (2004) Effects of different digestible carbohydrates on bile acid metabolism and SCFA production by human gut microflora grown in an *in vitro* semi-continuous culture. *Anaerobe*, **10**, 19–26.

Zapolska-Downar, D., Siennicka, A., Kaczmarczyk, M., Kolodziej, B. & Naruszewicz, M. (2004) Butyrate inhibits cytokine-induced VCAM-1 and ICAM-1 expression in cultured endothelial cells: the role of NF-κB and PPARα. *The Journal of Nutritional Biochemistry*, **15**, 220–228.

Zhang, D., Pan, Z. H., Zhang, X., Brideau, A. D. & Lipton, S. A. (1995) Cloning of a gamma-aminobutyric acid type C receptor subunit in rat retina with a methionine residue critical for picrotoxinin channel block. *Proceedings of the National Academy of Sciences of the United States of America*, **92**, 11756–11760.

10 Trouble Shooting

B.B.C. Wedding and H.C. Deeth

10.1 Introduction

The goal of any dairy processing operation is to assure uniform, high-quality products that satisfy market demand. Dairy processors are faced with a variety of quality issues from safety concerns about contaminants to less tangible issues about how production and storage methods affect the final product that reaches the consumer. Quality can be defined as 'fitness for purpose', the interpretation of which depends on the buyer, the user of the product and the agency that regulates the product. In relation to dairy products this generally involves availability, safety (chemical and microbiological), convenience, freshness, integrity and nutritional value.

The quality of all dairy products at the end-point of the handling chain depends on all stages along the chain from milking to processing and finally to the consumer. The end-point of the handling chain is where the consumer makes the final assessment of quality, judging the product, largely on its sensory properties of smell, sight, touch and taste. The aim of this chapter is to discuss the factors which may lead to faults or defects in fat-containing dairy products, namely cream, butter, dairy spreads and cream cheese.

Most of the difficulties experienced by manufacturers of dairy products are due to the poor quality of the raw milk supplied to them. Raw milk of the highest quality is required as an ingredient in the manufacture of good quality dairy products. Good-tasting milk is characterised as having a pleasing, slightly sweet taste with no unpleasant aftertaste (Bassette *et al.*, 1986; Rosenthal, 1991; Varnam & Sutherland, 2001). However, its bland nature makes it susceptible to flavour defects from a variety of sources. Off-flavours in milk are a major problem for dairy producers and processors. Potentially, these off-flavours can result in the final manufactured dairy product being rejected at great expense to the processor. An understanding of potential off-flavours in milk provides an insight into how they impact on the quality of manufactured dairy products.

10.2 Milk

10.2.1 *Transmitted flavours*

Off-flavours in milk caused by the transfer of substances from the cow's feed or environment into the milk while it is in the udder are known as transmitted flavours. They include flavours commonly described as 'feed', 'weed', 'cowy' and 'barny'. The compounds causing the off-flavours may enter the milk through the respiratory tract and/or digestive system and bloodstream (Bassette *et al.*, 1986; Walstra *et al.*, 2006).

Feed flavours

Undesirable feed flavours can enter milk and cause taints through the cow's consumption of certain types of feed or through the inhalation of the aroma of such feeds (Rosenthal, 1991; Varnam & Sutherland, 2001; Walstra *et al.*, 2006). Common causes of feed taint are lucerne (alfalfa), clover hay and silage. For instance, the chemical compounds responsible for off-flavours in milk from silage-fed cows include mixtures of methanethiol, aldehydes, ketones, alcohols and esters (Bassette *et al.*, 1986; Varnam & Sutherland, 2001). The problem can be controlled by withholding such feeds for 4 to 5 hours before milking (Varnam & Sutherland, 2001; Bruhn, 2002a). Thus, a feeding regime which restricts the intake of noxious and odorous weeds (e.g. wild garlic, onion weed, carrot weed) and roughage is essential for the production of milk free of undesirable feed taints.

Cowy, barny and unclean flavours

The term 'cowy' is often used to describe an 'unclean' flavour usually associated with barn odours and poor ventilation. The 'barny' or 'unclean' odours are transferred to the milk through the cow's respiratory system in the same manner as feed flavours (Bassette *et al.*, 1986; Bruhn, 2002a). The physical condition of the cow may also contribute to off-flavours in milk. Cows suffering from ketosis or acetonemia produce milk with a 'cowy' odour. They have a high concentration of acetone bodies in their blood from incomplete metabolism of fat (Bassette *et al.*, 1986; Bruhn, 2002a). Off-flavours in the milk may also result from the accumulation of gases in bloat, a condition caused by an upset of the cow's digestive system (Bassette *et al.*, 1986). Mastitis also adversely affects the flavour of milk, but the effects are variable. A flat flavour may result in mild cases, possibly due to a lower concentration of normal milk constituents while in more severe cases, the milk may exhibit 'cowy', 'unclean' and salty flavours (Munro *et al.*, 1984; Bruhn, 2002a).

10.2.2 *Chemical flavours*

Chemical flavours in milk may result from the contamination of milk through various environmental sources during milking and processing. These include sanitisers in improperly rinsed equipment, usually chlorine, iodine-based and quaternary ammonium salt compounds (Rosenthal, 1991; Varnam & Sutherland, 2001). Salve or phenolic residues from certain plastics, used in the treatment of teats, are possible sources of 'medicinal' flavours (Bruhn, 2002a). Some dairy producers and processors have had difficulties when using phenols and chlorine compounds in combination. These can react to form chlorophenols, which produce a flavour taint with much lower flavour thresholds than either phenols or chlorine alone (Kosinski, 1996; Bruhn, 2002a). Although transmitted flavours still exist, better farm practices have led to reduced incidence and the production of higher-quality raw milk.

10.2.3 *Flavours associated with oxidation*

The sensory quality of fat-containing dairy products is largely dependent on the chemical stability of milk lipids during storage. The major milk lipids, namely the glycerides, phosphoglycerides, glycolipids and cholesterol esters, contain fatty acids as common building moieties (Rosenthal, 1991). The unsaturated fatty acids are susceptible to oxidation which gives rise to off-flavours and limits the shelf-life of many fat-containing dairy products.

Auto-oxidation is a free radical chain reaction that, in milk, is primarily concerned with oleic (18:1, n9), linoleic (18:2, n6) and linolenic acids (18:3, n3) (Hamilton, 1983). The reaction results in the formation of peroxides which have no influence on flavour but are unstable and readily decompose to form unsaturated aliphatic ketones and aldehydes. These compounds exhibit off-flavours described as 'tallowy', 'fishy', 'metallic' and 'cardboard-like'. They are detectable by gas chromatography at very low levels, for example 10^{-3} mg g^{-1} (Goff & Hill, 1993; Joshi & Thakar, 1994; Varnam & Sutherland, 2001; Walstra *et al.*, 2006).

The fat in milk is naturally protected from auto-oxidation as it exists in the form of small fat globules enveloped in a biological membrane, known as the milk fat globule membrane (MFGM). Furthermore, the major proteins, especially casein, effectively bind pro-oxidant metal ions in the milk such as Cu^{2+} and Fe^{3+}, physically preventing contact between the metal ion and the fat (Hamilton, 1983). Pasteurisation causes the migration of naturally occurring copper ions to sites on the fat globule membrane sensitive to the deterioration and this renders the milk more prone to oxidation and the development of off-flavours (Rosenthal, 1991; Bruhn, 2002a). Higher heat treatments reverse this effect, possibly due to the formation of antioxidants, such as sulphydryl (SH) compounds (Rosenthal, 1991). Homogenisation of milk disrupts the fat globule membrane but leaves the fat droplets coated with casein micelles, thus maintaining the emulsion and protection from oxidation (Hamilton, 1983; Bruhn, 2002a). Natural antioxidants, such as tocopherol and β-carotene can block the chain reaction or prevent its initiation; however, they are consumed in this process thus exhausting their antioxidative effect (Walstra *et al.*, 2006).

The development of oxidised off-flavours is accelerated by oxygen and a beneficial effect is obtained by its removal from the product (Rosenthal, 1991). Pro-oxidants, especially copper and iron, also accelerate oxidation (Badings, 1970; Varnam & Sutherland, 2001). Conditions are more conducive to oxidation at the surface than inside the fat globule, as phospholipids, which are located in the MFGM on the surface of the fat globule, contain a higher percentage of unsaturated fatty acids than the remainder of the fat globule lipids (Bruhn, 2002a). Oxidation catalysed by copper is also concentrated at the surface of the fat globule; oxygen from the atmosphere dissolves in the skim portion of milk and also reaches the surface of the fat globule (Bruhn, 2002a). Since auto-oxidation is a chemical reaction with a low activation energy (4–5 kcal $mole^{-1}$ for the first step and 6–14 kcal $mole^{-1}$ for the second step), the rate of the reaction is not significantly diminished by lowering the temperature of storage (Hamilton, 1983).

Auto-oxidation can also be enhanced by exposure to light of short wavelengths (Rosenthal, 1991; Walstra *et al.*, 2006) and leads to light-induced flavours such as 'sunlight' or 'activated' flavours (Bassette *et al.*, 1986). These are commonly described as 'burnt feather', 'burnt protein', 'scorched', 'mushroom' and 'cabbage' flavours. As reported by Rosenthal (1991), the photochemical initiation leads to oxidation of not only lipids (unsaturated fatty acid residues and cholesterol) but also proteins. Thus, the initial 'activated' flavour is due to the oxidation of tryptophan and methionine of serum proteins and is followed by the oxidised off-flavour of oxidised lipids (Rosenthal, 1991). Riboflavin is an essential factor in light-induced auto-oxidation acting as a visible light-absorbing compound (Rosenthal, 1991; Jensen & Poulsen, 1992; Walstra *et al.*, 2006).

10.2.4 *Flavours associated with heat treatment*

The primary purpose of heat treatment of raw milk is to destroy pathogenic microorganisms including *Salmonella* spp., *Escherichia coli*, *Campylobacter* spp., *Yersinia enterocolitica*, *Listeria monocytogenes*, *Cryptosporidium parvum*, *Streptococcus* spp., *Staphylococcus aureus*, *Brucella abortus* and *Mycobacterium tuberculosis*, to make it safe for human consumption. At the same time, heat treatment prolongs shelf-life by destroying spoilage micro-organisms and enzymes. However, off-flavours may be produced by heat processing, the time and temperature of which govern the nature and intensity of the flavours. Heat-induced flavours have been described as 'cooked' or sulphurous', 'heated', 'scorched' and 'caramelised'.

The slight 'cooked' or 'sulphurous' flavour of pasteurised milk results from the formation of such volatile substances as ammonia, hydrogen sulphide, mercaptans, and volatile phosphorus compounds that occur with protein breakdown (Bassette *et al.*, 1986; van Boekel & Walstra, 1995; de Wit & Nieuwenhuijse, 2008). Hutton & Patton (1952) reported that β-lactoglobulin degradation accounts for practically all of the free SH groups in heated milk and is the source of the 'cooked' flavour. Bassette *et al.* (1986) reported that these compounds act as active antioxidants and appear responsible for inhibition of the 'tallowy' or 'oxidised' flavours in milk heated over 76°C. As these sulphydryls become oxidised, the cooked flavour decreases. The sweet caramel flavour of heat-treated milk has been attributed to non-enzymic browning while a scorched flavour is due to localised overheating in a heat exchanger creating excessive 'burn on'.

10.2.5 *Bacterial flavours*

The fresh milk from a healthy cow drawn aseptically contains only a few bacteria (10^2–10^3 cfu mL^{-1}) (Rosenthal, 1991). However all milk becomes contaminated by bacteria, the extent of which is influenced by the health of the animal, environmental conditions on the farm, equipment and personnel. Off-flavours in the milk from bacterial metabolism can occur at any stage of production or processing. They are not detectable until large numbers of bacteria are present, usually millions per millilitre (Bruhn, 2002a). Off-flavours of bacterial origin in raw milk are an indication of inadequate hygiene and sanitation, or of the milk having been stored for too long or at too high a temperature.

Pasteurisation destroys vegetative cells, but bacterial off-flavours that had developed prior to pasteurisation are affected little by this heat process and most cannot be removed by commercial vacuum treatment processes (Bassette *et al.*, 1986). The flavour defects from bacterial growth in both raw and pasteurised milk are described as 'acidic', 'sour', 'malty', 'fruity', 'unclean', 'bitter' and 'putrid'. The production of lactic acid or a combination of other acids by the principal acid-producing bacterial species in milk, such as *Streptococcus, Pediococcus, Leuconostoc, Lactobacillus*, and members of the Enterobacteriaceae family, cause an acidic flavour in raw and pasteurised milk. A 'malty' flavour in milk, sometimes described as 'cooked', 'caramelised' and 'burnt' can be associated with the metabolism of *Lactococcus lactis* biovar *maltigenes* (Bassette *et al.*, 1986; Deeth, 1986), while a fruity flavour in pasteurised and processed dairy products is associated with post-pasteurisation contamination by *Pseudomonas fragi* producing ethyl esters (Bassette *et al*., 1986; Deeth, 1986).

Spoilage by psychrotrophic organisms often produces 'unclean', 'bitter', 'putrid' and 'rancid' flavours. The maximum keeping quality of raw milk in refrigerated storage tanks is determined by the growth of these organisms. Prior to processing, bacterial numbers greater than 10^6 or 10^7 cfu mL^{-1} in milk imply a risk that psychrotrophs have produced heat-stable extracellular enzymes, particularly lipases and proteinases (Downey, 1980; Walstra *et al*., 1999). These can act directly on micellar casein or on the fat in milk, thus producing 'unclean', 'bitter','putrid' and 'rancid' flavours.

10.2.6 *Lipolysed flavour*

Hydrolysis of milk fat, or lipolysis, catalysed by lipases, is a major dairy industry problem because of the production of flavour defects in milk, cream and manufactured dairy products. The flavours produced are often described as 'soapy', 'rancid', 'butyric', 'bitter', 'unclean' and 'astringent'. Lipolysis produces free fatty acids (FFA) and mono- and diacylglycerols. Short-chain free fatty acids, particularly butyric acid (C4) and caprylic acid (C6), are mostly responsible for the unpleasant rancid flavour (Couvreur *et al*., 2006; Deeth & Fitz-Gerald, 2006). The lipases responsible for lipolysis in milk and milk fat products are of two types, the endogenous milk lipase and exogenous bacterial lipases.

Endogenous milk lipase

The major endogenous milk lipase is a lipoprotein lipase (LPL) (Deeth & Fitz-Gerald, 2006). Cow's milk contains sufficient LPL to release about 2 μmol of FFA min^{-1} at 37°C, but the actual activity during storage of raw milk at 4°C may be as low as 0.002 μmol of FFA min^{-1}. Of the total LPL present in milk, about 80% is bound to micellar caseins, 10 to 20% is present in the serum, and only 0 to 10% is associated with the fat globule (Goff & Hill, 1993). The lipases are loosely bound to the casein micelle and are not released until either the milk is cooled or an activation treatment is applied (Bassette *et al*., 1986). In general, LPL is ineffective in milk unless the MFGM is damaged or weakened in some way (Driessen, 1989; Goff & Hill, 1993; Kon & Saito, 1997). However, the addition of blood serum facilitates the interaction between the enzyme and the fat globule and lipolysis ensues, possibly as a result of the activating apolipoprotein activator, apo-LP CII, enhancing the binding of LPL to the fat globule and its catalytic efficiency (Deeth & Fitz-Gerald, 2006).

Induced lipolysis

Induced lipolysis results when the milk lipase system is activated by physical or chemical means. For instance, lipolysis can be initiated by homogenisation, agitation (e.g. through milking machines, pumps and other equipment) and temperature manipulation (cooling–warming–cooling) (McDowell, 1969; Downey, 1975; Kon & Saito, 1997; Collins *et al*., 2003; Deeth, 2006; Deeth & Fitz-Gerald, 2006). Excessive mechanical agitation and turbulence disturb the natural MFGM, and homogenisation increases the number of small fat globules which are not surrounded by the normal MFGM, but are covered by a film of protein which is more permeable to LPL (Griffiths, 1983). This greatly increases the area of the fat–water interface and makes the milk susceptible to the action of lipase. The amount of lipolysis that occurs is dependent on both the degree of activation resulting from mechanical treatment and the duration and conditions of subsequent storage (Deeth & Fitz-Gerald, 1977; Downey, 1980).

Spontaneous lipolysis

Milk that undergoes lipolysis without being subjected to any mechanical treatment other than cooling soon after milking is referred to as 'spontaneously lipolytic' or 'spontaneous' (Downey, 1980; Deeth & Fitz-Gerald, 2006). The sooner the milk is cooled and the lower the temperature to which it is cooled, the more lipolysis occurs (Bachman & Wilcox, 1990). A characteristic of spontaneous lipolysis is its inhibition by mixing normal milk and spontaneous milk before cooling, thus preventing lipolysis (Deeth & Fitz-Gerald, 1977; Bassette *et al.*, 1986; Deeth, 2006). The milks from individual cows show a wide variation in their susceptibility to spontaneous lipolysis. The hormonal balance (stage of lactation, pregnancy), and the quality and quantity of the feed significantly influence a cow's tendency to produce spontaneous milk (Kon & Saito, 1997; Deeth, 2006; Deeth & Fitz-Gerald, 2006). However, lipolysis caused by natural milk lipase enzyme(s) accounts for most of the rancidity problems that occur in raw milk and cream. The LPL, which causes spontaneous lipolysis, is a relatively unstable enzyme, being inactivated by acid, heat, ultraviolet (UV) light, oxidising agents and prolonged freezing. High-temperature short-time (HTST) treatment (72°C for 15 s) of milk almost completely inactivates the enzyme (Fox & Stepaniak, 1993; Collins *et al.*, 2003; Deeth & Fitz-Gerald, 2006). However, in stored milk products, lipolysis by heat-stable microbial lipases is of most significance. Once lipolysis has occurred there is little that can be done to reduce its effects on quality.

Exogenous bacterial lipases

The principal bacterial lipases that occur in milk are the heat-stable extracellular lipases produced by psychrotrophic bacteria (Goff & Hill, 1993). The bacteria most commonly responsible are *Pseudomonas*, particularly *P. fluorescens*, and *P. fragi, Enterobacteriaceae*, such as *Serratia* spp. and *Acinetobacter* spp., and other organisms including *Aeromanas, Bacillus, Achromobcter, Micrococcus, Flavobacterium* and *Moraxella* species (Chen *et al.*, 2003; Deeth & Fitz-Gerald, 2006). The heat stability of the lipases varies with species and strain, but many are sufficiently stable to retain some activity after pasteurisation (Chen *et al.*, 2003; Muir & Banks, 2003) and even after ultra-high temperature (UHT) treatment (Fitz-Gerald *et al.*, 1982; Hoffmann, 2002b). In fact, heating at temperatures >70°C (up to ~120°C) can cause activation of some lipases (Fitz-Gerald *et al.*, 1982). However, phospholipases can degrade the phospholipids in the MFGM and increase the susceptibility of the milk fat to lipolytic attack (Griffiths, 1983). Several psychrotrophic bacteria produce extracellular phospholipases, those prevalent in milk being *Pseudomonas, Alcaligenes, Bacillus* and *Acinetobacter* species (Deeth & Fitz-Gerald, 2006). Like the lipases, many of these enzymes have considerable heat stability and are not destroyed by pasteurisation (72°C for 15 s) (Koka & Weimer, 2001). The phospholipases of *Bacillus cereus* are particularly well known because of their association with the 'bitty cream' defect in milk (Deeth, 1986).

10.2.7 *Proteolysis*

Proteinases act on milk proteins, particularly the caseins, to form peptides and amino acids. Certain small peptides impart a bitter or astringent taste to milk and milk products (Driessen, 1989). The action of proteinases may also change rheological properties by destabilising the casein micelles resulting in defects such as gelation (Driessen, 1989; Kosniski, 1996; Datta & Deeth, 2001). As with lipolytic bacteria, flavour problems resulting from proteolytic bacteria are not usually observed until the bacterial count reaches about 10^7 cfu mL^{-1} (Deeth, 1986). Indigenous milk proteolytic enzymes of concern are the milk alkaline proteinase, plasmin, and to a lesser extent, the acidic proteinase cathepsin D (Chen *et al.*, 2003). Mastitis can lead to an increase in plasmin activity in milk, which corresponds to an increase in somatic cell count, and flavour defects (Chen *et al.*, 2003; Fox & Kelly, 2006).

10.2.8 *Antibiotics*

Antibiotics and antimicrobial drugs are widely used in the treatment of mastitis and prophylactic treatment of non-lactating cows (Rosenthal, 1991; Varnam & Sutherland, 2001). Residues might be excreted in the milk for 4 or more days depending on the drug being used and the condition being treated. Antimicrobial drugs may also enter milk through improper use of drugs and through medicated feed. From a processing viewpoint, the presence of antibiotic residues is undesirable because of interference with growth of starter micro-organisms during the manufacturing of cheese and fermented milks (Robinson & Wilbey, 1998). Antibiotic residues may result in partial or total inhibition of bacterial starters employed in the manufacture of dairy products. In addition, some public health problems may be associated with the presence of antibiotics in milk. Antibiotic allergic reactions might be triggered in people with hypersensitivities after consuming the dairy product (Rosenthal, 1991; Rinken and Riik, 2006). Therefore milk contaminated with antibiotics should be rejected.

10.3 Cream

Cream is produced with different fat contents for use in different applications. The fat content (g 100 g^{-1}) may range from 10 to 18 for half cream or coffee cream, to 35 to 48 for whipping and double cream. The yield and the physical, microbiological and chemical properties are controlled by the quality of the raw materials, the effectiveness of processing, packaging and storage operations, and subsequent distribution and marketing conditions (Walstra *et al.*, 2006). To meet the quality attributes demanded by the consumer, the first consideration for producing a desirable cream with no defects, is to ensure the quality of the raw milk.

Cream must have a clean smell, free of any foreign odour, and taste sweet, fresh and free from taints. The texture must be smooth and rich. It should contain no sediment, have no discolouration or visible phase separation. Before processing, the raw product must be held at $< 5^\circ C$ for as short a time as possible to limit the growth of spoilage micro-organisms and prevent the formation of enzymes such as lipases, proteinases and phospholipases. Provided post-heat treatment microbial contamination is low, the shelf-life of cream should be at least 8 to 10 days at 6 to $8^\circ C$ (Muir & Banks, 2003).

10.3.1 *Transmitted flavours*

The quality of raw milk should be the same as that of milk destined for the liquid milk market. The milk should be free of feed taints as these often partition into the fat phase, and any off-flavours of fat become concentrated in the cream (Nelson and Trout, 1965; Walstra *et al.*, 2006). The cream may be deodorised in an open vat at separation temperatures (i.e. 62–64°C) (Anonymous, 2003) or by vacuum treatment (vacreation) with or without heat (Varnam & Sutherland, 2001). However, vacreation may result in fat globule damage (Walstra *et al.*, 2006).

The fat in cream has an enormous surface area in relation to its volume and readily absorbs odours from the atmosphere (Davis & Wilbey, 1990). Therefore it should not be stored near odoriferous material or in packaging material that may result in migratory contamination; for example, styrene monomer migrating from polystyrene containers imparts an undesirable flavour to the product (Davis & Wilbey, 1990; Bull, 1992).

10.3.2 *Microbiological defects*

Milk used in the manufacture of cream should have a low bacterial count. As some spoilage bacteria preferentially associate with the fat phase, problems due to heat-resistant lipolytic and proteolytic enzymes of these bacteria are exacerbated in the cream after separation. These enzymes may cause a rancid or bitter flavour, or thickening and gelation of the cream (Lewis, 1989). Thus the storage time and temperature of the raw milk before separation must be carefully controlled (Varnam & Sutherland, 2001).

Pasteurisation (72°C for 15 s) of cream destroys vegetative organisms, including pathogens, but in some cases may activate spores to germinate and grow. The cream should be cooled quickly after the heat treatment to prevent possible growth of any surviving organisms. These bacteria together with those introduced post-heat treatment, cause spoilage of cream, resulting in the production of off-flavours. The microflora in pasteurised cream stored at 5°C, which had not been contaminated post-heat processing, have been reported to be *Bacillus cereus, Bacillus licheniformis*, *Bacillus coagulans*, *Bacillus subtilis* and *Bacillus stearothermophilus* (now *Geobacillus stearothermophilus*); these can cause bitter off-flavours and cream thinning due to lipolysis, proteolysis and acid production. The post-pasteurisation contaminants are similar to those in pasteurised milk, mostly *Pseudomonas* spp. (Lewis, 1989; O'Donnell, 1989).

Heating cream above 80°C may cause bacterial spores to germinate, and an increase in the number of vegetative organisms and bacterial activity (Rothwell, 1989b). Higher heat treatment, for example, at 125°C, can destabilise the MFGM, and lead to a marked separation of free fat during storage (Kessler & Fink, 1992). Furthermore, if the heat treatment is excessive, either in temperature or time, sulphurous off-flavours may be produced (Rothwell *et al.,* 1989a).

The production of acid through the growth of micro-organisms in cream can be detected easily by a characteristic sour, slightly unclean odour and taste. Acidity developed in cream is an objectionable defect that consumers criticise severely. This defect in cream also favours feathering, a phenomenon associated with protein denaturation, when the cream is added to a hot acidic solution such as coffee (Nelson & Trout, 1965). Ropiness or sliminess in cream may be caused by some coliforms or lactic streptococci (Davis & Wilbey, 1990). In extreme cases of spoilage in cream, gas may be formed, usually by lactose-fermenting yeasts, and mould growth may be visible on the product's surface (Davis & Wilbey, 1990).

It is therefore important that bacteria which are able to spoil cream, principally the gram-negative bacteria, are destroyed by pasteurisation. A coliform count in the final cream can be used as an indication of either faulty processing or contamination after processing. The presence of these micro-organisms is an indication of poor manufacturing practice and their presence is often associated with impaired shelf-life (Rothwell *et al.*, 1989a).

From a food safety aspect, cream (pasteurised) has been implicated on epidemiological grounds as a source of *L. monocytogenes* and *B. cereus* infection. Cream-based products have also been associated with a number of outbreaks of staphylococcal food poisoning (Varnam & Sutherland, 2001). In general, it is recommended that cream must be pasteurised at a slightly higher temperature ($\sim$75°C rather than 72°C for 15 s) than milk because of the protective effect of the fat on the micro-organisms (Enright *et al.*, 1956).

10.3.3 *Defects associated with oxidation*

The contamination of cream by copper or iron through poor manufacturing practices and use of equipment containing these metals result in rapid oxidation of lipids producing off-flavours, typically 'cardboardy', 'metallic', 'tallowy', 'oily' and 'fishy' (Nelson & Trout, 1965; Jensen & Poulsen, 1992). This defect is now comparatively rare because of the widespread use of stainless steel and elimination of copper from dairy equipment (Deeth, 1986). Protection from light and/or oxygen is most important for cream as these factors may induce oxidation of unsaturated fatty acids. This leads to flavour degradation and reduced biological safety, due to the formation of free radicals, peroxides and oxidised cholesterols (Eyer *et al.*, 1996).

Light-induced flavours can develop when cream is exposed to sunlight, fluorescent light or even diffused daylight (Jensen & Poulsen, 1992). The most damaging wavelengths are in the UV range between 440 and 490 nm, while 310 to 440 nm and 490 to 500 nm also contribute to accelerated degradation (Bull, 1992; Robertson, 2005). Homogenisation and freezing increases the light sensitivity of cream through the breakdown of the fat globules and the increase in surface area available to light and oxygen (Faulks, 1989b; Bull, 1992). Too vigorous agitation during processing may also lead to the introduction of excessive air, which may increase oxidation and off-flavour production (Rothwell *et al.*, 1989a).

Cardboard containers, waxed paper and pigmented packaging give more protection than natural un-pigmented polyethylene (PE), polystyrene (PS), polypropylene (PP) and polyvinylchloride (PVC) containers (Bull, 1992). 'Fillers' can be incorporated into un-pigmented plastic to reduce the transmittance of light (Bull, 1992). Creams designed for a longer shelf-life, such as UHT cream, should be packed in containers with additional protection against light, such as paper board/foil/plastic cartons (Robertson, 2005). To some extent, the strong reducing activity of SH groups released from β-lactoglobulin during UHT treatment inhibits the development of oxidised flavours more than in pasteurised products (Bull, 1992; Bosset *et al.*, 1994).

10.3.4 *Physical defects and stability*

Fat lumps, fat deposition and 'cream plug' result from unsatisfactory emulsion, which is typically caused by mechanical and/or thermal damage of the MFGM during manufacture and storage. Pumping of fresh cooled cream and cream crystallised at 10 to 30°C should be avoided as the fat globules have a lower specific volume than liquid fat and are very sensitive to mechanical treatment (Jensen & Poulsen, 1992; Kessler & Fink, 1992; Muir & Kjaerbye, 1996; Anonymous, 2003). Therefore, cream should not be processed or pumped unless the fat is completely solid or liquid (i.e. at temperatures below 5°C or above 40°C) (Walstra *et al.*, 2006).

Incorporation of air into cream makes the emulsion less stable, as air bubbles act as centres for fat globule aggregation (Jensen & Poulsen, 1992). Storage temperatures also impact emulsion stability. Too high temperatures (10–13°C) make cream layers very firm and accelerate fat deposition (Jensen & Poulsen, 1992). Fluctuations in cream storage temperature can lead to partial melting and re-crystallisation of the fat inside the globules. This leads to rupture of the MFGM, resulting in more cream plug defects than storage at constant temperatures (Børgh-Sørensen, 1992). Direct steam injection also causes disruption of the fat globules to much the same degree as in homogenisers (van Boekel & Walstra, 1995).

A wide range of processing conditions will alter the final viscosity of cream, the most common being homogenisation and cooling. In addition, the higher the fat content of the cream the greater is its viscosity (Rothwell *et al.*, 1989b). Ageing increases the viscosity of most creams due to crystallisation and hardening of the fat at 5°C. Separation temperature may also influence the viscosity of cream. The majority of separators are designed to function most efficiently above 38°C (Faulks, 1989a; Rothwell *et al.*, 1989b). However, cream has a higher viscosity at this temperature than at higher temperatures favoured for cream, for example, 55 to 71°C, and may suffer excessive damage to the fat globules making the cream more physically unstable (Faulks, 1989a; Kosinski, 1996; Hoffmann, 2002a). Another disadvantage of low-temperature cream separation is that natural lipases present in the original milk may cause the production of rancid off-flavours. For this reason, it is advisable to use a separation temperature of not less than 45°C, as this temperature minimises the problem (Faulks, 1989a).

The method of cooling has considerable effect on the viscosity of the final cream (Rothwell, 1989b). In general, cream for direct consumption is cooled to about 5°C in a heat exchanger before being packed (Rothwell *et al.*, 1989b). A two-stage cooling process may be employed to produce cream of a higher viscosity. The cream is first cooled to approximately 20 to 25°C in a plate heat exchanger, then packaged and placed in cold storage for final cooling (Rothwell *et al.*, 1989b). This practice may result in microbiological defects as spoilage bacteria can multiply rapidly and reduce the quality of the cream.

The basic aim of homogenisation of cream is to stabilise the product during storage and prevent or at least minimise creaming and phase separation (Hinrichs & Kessler, 1996). However, splitting of the fat globules modifies the MFGM and greatly increases the surface area of the fat, thus favouring the action of lipases which may be present (Davis & Wilbey, 1990). Homogenised cream has poor stability towards heat coagulation, although its stability can be improved by adjusting the pH and by adding stabilising salts (e.g. citrate) (Walstra *et al.*, 2006).

Clustering of fat globules due to homogenisation increases the viscosity by effectively increasing the volume fraction of the fat through entrapment and immobilisation of plasma. Furthermore, the clusters effectively occupy a greater volume due to their irregular shape.

Clustering is more extensive for creams with high fat content. A second homogenisation at a lower pressure can be used to reduce the viscosity, as the clusters are reduced in size and become more rounded (Faulks, 1989b; Rothwell, 1989b; Walstra *et al.*, 2006).

For all types of lower-fat creams (e.g. half and single creams) homogenisation is essential. Otherwise, they quickly separate out on standing and give a fatty layer floating on a very thin milk serum. A relatively high pressure is used to homogenise this product to avoid such separation. An increase in homogenisation temperature may further reduce the amount of serum, and will also reduce the viscosity, so the homogenisation pressure will also have to be increased. Homogenisation of double cream, even at very low pressures of not more than ~3 MPa, results in a very thick cream. This thick cream may be relatively unstable during storage, sometimes even 'buttering' (Rothwell, 1989b; Rothwell *et al.*, 1989a, 1989b).

10.3.5 *Lipolysis*

Lipolysed flavour defects are among the most common flavour defects in pasteurised market cream (Jensen & Poulsen, 1992). Adequate heat treatment, that is, 75°C for 15 s, for 18 g 100 g^{-1} cream and 80°C for 15 s for ≥ 35 g 100 g^{-1} fat (Castberg, 1992; Davis & Wilbey, 1990; Deeth, 2006), is required for the inactivation of the naturally occurring LPL. Anderson *et al.* (1984) suggested that the ultimate amount of FFA in cream depends not only on the effect of temperature but also on mechanical damage to the fat during separation and the length of time during which warm raw cream is held in the standardising tank before pasteurisation.

10.3.6 *Defects associated with whipped cream*

Processing conditions, chemical composition of cream and seasonal variations influence the whipping ability of cream (Jensen & Poulsen, 1992). The cream (35–40 g 100 g^{-1} fat) should be quickly and easily whipped to form a firm homogenous product containing ~50 mL 100 mL^{-1} of air (Walstra *et al.*, 2006). Whipping cream should not be homogenised, as most homogenised cream cannot be whipped due to the damage to the MFGM. Direct steam injection UHT heating causes homogenisation and the cream should then be aseptically homogenised under low pressure to ensure stability (Walstra *et al.*, 2006). Temperature fluctuations during storage of UHT cream can cause 'rebodying', an increase in viscosity. Attempted whipping of such cream results in churning rather than whipping (Walstra *et al.*, 1999).

With whipping cream, the harder the fat, the better the whip. However, during cooling to solidify the fat, the cream must not be frozen as MFGM damage may result causing clumping and giving the cream a high viscosity and poor whipping properties (Faulks, 1989b; Anonymous, 2003). Stabilisers, including sodium alginate, carrageenan, gelatin, sodium bicarbonate, tetrasodium pyrophosphate and alginic acid, have been used to improve the whipping properties of cream. Faulks (1989b) suggests that the improvement with alginate is due to the stabiliser forming a precipitate of calcium alginate with the calcium ions in cream. This adsorbs onto the surface of each bubble and augments the fat in stabilising the system. It may assist by merely increasing the viscosity of the cream as does gelatine (Vanderghem *et al.*, 2007).

10.3.7 *Defects associated with coffee cream*

It is important for cream used in coffee not to feather or form large droplets of oil on the surface (Walstra *et al.*, 2006). Feathering is a term used to describe the formation of an unpleasant scum or curd on the surface of coffee when cream is added. This is a calcium-induced aggregation of the fat globules and occurs when the milk protein is denatured. The susceptibility of a cream to feathering tends to parallel its heat stability and thus UHT cream is rather susceptible to feathering (Muir & Banks, 2003; Walstra *et al.*, 2006). Feathering depends on temperature, pH and Ca^{2+} activity (Deeth, 1986; Varnam & Sutherland, 2001). Most coffee solutions have a pH of about 5.0, which is near the isoelectric point of casein (Hoffmann, 2002b); this pH, in combination with high temperatures, can cause coagulation and aggregation of unstable creams.

Walstra *et al.* (2006) reported that stability of cream in coffee may be improved by increasing the solids-non-fat content of the cream, which increases its buffer capacity. According to Muir & Kjaerbye (1996), partial demineralisation of cream followed by addition of small amounts of sodium caseinate improves the stability of UHT cream. Demineralisation is expensive and difficult to incorporate into the manufacturing process, while the addition of sodium caseinate is easily accomplished depending on legislative restrictions. Stabilising salts, such as phosphates and citrates, complex Ca^{2+} and may affect the pH value, thereby decreasing aggregation of casein micelles during heat treatment and in hot acidic (e.g. coffee) beverages (Hoffmann, 2002a).

10.3.8 *Defects associated with UHT cream*

UHT cream has an initial cooked, cabbage-like flavour due to the formation of volatile SH compounds and hydrogen sulphide. These compounds will be oxidised at a rate depending on the temperature and oxygen content, and the initial cooked flavour will disappear, usually within days (Lewis, 1989). Homogenisation of UHT cream reduces the fat particle size inhibiting creaming and leading to the creation of a new fat surface, which is stabilised by the adsorption of milk protein (Muir & Kjaerbye, 1996). However, this layer of milk protein is susceptible to proteolysis by plasmin and heat-stable proteinases from psychrotrophic bacteria. During extended storage, the heat-stable enzymes slowly degrade milk protein, resulting in a loss of product quality. This may include the development of a bitter flavour, a thickening of the product, and the gel or emulsion stability may be lost (Muir & Kjaerbye, 1996).

The characteristics of UHT cream are influenced greatly by the temperature and pressure of homogenisation and the position of the homogeniser in the process. In order to achieve good emulsion stability and optimal dispersion, the mean fat globule diameter should be as small as possible and the number of aggregates as few as possible. The most favourable homogenisation temperature for UHT cream is 65 to 75°C. Deviations in temperature in either direction may lead to a loss in stability during storage, resulting in creaming and settling of the serum. Rothwell *et al.* (1989b) suggested using a pressure of $\sim$10 MPa for single cream or for a two-stage homogenisation, $\sim$17 MPa in the first stage and $\sim$3.5 MPa in the second. For half cream, homogenisation pressures of up to 30 MPa may be necessary to produce a cream with reasonable viscosity and stability (Hinrichs & Kessler, 1996).

10.3.9 *Defects associated with sterilised cream*

Owing to the intense heat treatment designed to achieve 'commercial sterility' (e.g. 108°C for 45 min; Davis & Wilbey, 1990) lipolytic and oxidative deterioration rarely occurs in in-container sterilised cream (O'Donnell, 1989; Walstra *et al.*, 2006). It is important not to over-sterilise and impart a caramelised flavour to the cream. Over-processed cream has a dark colour due to Maillard browning, and poor flavour acceptability. Heat penetration is slow, and to avoid a cooked flavour the heat treatment should be kept to the minimum necessary for bacteriological safety. However, there is a risk that such a product may not be completely sterile. Growth of residual heat-resistant spore-formers such as *B. subtilis* can lead to thinning and the development of rancid and bitter flavours due to the action of lipolytic and proteolytic enzymes (Lewis, 1989; Davis & Wilbey, 1990). To minimise this risk, particular attention should be paid to the bacteriological quality, particularly the spore count of the raw material before processing, plant hygiene and processing conditions. An efficient method that has been widely adopted to reduce the risk from *Bacillus* spp. spoilage is an initial UHT treatment at 140°C for 2 seconds to kill any spores prior to in-container sterilisation (Davis & Wilbey, 1990).

Maintaining careful control of retort rotation during sterilisation is important as speeds higher than 1 rev min^{-1} result in broken texture or grainy cream (Smith, 1989). Very strict bacteriological and chemical control must be applied to the cooling water, especially for pressure cooling, since there is a tendency for water to be forced through the seams (Smith, 1989). Entry of non-spore-forming micro-organisms into defective cans may result in acid curdling with or without gas (Lewis, 1989). Similarly, when steam is used for direct cream sterilisation, it has to be of satisfactory quality, that is, derived from potable water.

Defects from corrosion of cans occur more with plain cans than internally lacquered cans. Such defects have been described as 'blackening' and 'purpling', and are caused by degradation of the cream during heat processing, releasing free sulphide or hydrosulphide ions, which convert to hydrogen sulphide gas in the headspace (Smith, 1989).

Sterilised cream may exhibit a 'broken' or 'grainy' appearance, which may be apparent immediately after sterilisation or may only become apparent several weeks after manufacture. Graininess is the presence of pin-point grains of varying sizes that produce an undesirable appearance. These defects can be controlled to some extent by the addition of stabilising salts such as sodium bicarbonate, tri-sodium citrate and di-sodium hydrogen phosphate, which inhibit calcium–protein interaction (Smith, 1989; Muir & Banks, 2003). However, the use of sodium bicarbonate should be carefully controlled as it tends to darken the product (Smith, 1989). If the use of the stabiliser is excessive, undesirable bitter or salty flavours may develop in the cream.

Creaming and (partial) coalescence of fat globules can also occur in sterilised cream. This can be overcome by homogenising the cream before sterilisation (Walstra *et al.*, 2006).

10.4 Butter

Butter is a natural dairy product consisting mainly of milk fat, water and non-fat solids (proteins, lactose, vitamins and minerals). Sweet cream butter is produced by churning fresh pasteurised uncultured cream, while cultured cream butter is made from cream that has been cultured by lactic acid and flavouring micro-organisms, or cultured after churning using the NIZO method (Varnam & Sutherland, 2001). Butter may be salted or unsalted, with unsalted butter having a shorter shelf-life than salted butter (Manners *et al.*, 1987).

The manufacture of high-quality butter requires high-quality cream from high-quality milk. A number of flavour defects associated with butter can be attributed to off-flavours in the cream. Thus, the cream must be of good microbiological quality and be free of any transmitted taints. The cream may be deodorised in an open vat or by vacuum (vacreation) to remove some of the off-flavours that may have developed in raw cream as a result of feed or bacterial growth. However, this process may affect the flavour of butter by removal of desirable butter flavour volatiles (Varnam & Sutherland, 2001). Also, intense heat treatment of the cream may give the butter a cooked flavour (Walstra *et al.*, 2006).

10.4.1 *Microbiological defects*

The two principal types of microbiological spoilage of butter are surface taint and hydrolytic rancidity. The microbiology of butter reflects the microflora present in the cream from which it was made, the wash water, salt slurry, sanitary conditions of the processing environment and equipment, process hurdles and conditions of storage (Kornacki & Flowers, 1998). Micro-organisms already present in the butter, and any entering during reworking and packaging, are potentially capable of growth during retail and domestic storage at temperatures greater than 0°C (Varnam & Sutherland, 2001).

Hydrolytic rancidity in butter is characterised by off-flavours described as 'bitter', 'butyric', 'rancid', 'unclean' or simply 'lipase'. The defect may be evident at manufacture but often develops during storage (Deeth *et al.*, 1979). The type of cream employed, the salt concentration and the pH can affect both the actual and perceived lipolytic rancidity (Allen, 1983). Difficulties can often be experienced during the manufacture of butter from lipolysed cream. Churning durations up to five times longer than normal can occur with rancid cream. The high levels of free fatty acids of lipolysed cream result in the butter being more susceptible to oxidation than butter made from good quality cream (Deeth & Fitz-Gerald, 2006). The risk of post-manufacture lipolysis and associated microbial growth are both greatly enhanced if the butter is not packaged immediately following manufacture, but is done following cold storage in bulk (Downey, 1980).

High free fatty acid (FFA) levels in freshly churned butter may not impart a rancid flavour to butter since the most flavoursome, short-chain acids are removed with the buttermilk and wash water. However, if FFAs are produced during storage by either bacterial growth or residual heat-resistant bacterial lipases, all FFAs remain in the butter and cause a rancid flavour at a relatively low total FFA level (Deeth *et al.*, 1979).

Psychrotrophic micro-organisms, particularly *Pseudomonas* and *Micrococcus* species, have been implicated as important spoilage micro-organisms of butter (Varnam & Sutherland, 2001). *Pseudomonas putrifaciens* is able to grow on the surface of butter and produce a putrid odour through the liberation of certain organic acids (e.g. isovaleric acid) within 7 to 10 days. Pathogenic micro-organisms that have been implicated with butter include *Listeria* spp., *Salmonella* spp., *S. aureus, Streptococcus* spp. and *Mycobacterium* spp. (Kornacki & Flowers, 1998). However, if butter is made from pasteurised cream under good hygienic conditions, the risk from food-borne pathogens is very low.

Yeasts, particularly *Candida lipolyticum*, *Torulopsis* spp., *Cryptococcus* spp. and *Rhodotorula* spp., are capable of growth and can cause lipolysis in butter at low temperatures (Varnam & Sutherland, 2001). Spoilage of butter by mould, including discolouration, typically involves lipolytic genera, such as *Aspergillus, Cladosporium, Geotrichum* and *Penicillium* (Varnam & Sutherland, 2001). In cultured cream butter, microbial spoilage usually involves yeast and moulds due to the low pH ($\sim$4.6) (Walstra *et al.*, 2006). Moulds can grow on parchment packaging material, containing a high proportion of water-soluble organic matter, if the humidity conditions are favourable (Murphy, 1990). Thus, special precautions are required for parchment packaging. Generally, the extent of mould growth depends partly on the degree of contamination of the air (processing environment) with mould spores, and partly on the cleanliness of the manufacturing equipment.

Moisture dispersion

Milk is an emulsion of oil (milk fat) in water, whereas butter is an emulsion of water in oil ($\sim$80 g fat 100 g^{-1}) (Manners *et al.*, 1987; Varnam & Sutherland, 2001; Walstra *et al.*, 2006). Well-worked butter has small water droplets ($\sim$5 μm diameter) dispersed between the fat globules; every gram contains 10 to 18 billion tiny water droplets (Manners *et al.*, 1987; Kornacki & Flowers, 1998). The extent of microbial growth is severely restricted by the physical size of the droplets. Also, if there are about 10^3 cfu micro-organisms per millilitre and 10^{10} moisture droplets per mililitre, some of the water droplets will actually be sterile. If butter is not properly worked, there will be fewer and larger water droplets, and the bacterial cells will have a better chance of multiplying and causing defects (Manners *et al.*, 1987; Walstra *et al.*, 2006).

Butter from neutralised cream

Lactic acid can be produced in raw cream due to the growth of contaminating bacteria. This must be neutralised with sodium hydroxide or sodium bicarbonate before butter manufacture; generally, butters have a titratable acidity of $\leq$ 0.2 mL 100 mL^{-1} (Ronsivalli & Vieira, 1992; Ogden, 1993). Caution is required, as the addition of excessive quantities of neutraliser results in a 'bitter', 'soda' or 'soapy' flavour defect (Ogden, 1993). High acidity promotes the development of oxidation defects in cream and butter and may also hinder the churning process, as acidity has a profound effect on both cream viscosity and fat globule stability (Badings, 1970; Kessler, 1981; Manners *et al.*, 1987; Murphy, 1990). In many countries, the problem of high acidity cream has greatly diminished due to the use of refrigeration on farms and separation of milk to produce cream in the factory rather than on the farm.

10.4.2 *Cultured butter*

Cultured butter (salted or unsalted) is produced using lactic acid-producing starters that impart a distinctive diacetyl smell and characteristic lactic acid flavour (Manners *et al.*, 1987; Varnam & Sutherland, 2001; Walstra *et al.*, 2006). The starters used typically include *Lactococcus lactis* subsp. *lactis* or *Lactococcus lactis* subsp. *cremoris* in combination with

Lactococcus lactis subsp. *lactis* biovar *diacetylactis* or *Leuconostoc mesenteroides* subsp. *cremoris* (Tamime & Marshall, 1997; Varnam & Sutherland, 2001; Frede, 2002a; Lyck *et al.*, 2006). Due to both competition of lactic acid starter bacteria with other bacteria and the inhibitory effect of lactic acid (reduced pH), cultured butter is less susceptible to microbiological spoilage than sweet cream butter (Manners *et al.*, 1987; Varnam & Sutherland, 2001).

Low acid production or aroma formation may be due to slow cultures, or contamination by other micro-organisms, bacteriophages or antibiotics (Kessler, 1981; Surono & Hosono, 2002). Generally, during pasteurisation, those micro-organisms that might interfere with the starter culture and the production of acid are destroyed (Kessler, 1981). However, infection of starter cultures with bacteriophages can result in the failure of the fermentation and loss of product (Frank & Hassan, 1998). Despite the implementation of control measures, bacteriophage infection still causes production problems in the modern dairy fermentation industry. The main growth niches for bacteriophages are raw milk, whey, spilled product, floor drains, equipment and water (Frank & Hassan, 1998).

As discussed in Section 10.2, antibiotics can be present in raw milk. Sensitivity of starter cultures is highly strain and species dependent, and low-level antibiotic contamination may be sufficient to inhibit starter micro-organisms (Frank & Hassan, 1998). Chemical sanitisers contamination can also result in starter culture inhibition, depending on the type and quantity of contaminating sanitisers (Frank & Hassan, 1998).

10.4.3 *Butter churning defects*

In the batch production process, an over-filled churn will take longer to produce butter, and may result in an increase in temperature that will cause the butter to become sticky during working. This moist open-textured butter is often referred to as 'leaky butter' and can cause problems with packaging and printing. High temperatures resulting in sticky butter may also be attributed to churning the cream too soon after the heat treatment. Also, an under-filled churn will churn too rapidly and may produce high-moisture butter and create fat losses in the buttermilk (Manners *et al.*, 1987).

Ideally, at the end of the churning process, the butter granules should be 5 to 10 mm in diameter. If the butter granules are too small, high fat losses occur in the buttermilk and there will be a lower curd content due to less buttermilk being trapped in the granules. On the other hand, if the butter granules are too large, a high moisture level may occur due to too much buttermilk being trapped in the granules; salt incorporation problems may also occur. To achieve the optimum moisture level in butter (13.5–15.5 g 100 g^{-1}), it may be necessary to add potable water to low-moisture butter. However, the addition of too much water should be avoided as it may require so much working that the butter becomes sticky (Manners *et al.*, 1987).

10.4.4 *Oxidative defects*

Even if butter is made from good-quality milk and does not contain heat-resistant lipolytic enzymes, it will eventually deteriorate due to auto-oxidation, after a period of 1 month to 2 years (Walstra *et al.*, 2006). Oxidation is a major cause of spoilage of butter resulting in metallic, fatty, oily, trainy (fishy) and tallowy flavour defects (Badings, 1970; Kessler,

1981). The rate of spoilage is increased by the presence of light, oxygen, heat and traces of metal (Kessler, 1981; Manners *et al.*, 1987; Varnam & Sutherland, 2001). It is recommended that the air content of butter should be less than 1 g 100 g^{-1}, as this reduces the risk of oxidation and lamination (or cracking), and ensures that the texture and colour are not affected (Manners *et al.*, 1987).

Packaging materials for butter should be impermeable to light, fat, moisture and oxygen. Further, any materials in direct contact with butter, which contains >5 mg L^{-1} copper, will act as a catalyst for oxidation. This may also result in bleached spots and/or a tallowy flavour (Manners *et al.*, 1987). Impermeability to water vapour should prevent oiling out on the surface of the butter due to the loss of moisture (Kessler, 1981). The packaging material must also be impermeable to odours, as butter readily takes up foreign odours.

During acidification of cream or milk, a considerable proportion (30–40 g 100 g^{-1}) of added copper (e.g. from contamination) moves to the fat globules. Consequently, butter from sour cream is much more affected by auto-oxidation than sweet cream. The heating of cream largely prevents the migration of copper during souring, as Cu^{2+} becomes bound to low-molecular-mass sulphides, especially H_2S (Walstra, *et al.*, 2006).

The effect of the free fatty acids in ripened cream butter is greater than in sweet cream butter because of the lower pH and this can lead to a greater risk of rancid or soapy flavours in this product (Varnam & Sutherland, 2001). However, because of its characteristic lactic flavour, cultured butter can tolerate higher free fatty acid levels than sweet cream butter without equivalent flavour deterioration (Downey, 1980). The addition of salt also promotes oxidation of milk fat, especially in cultured butter, because of the combined effects of low pH and catalytic action of the salt (Murphy, 1990).

10.4.5 *Physical defects*

The pumping of butter through large stainless steel pipes may result in structure defects and may result in a change in the size and distribution of the water droplets. Pumping of butter may also result in an increase in air content, leading to increased oxidation, and 'flaky' or 'split' butter due to cohesion and adhesion of the butter to the walls of the pipes (Manners *et al.*, 1987; Varnam & Sutherland, 2001). The flaky butter defect can be prevented by maintaining a low air content in the butter, generally below 0.2% (Manners *et al.*, 1987).

Salting

Salt is added for its preservation effects and flavour enhancement. It is generally added to butter in the form of a salt-water slurry, typically aiming for a final salt concentration of 2 g 100 g^{-1} (Manners *et al.*, 1987). The salt slurry should also be of high purity, with less than 1, 10 and 2 mg L^{-1} for lead, iron and copper, respectively (Kornacki & Flowers, 1998). With continuous butter making, the salt slurry is injected and the butter is only given a short working time. It is therefore important to only use fine salt since, if the salt is not thoroughly worked and dissolved into the water phase of the butter, defects will develop during storage. A gritty texture may result from undissolved salt, which will attract surrounding moisture causing free moisture, discolouration and possibly a mottled colour defect

(Manners *et al.*, 1987; Murphy, 1990; Kornacki & Flowers, 1998; Varnam & Sutherland, 2001). Manners *et al.* (1987) suggested that buttermaking machines with a two-stage working section have the advantage of thorough working of the butter, which gives very fine moisture droplet distribution and complete dissolution of the salt crystals.

Milk fat crystallisation

Crystallisation of the fat phase occurs during the related processes of ageing and ripening. Many of the physical properties of butter are determined by the ratio of solid to liquid fat present and the shapes and sizes of the fat crystals (Varnam & Sutherland, 2001; Frede, 2002a). Butter hardens due to the gradual formation of a tighter crystalline structure. This thixotropic hardening may continue for days or weeks; it is fastest at 13°C and stops completely at −4°C, as the butter freezes. Butter and other yellow fats should be cooled rapidly, usually to 3 to 7°C, to produce the desired large numbers of small fat crystals. However, the final temperature and length of holding time are more important than cooling rate, and a minimum of 4 hours holding is necessary for the development of an extensive crystalline network (Varnam & Sutherland, 2001). The physical consistency of the butter can be improved through pre-heat treatment of the cream to control crystallisation of the fat. A 'cold–warm–cold' process will produce a soft butter, while a 'warm–cold–cold' process will produce a firmer butter (Frede, 2002a). However, extremely rapid cooling may result in the formation of impure crystals, or a mixture of middle- and high-melting triacylglycerols with higher levels of solid fat, thus resulting in hard and crumbly butter (Manners *et al.*, 1987). This fault can be minimised through tempering. Tempering involves a 'cold (6–8°C) – warm (14–21°C) – cold (8–13°C)' cycle, which increases the relative quantity of liquid fat and improves the spreadability and consistency of the finished butter (Varnam & Sutherland, 2001). To maintain the spreadability of butter during extended storage of up to 8 months, butter should be kept at −12 to −20°C (Manners *et al.*, 1987). If the storage temperature increases to a point where many of the crystals melt, the network becomes less dense and coarser, and destabilisation can eventually occur (i.e. oil separates from the butter) (Walstra *et al.*, 2006). The spreadability of butter can be improved by whipping nitrogen gas into the butter, which also reduces the hardness of the butter (Bruhn, 2002b). However, the structure is coarse, spongy and crumbly, and differs in appearance and aroma from conventional butter (Varnam & Sutherland, 2001).

Working of butter

Working of the butter is an extremely important process in buttermaking. It has a major influence on the colour, appearance, consistency, spreadability and microbial stability of the butter (Varnam & Sutherland, 2001). Under-working and incorrect churning temperatures may contribute to a hard and crumbly texture (Manners *et al.*, 1987). Over-working may result in a weak body resembling thick cream (Varnam & Sutherland, 2001). The working speed also influences the butter's texture. High working speeds produce small water droplets, and the butter may become dry, while low working speeds result in larger water droplets, especially at low temperatures, and the butter becomes 'wet' (Walstra *et al.*, 2006).

Colour

A mottled colour defect is generally due to improper salt incorporation while a 'streaky' effect, that is, different shades of yellow in the butter, can be attributed to uneven working, working-in over-moist butter from previous churnings, and/or the evaporation of surface moisture before packaging. Speckled butter can arise due to the inclusion of curd particles that have precipitated during heat treatment; poor neutralisation of acid cream can accentuate this defect. The presence of copper may result in green spots, while iron causes brown or black spots. Localised oxidation around particles of iron-containing additives, such as parsley, can result in white spots due to oxidative bleaching of the β-carotene. The colour resulting from surface moisture loss from evaporation is described as a 'primrose' defect and can be prevented by good manufacturing procedures and the use of packaging with good moisture barrier properties (Manners *et al*., 1987).

10.5 Dairy spreads

Consumer demand for a cold-spreadable butter-based table spread has led to the development of dairy spreads made with a reduced milk fat content, blends of butter (55–80 g 100 g^{-1}) and vegetable oil (20–35 g 100 g^{-1}), fractionated milk fat and chemically interesterified milk fat (Tuorila *et al*., 1989; Rousseau & Marangoni, 1999; Frede, 2002b). These water-in-oil emulsion spreads have been developed further to include low-fat (20–25 g 100 g^{-1}) and cultured dairy blends (Manners *et al*., 1987). The manufacturing technology usually involves the buttermaking process, although margarine-based technology can also be used. The vegetable oil may be blended with butterfat at any stage from the milk before separation to the finished butter. Vegetable oil, such as un-hydrogenated soya bean oil, sunflower oil and canola oil may be used, but to be effective it must be present at levels of 15 to 35 g 100 g^{-1} (Klapwijk, 1992; Varnam & Sutherland, 2001).

10.5.1 *Fat phase structure*

As with butter, dairy spreads must be carefully controlled during crystallisation and working to obtain the desired body and consistency. The metastable β' crystals are desired for butter and dairy spreads, and the 'graininess' defect may result from the formation of large coarse triclinic β-type crystals (Frede, 2002b). The graininess defect may be overcome by using β-type crystal fats as the 'soft' component of the fat blend. In addition, some types of spreads contain globular fat, which limits the tendency of higher melting point crystals to form an over-rigid network, resulting in a brittle texture. Hydrogenation may be used to raise the melting point of the fat and help stabilise the emulsion. This is generally not applied to milk fat because of its existing low degree of unsaturation and high costs as a raw material (Varnam & Sutherland, 2001).

Apart from size, shape and arrangement of fat crystals, their amount measured as solid fat content and expressed in relation to temperature is important in controlling the physical properties of spreads (Frede, 2002b). To attain plasticity it is necessary to have the liquid and solid fat present in the correct ratio. Sufficient crystals must be present to retain the liquid phase, but an excess leads to a very rigid structure, lack of plasticity, and brittleness. Some spreads contain in excess of 35 g 100 g^{-1} vegetable oil and must contain some hydrogenated oil to impart plasticity (Varnam & Sutherland, 1994).

With reduction in the milk fat content, the aqueous phase becomes increasingly important and technical difficulties in maintaining a stable water-in-oil emulsion increase. Mono- and diacylglycerols (mono- and di-glycerides) can stabilise the emulsion by nucleating crystallisation of triacylglycerols and crystallising at the oil–water interface (Frede, 2002b). However, changes in emulsion often occur during production, process, storage or transport, resulting in a loss of stability, meltdown and flavour release (Tossavainen *et al.*, 1996).

The lower-fat, higher-moisture spreads have a tendency towards larger water droplets (e.g. 4–90 μm diameter) and coalescence. Emulsifiers and aqueous phase structuring agents are often added to assist churning, to stabilise the final product and to minimise the coalescence of water droplets during processing and spreading (Frede, 2002b). Sodium caseinate, butter milk powder, carrageenan, alginates, gelatine and pectins have been used for this purpose, although excess quantities can lead to undesirably high residual viscosity after the fat has melted during consumption (Tossavainen *et al.,* 1996; Varnam & Sutherland, 2001).

Smaller droplet size results in a paler colour and less flavour release in the mouth, and leads to a higher fat oxidation rate (Frede, 2002b). Dairy spreads exhibit body defects and oiling-off at temperatures above 16°C, particularly the spreads whose spreadability ex-refrigerator approaches that of soft margarines. Milk fat levels around 50 g 100 g^{-1} are still too high to combine good spreadability at about 6 to 8°C with acceptable stand-up properties at ambient temperatures. Furthermore, reducing the milk fat level below 50 g 100 g^{-1} leads to a rapid loss of butter flavour and taste (Frede, 2002b).

Inter-esterification is a chemical process used to modify the structure of triacylglycerols by redistribution of fatty acid moieties on the glycerol backbone. This causes changes in physical properties of fats and fat blends, including spreadability (Rousseau & Marangoni, 1999). However, random inter-esterification of milk fat results in an increase in the solid fat content at 35°C, which produces an undesirable 'tallowy' mouth feel (Varnam & Sutherland, 2001) and a loss of the characteristic butter flavour (Rousseau & Marangoni, 1999).

10.5.2 *Microbiological defects*

There is an interrelationship between the water droplet size and quality parameters, such as microbiological, chemical and physical, as well as organoleptic properties (Rønn *et al.*, 1998; Frede, 2002b). Compared to butter, dairy spreads contain greater amounts of water with increased droplet size, and as a result the microbiological stability of the products is reduced (Klapwijk, 1992; Rousseau & Marangoni, 1999; Varnam & Sutherland, 2001). Preservatives including potassium sorbate and benzoate are commonly added to spreads to inhibit microbial growth (Nichols, 1989; Klapwijk, 1992). However, preservatives are not permitted in some countries.

The shelf-life of refrigerated dairy spreads is very short compared with similar products of higher fat content. Storing dairy spreads at elevated temperatures may destabilise the emulsion, compromising the preservative effects of their compartmentalised structure (Holliday *et al.*, 2003). The combining of ingredients for low-fat spreads occurs at 45°C in an emulsifying unit. This temperature may also allow growth of thermoduric micro-organisms (e.g. *Enterococcus faecium* and *Enterococcus faecalis*) (Kornacki & Flowers, 1998).

There are few published reports of the spoilage of spreads, but spoilage by yeasts and moulds is a common problem (Corradini *et al.*, 1992). Yeast, particularly *Yarrowia lipolytica*, was found to be an important spoilage organism in low-fat dairy spreads; *Bacillus polymyxa* and *E. faecium* have also been implicated. Unlike bacteria and yeasts, moulds have the ability to transit through the oil phase and may generate free fatty acids, which result in off-flavours and breakdown of the emulsion. Psychrotrophic lipolytic bacteria may also cause breakdown of the emulsion due to the production of extracellular enzymes, and rancid off-flavours may develop (Delamarre & Batt, 1999). For the manufacture of cultured dairy spreads, controlled hydrolysis of milk fat is used for flavour development – low levels of lipolysis produce butter flavours, while higher levels produce cheese flavours (Varnam & Sutherland, 2001).

The control of cross-contamination during packaging is more critical in dairy spreads than in butter manufacture because of the high potential for microbial growth in spreads. Thus rigorous adherence to hygienic standards, achievement of proper water dispersion with appropriate sampling and hazard analysis and critical control points (HACCP) are required to ensure the desired quality, safety and shelf-life of dairy spreads (Murphy, 1990; Kornacki & Flowers, 1998).

10.5.3 *Oxidative defects*

As with butter, auto-oxidation of dairy spreads, involving reaction with molecular oxygen and photosensitised oxidation, is important and may result in the development of off-flavours. Light, oxygen, heat and metals such as copper play an important role as catalysts. Low concentrations of antioxidants (e.g. tocopherols) in milk fat provide some protection, while low-fat spreads with high levels of polyunsaturated fatty acids are additionally protected by natural or synthetic antioxidants, for example, tocopherols or butylated hydroxytoluene (Frede, 2002b).

10.6 Cream cheese

Cream cheese is a fresh acid curd cheese produced by the acidification of milk (Lucey, 2002). It has a mild to acidic flavour, with a minimum fat content of 33 g 100 g^{-1} and a maximum moisture content of 55 g 100 g^{-1} (Charley, 1982; Rosenthal, 1991; Sanchez *et al.*, 1996). When the milk is clotted, it is stirred and the mixture is centrifuged to remove the whey. The curd is then either packaged as a cold-packed cream cheese or used as the major ingredient in a mixture with cream, non-fat dry milk or condensed milk, stabilisers and salt to make hot-packed cream cheese (Kessler, 1981; Johnston & Law, 1999). In this process, the mixture is homogenised, heated and then packaged (Johnston & Law, 1999).

Any flavour taint in the raw milk will result in the cheese exhibiting the flavour taint, and in some cases because of concentration, becoming more noticeable (Scott, 1981). Similarly, oxidation and rancid flavours associated with the raw milk may result in the final cheese product being unacceptable. Major faults in these products arise from excess moisture, off-flavours due to contamination, high proteolytic activity in the curd and uneven quality of product from batch to batch (Robinson & Wilbey, 1998).

10.6.1 *Microbiological defects*

Insufficient heat treatment of the milk used to make the cream cheese may allow lactic acid bacteria present in the milk to cause undesirable spontaneous acidification and flavour defects (Høier *et al*., 1999). In addition, both pathogenic and spoilage micro-organisms can contaminate cheese at any stage of manufacture. Production hygiene is therefore of great importance. The production of lactic acid by starter cultures reduces the pH and prevents the growth of pathogens and many spoilage micro-organisms.

A major spoilage micro-organism of cream cheese is mould, which causes the development of undesirable off-flavours (Varnam & Sutherland, 2001; Lucey, 2002). The high temperature of hot-packed cream cheese when packaged minimises contamination, and the absence of air in the packages prevents the growth of mould (Van Slyke, 1949). For these reasons the shelf-life of hot-packed cream cheese in an air-tight, high-moisture-barrier package is generally superior (3–4 months in refrigeration) to that of cold-packed cream cheese (a few weeks) (Johnston & Law, 1999; Lucey, 2003; Hinrichs *et al*., 2004).

Generally, cream cheese curds are cooked at temperatures around 51.5 to 60°C to promote syneresis and provide product stability (Scott, 1981; Lucey, 2003). These temperatures are sufficient to inactivate the culture and prevent further acid development (Rajinder Nath, 1993). Insufficient inactivation may lead to an over-acid product with a sharp acid taste. The acidity may be reduced by washing the curd before salting (Anonymous, 2003).

Starter cultures

The lactic acid starters used in the manufacture of cream cheese include *L. lactis* subsp. *lactis*, *L. lactis* subsp. *cremoris* and *L. lactis* subsp. *lactis* biovar *diacetylactis* (or *Leuconostoc* spp.) (Kessler, 1981; Marshall & Law, 1984; Rosenthal, 1991; Tamime & Marshall, 1997; Tamime, 2006). However, *L. lactis* subsp. *lactis* biovar *diacetylactis* may produce a 'yoghurt flavour' defect due to high acetaldehyde production (Chapman & Sharpe, 1990). The fermentation process of any cultured dairy product relies on the purity and activity of the starter culture, provided that the milk or growth medium is free from inhibitory agents, such as sanitisers, antibiotics or bacteriophage. Milk and cream intended for cream cheese production should be free from these agents. Some compounds naturally present in milk, which have been identified as inhibiting the growth of starter culture bacteria, are lactenins and the lactoperoxidase/hydrogen peroxide/thiocyanate system (LPS) (Tamime, 1990, 2002). The setting temperature of the cheese milk is an important parameter for the growth of the starter culture. It usually matches the optimum growth temperature of the culture used, allowing the starter bacteria to reach their exponential growth phase in the shortest time possible (Broome *et al*., 2002). Incorrect temperature setting (i.e. too low or high) may result in slow starter growth or possibly complete inactivation. Failure of a starter culture to produce acid may result in a gassy, bitter, unclean-flavoured cheese due to unwanted fermentations (Rosenthal, 1991). The lactic acid promotes the exudation of whey from the curd in cheese making, and a poor or sluggish starter generally results in a cheese with a higher than normal moisture content (Rosenthal, 1991). Spoilage is more likely to occur under conditions of high moisture, low acid and excess lactose (Linklater, 1968).

Bacteriophage

Infection by bacteriophage is one of the major problems during cheese manufacture (Rajinder Nath, 1993; Robinson & Wilbey, 1998; Høier *et al.*, 1999). Bacteriophage can cause a complete vat failure due to reduced or no acid production. It is, therefore, important that the cultures used have the best possible phage insensitivity and are produced phage free. Infection within the dairy cannot be completely avoided as raw milk itself may contain bacteriophage, which may not be completely inactivated by pasteurisation. In general, sanitisation of equipment and the processing environment and the use of closed fermentation vessels can minimise the risk of phage attack (Høier *et al.*, 1999).

10.6.2 *Emulsion stability*

Consumer acceptance of cream cheeses requires good product stability and the desired flavour. Instability of dairy emulsions may lead to creaming, flocculation and coalescence. Due to the high fat content of cream cheeses, especially double cream cheese (fat in dry matter >60 g 100 g^{-1}), the structure can be damaged by treatment in processing machinery, for example, shearing and temperature fluctuations (Sanchez *et al.*, 1994b).

Curd homogenisation and cooling rate are important parameters affecting the texture of cream cheese. Prior to curd homogenisation, fat globule damage and coalescence are promoted by centrifugal separation and thermal treatment in a heat exchanger (Sanchez *et al.*, 1994b). Curd homogenisation >15 to 20 MPa allows emulsification of the curd, and the extensively damaged fat globules undergo clustering, forming new aggregates (Sanchez *et al.*, 1994a). Static cooling at 5°C results in strengthening of ionic and hydrogen bond interactions leading to more protein aggregation forming a stronger network (Sanchez *et al.*, 1994a, 1994b). Conversely, during dynamic cooling in heat exchangers, cream cheese undergoes shearing forces, which rupture hydrogen bonds and other weak interactions, leading to aggregate dispersal (Sanchez *et al.*, 1994b). Another fat globule destabilisation parameter is the presence of fat crystals. For instance, hot-packed cheese (70°C) undergoes slow cooling during storage at 5°C, and thus relatively large fat crystals can be formed. These large crystals in conjunction with the additional shearing treatments enhance the brittle texture of the cream cheese (Sanchez *et al.*, 1994b; Lucey, 2003).

10.6.3 *Flavour defects*

Undesirable rancidity in cheese is largely the result of contaminating psychrotrophic bacteria and their heat-resistant enzymes (in milk or cheese) or native milk lipase action in the raw milk (Kessler, 1981; Johnston & Law, 1999). As previously explained, while the activity of native milk lipase is greatly decreased by pasteurisation, the activity of lipase produced by certain psychrotrophic bacteria, typically *Pseudomonas* spp., is not (Kessler, 1981). High levels of free fatty acids caused by excessive lipase-catalysed fat hydrolysis impart an unclean, rancid flavour to cheese (Deeth & Fitz-Gerald, 1975). Proteolysis by enzymes in the milk tends to cause soft curds, which do not drain well and give poor yield, and can cause bitter off-flavours in the cheese (Scott, 1981; Law, 1984).

Rapid coagulation as a result of too much starter or excess rennet may cause a bitter acid flavour (Robinson & Wilbey, 1998). Bitterness from the accumulation of hydrophobic peptides formed by the breakdown of milk proteins (caseins) by the action of starter bacteria may be prevented by careful starter selection and process control (Powell *et al.*, 2002). However, selection of starters for fast acidification and bacteriophage resistance can result in an increased likelihood of bitter cheese.

10.6.4 *Texture defects*

Texture and physical characteristics of cream cheese can be influenced by the source and quantity of fat, quantity of non-fat milk solids, heat treatment at various stages of manufacture, homogenisation pressure, pH, agitation, starter culture and the coagulating enzyme (Rajinder Nath, 1993; Sanchez *et al.*, 1996). Homogenisation (10–15 MPa at 50°C) of the mix is required to produce a smooth cheese and to minimise fat losses in the whey (Lucey, 2003). Efficient homogenising distributes the fat of the mix so that it is retained in the curd and not lost in the whey, and also imparts smoothness to the cheese (Van Slyke, 1949). Salting promotes syneresis, regulates moisture content and helps control microbial growth and activity (Rajinder Nath, 1993). Small amounts of rennet may also be used to promote gelation at a higher pH (Lucey, 2002) to assist in syneresis and to give the desired texture (Van Slyke, 1949; Hinrichs *et al.*, 2004). Excessive moisture can cause leakage of the cheese in the package, reducing the keeping quality of the product and resulting in the loss of visual appeal for the consumer (Van Slyke, 1949).

Heat treatment of cheese milk increases the firmness of milk gels formed by acidification although these gels are more susceptible to wheying-off as the gel may undergo greater rearrangement (Lucey, 2002). Manufacturers try to prevent whey separation by increasing the total solids content of the milk and by adding stabilisers (Lucey *et al.*, 2001). The major problem, that is, with the addition of solids (fat and casein) during standardisation of the cheese milk, is the carry-over to the cheese of any off-flavours from these solids, particularly stale or oxidised flavours (Johnston & Law, 1999). Also, the use of powders to increase the solids may result in a grainy texture if they are not correctly hydrated. Stabilisers, typically locust bean gum, guar gum, xanthan gum, carrageenan, gelatine and whey protein concentrates, can be added to reduce syneresis and the subsequent release of free moisture on the surface of the product during storage (Van Slyke, 1949; Johnston & Law, 1999; Lucey *et al.*, 2001; Hinrichs *et al.*, 2004; Lucey, 2004).

Cream cheese may be dry, crumbly, mealy and grainy if made from less than 14 g fat 100 g^{-1}. If the pH of the cheese is too high (i.e. >4.8), the texture will be soft and the cheese will lack flavour (Lucey, 2003). At very low pH (i.e. <4.6), the texture may become too grainy and crumbly, and the flavour too acid (Van Slyke, 1949; Lucey, 2003).

10.6.5 *Oxidative defects*

Cheeses are exposed to varying degrees of light during processing, distribution and retail sale. Light with high quantum energy (e.g. lower wavelength light in the visible/UV spectrum) has the potential for the most severe effects as it can be absorbed by a variety of molecular structures (Mortensen *et al.*, 2004). For this reason cheeses should never be exposed to direct sunlight or UV light during manufacturing. Fresh cheeses are not as light-sensitive as milk, cream and butter (Mortensen *et al.*, 2004), but still require protection from light by the appropriate packaging. Packaging of cream cheeses must be sterile, of low permeability to water vapour (preventing moisture loss), oxygen, odours, flavours and light (Robertson, 2005). The higher the fat content of the cheese, the more important it is for the packaging material to protect the cheese from light and subsequent auto-oxidation (Kessler, 1981).

Light-induced degradation of lipids, proteins and vitamins in cheeses causes both the formation of off-flavours and colour changes, which rapidly impair product quality and marketability and eventually may lead to loss of nutritional value and formation of toxic products (e.g. cholesterol oxides). Light-induced oxidation requires oxygen, a light source and a photosensitiser (Pettersen *et al.*, 2005). Riboflavin (vitamin B2) in cheeses is an efficient photosensitiser, and, on absorbing light, is excited to a higher energy level (Pettersen *et al.*, 2005). Reverting to its ground state, the photosensitiser may react with fatty acids or oxygen, producing either free radicals or reactive oxygen species. Both of these pathways may lead to formation of lipid free radicals, which, once formed, result in autocatalytic oxidative processes. The off-flavours associated with light-induced oxidation, including metallic, oily, fishy, tallowy, rancid and oxidised, become apparent when the lipid hydroperoxides decompose to alkanes, alkenes, alcohols, aldehydes, esters, ketones and acids (Mortensen *et al.*, 2004; Pettersen *et al.*, 2005).

Pro-oxidants such as copper and iron may enter the product via processing equipment. Chlorophyll may be added with herbs and fruit to cream cheese, initiating immediate oxidation of the unsaturated lipids of the herbs or fruit, and leading to oxidation of milk lipids (Mortensen *et al.*, 2004). Limiting the availability of oxygen in the package and reducing light penetration with the use of innovative packaging will minimise light-induced oxidation.

10.7 Conclusion

The quality of any dairy product at the point of sale depends on the way it is handled at all stages along the production and distribution chains. To maintain the quality of fat-containing dairy products, emphasis must be placed on temperature control to reduce microbial, biochemical and chemical spoilage in the raw milk and subsequently derived products. However, the quality and shelf-life of these products are often compromised by flavour and functional defects.

Undesirable flavours in fat-containing products can be transmitted to the milk and to the final products from the environment. High-fat products such as butter readily absorb extraneous flavours. Flavour defects can also arise from the activities of enzymes, principally lipases and proteinases, which may be indigenous to milk or derived from contaminating bacteria, either in the raw milk prior to processing or in the final product. Another major cause of flavour defects related to fat is oxidation, which gives rise to oxidised flavours, such as stale, cardboardy and metallic; yeasts and moulds may also cause flavour problems in some products, particularly as surface contaminants.

Functional defects are particularly significant for products with fat contents higher than that of milk since the physical stability of the product usually depends on the stability of an oil/fat-in-water or a water-in-oil/fat emulsion. In this respect, product handling, for example, pumping, and homogenisation conditions and, in some cases, additives are critical.

This is important for minimising phase separation during storage, which is unacceptable to consumers.

In the end, the consumer is the final judge of quality and all precautions must be taken to maintain the quality throughout the production chain. It is impossible to generalise about the prevention and control of problems during the manufacture of these dairy products. Instead, the HACCP approach, in conjunction with good manufacturing practices, is recommended for the maintenance of product quality and prevention of spoilage (Neaves & Williams, 1999).

References

Allen, J. C. (1983) Rancidity in dairy products. *Rancidity in Foods*, (eds. J. C. Allen & R. J. Hamilton), pp. 169–178, Applied Science Publishers, London, England.

Anderson, M., Needs, E. C. & Price, J. C. (1984) Lipolysis during the production of double cream. *Journal of the Society of Dairy Technology*, **37**(1), 19–22.

Anonymous (2003) *Dairy Processing Handbook*, 2nd and revised edn. of G. Bylund (1995), Tetra Pak Processing Systems AB, Lund.

Bachman, K. C. & Wilcox, C. J. (1990) Effect of time of onset of rapid cooling on bovine milk fat hydrolysis. *Journal of Dairy Science*, **73**, 617–620.

Badings, H. T. (1970) Cold storage defects in butter and their relation to the auto-oxidation of unsaturated fatty acids. *Netherlands Milk and Dairy Journal*, **24**, 145–246.

Bassette, R., Fung, D. Y. C. & Mantha, V. R. (1986) Off-flavours in milk. *CRC Critical Reviews in Food Science and Nutrition*, **24**, 1–52.

van Boekel, M. A. J. S. & Walstra, P. (1995) Effects of heat treatment on chemical and physical changes to milk fat globules. *Heat-induced Changes in Milk*, (ed. P. F. Fox), 2nd edn, Document No. 9501, pp. 22–50, International Dairy Federation, Brussels.

Børg-Sørensen, T. (1992) Cream pasteurisation technology. *Pasteurisation of Cream*, Document No. 271, pp. 32–39, International Dairy Federation, Brussels.

Bosset, J. O., Gallman, P. U. & Sieber, R. (1994) Influence of light transmittance of packaging materials on the shelf life of milk and dairy products – a review. *Food Packaging and Preservation*, (ed. M. Mathlouthi), pp. 222–268, Blackie Academic & Professional, Glasgow.

Broome, M. C., Powell, I. B. & Limsowtin, G. K. Y. (2002) Starter cultures: specific properties. *Encyclopedia of Dairy Sciences*, (eds. H. Roginski, J. W. Fuquay & P. F. Fox), Vol. 1, pp. 269–275, Academic Press, London.

Bruhn, J. C. (2002a) *Dairy Processing; Milk Flavours*. Dairy Research and Information Centre, Department of Food Science and Technology, University of California, Davis.

Bruhn, J. C. (2002b) *Dairy Processing; Butter: Some Technology and Chemistry*. Dairy Research and Information Centre, Department of Food Science and Technology, University of California, Davis.

Bull, M. (1992) Packaging requirements for cream. *Pasteurisation of Cream*, Document No. 271, pp. 40–44, International Dairy Federation, Brussels.

Castberg, H. B. (1992) Lipase activity. *Pasteurisation of Cream*, Document No. 271, pp. 18–20, International Dairy Federation, Brussels.

Chapman, H. R. & Sharpe, M. E. (1990) Microbiology of cheese. *Dairy Microbiology*, Vol. 2: *The Microbiology of Milk Products*, (ed. R. K. Robinson), 2nd edn., pp. 203–290, Elsevier Science Publishers, London.

Charley, H. (1982) *Food Science*, 2nd edn., John Wiley & Sons, New York.

Chen, L., Daniel, R. M. & Coolbear, T. (2003) Detection and impact of proteinase and lipase activities in milk and milk powders. *International Dairy Journal*, **13**, 255–275.

Collins, Y. F., McSweeney, P. L. H. & Wilkinson, M. G. (2003) Lipolysis and free fatty acid catabolism in cheese: a review of current knowledge. *International Dairy Journal*, **13**, 841–866.

Corradini, C., Pittia, P., Sensidoni, A. & Innocente, N. (1992) Improvement of emulsion characteristics in low-fat butter. *Protein and Fat Globule Modifications by Heat Treatment, Homogenization and other Technological Means for High Quality Dairy Products,* Special Issue 9303, pp. 25–28, International Dairy Federation, Brussels.

Couvreur, S., Hurtaud, C., Lopez, C., Delaby & Peyraud, J. L. (2006) The linear relationship between the proportion of fresh grass in the cow diet, milk fatty acid composition, and butter properties. *Journal of Dairy Science*, **89**, 1956–1969.

Datta, N. & Deeth, H. C. (2001) Age gelation of UHT milk – a review. *Transactions of the Institute of Chemical Engineers C. Food and Bioproducts Processing*, **79**, 197–210.

Davis, J. G. & Wilbey, R. A. (1990) Microbiology of cream and dairy desserts. *Dairy Microbiology*, Vol. 2: *The Microbiology of Milk Products*, (ed. R. K. Robinson), 2nd edn., pp. 41–108, Elsevier Science, New York.

Deeth, H. C. (1986) The appearance, texture, flavour and defects of pasteurised milk. *Pasteurized Milk*, Document No. 200, pp. 22–26, International Dairy Federation, Brussels.

Deeth, H. C. (2006) Lipoprotein lipase and lipolysis in milk. *International Dairy Journal*, **16**, 555–562.

Deeth, H. C. & Fitz-Gerald, C. H. (1975) The relevance of milk lipase activation to rancidity in Cheddar cheese. *Australian Journal of Dairy Technology*, **30**, 74–76.

Deeth, H. C. & Fitz-Gerald, C. H. (1977) Some factors involved in milk lipase activation by agitation. *Journal of Dairy Research*, **44**, 569–583.

Deeth, H. C. & Fitz-Gerald, C. H. (2006) Lipolytic enzymes and hydrolytic rancidity. *Advanced Dairy Chemistry*, Vol. 2: *Lipids*, (eds. P. F. Fox & P. McSweeney), pp. 481–556, Springer, New York.

Deeth, H. C., Fitz-Gerald, C. H. & Wood, A. F. (1979) Lipolysis and butter quality. *Australian Journal of Dairy Technology*, **34**, 146–149.

Delamarre, S. & Batt, C. A. (1999) The microbiology and historical safety of margarine. *Food Microbiology*, **16**, 327–333.

Driessen, F. M. (1989) Inactivation of lipases and proteinases (indigenous and bacterial). *Heat-induced Changes in Milk*, (ed. P. F. Fox), Document No. 238, pp. 71–93, International Dairy Federation, Brussels.

Downey, W. K. (1975) Lipolysis in milk and dairy products. *Proceedings of the Lipolysis Symposium, Cork (Ireland)*, Annual Document No. 86, pp. 2–4, International Dairy Federation, Brussels.

Downey, W. K. (1980) Risks from pre- and post-manufacture lipolysis. *Flavour Impairment of Milk and Milk Products due to Lipolysis*, Document No. 118, pp. 4–18, International Dairy Federation, Brussels.

Enright, J. B., Sadler, W. W. & Thomas R. C. (1956) Observations on the thermal inactivation of the organism of Q fever in milk. *Journal of Milk and Food Technology*, **19**, 313–318.

Eyer, H. K., Rattray, W. & Gallmann, P. U. (1996) The packaging of UHT cream. *UHT Cream*, Document No. 315, pp. 23–24, International Dairy Federation, Brussels.

Faulks, B. (1989a) Separation and standardisation. *Cream Processing Manual*, (ed. J. Rothwell), pp. 12–25, The Society of Dairy Technology, Huntingdon.

Faulks, B. (1989b) The whipping of cream. *Cream Processing Manual*, (ed. J. Rothwell), pp. 88–98, The Society of Dairy Technology, Huntingdon.

Fitz-Gerald, C. H., Deeth, H. C. & Coghill, D. M. (1982) Low temperature inactivation of lipases from psychrotrophic bacteria. *Australian Journal of Dairy Technology*, **37**, 51–54.

Fox, P. F. & Kelly, A. L. (2006) Indigenous enzymes in milk: overview and historical aspects – part 1. *International Dairy Journal*, **16**, 500–516.

Fox, P. F. & Stepaniack, L. (1993) Enzymes in cheese technology. *International Dairy Journal*, **3**, 509–530.

Frank, J. F. & Hassan, A. N. (1998) Starter cultures and their use. *Applied Dairy Microbiology*, (eds. E. H. Marth & J. L. Steele), pp. 131–172, Marcel & Dekker, New York.

Frede, E. (2002a) Butter. *Encyclopedia of Dairy Sciences*, (eds. H. Roginski, J. W. Fuquay & P. F. Fox), Vol. 1, pp. 220–226, Academic Press, London.

Frede, E. (2002b) Milk-fat based spreads. *Encyclopedia of Dairy Sciences*, (eds. H. Roginski, J. W. Fuquay & P. F. Fox), Vol. 3, pp. 1859–1868, Academic Press, London.

Griffiths, M. W. (1983) Synergistic effects of various lipases and phospholipase C on milk fat. *Journal of Food Technology*, **18**, 495–505.

Goff, H. D. & Hill, A. R. (1993) Chemistry and physics. *Dairy Science and Technology Handbook*, Vol. 1, *Principles and Properties*, (ed. Y. H. Hui), pp. 1–82, VCH Publishers, New York.

Hamilton, R. T. (1983) The chemistry of rancidity in foods. *Rancidity in Foods*, (eds. J. C. Allen & R. J. Hamilton), pp. 1–20. Applied Science Publishers, London.

Hinrichs, J., Götz, J., Noll, M., Wolfschoon, A., Eibel, H. & Weisser, H. (2004) Characterisation of the water-holding capacity of fresh cheese samples by means of low resolution nuclear magnetic resonance. *Food Research International*, **37**, 667–676.

Hinrichs, J. & Kessler, H. G. (1996) Processing of UHT cream. *UHT Cream*, Document No. 315, pp. 17–22, International Dairy Federation, Brussels.

Hoffmann, W. (2002a) Cream. *Encyclopedia of Dairy Sciences*, (eds. H. Roginski, J. W. Fuquay & P. F. Fox), Vol. 1, pp. 545–550, Academic Press, London.

Hoffmann, W. (2002b) Products. *Encyclopedia of Dairy Sciences*, (eds. H. Roginski, J. W. Fuquay & P. F. Fox), Vol. 1, pp. 551–557, Academic Press, London.

Høier, E., Janzen, T., Henrisen, C. M. Rattray, F., Brockmann, E. & Johansen, E. (1999) The production, application and action of lactic cheese starters. *Technology of Cheesemaking*, (ed. B. A. Law), pp. 99–131, Academic Press, Sheffield.

Holliday, S. L., Adler, B. B. & Beuchat, L. R. (2003) Viability of *Salmonella, Escherichia coli* 0157:H7, and *Listeria monocytogenes* in butter, yellow fat spreads, and margarine as affected by temperature and physical abuse. *Food Microbiology*, **20**, 159–168.

Hutton, J. T. & Patton, S. (1952) The origin of sulphydryl groups in milk proteins and their contribution to cooked flavour. *Journal of Dairy Science*, **35**, 699.

Jensen, G. K. & Poulsen, H. H. (1992) Sensory aspects. *Pasteurisation of Cream*, Document No. 271, pp. 26–31, International Dairy Federation, Brussels.

Johnston, M. & Law, B. A. (1999) The origins, development and basic operation of cheesemaking technology. *Technology of Cheesemaking*, (ed. B. A. Law), pp. 1–32, Academic Press, Sheffield.

Joshi, N. S. & Thakar, P. N. (1994) Methods to evaluate deterioration of milk fat – a critical appraisal. *Journal of Food Science and Technology*, **31**, 181–196.

Kessler, H. G. (1981) *Food Engineering and Dairy Technology*, Publishing House Verlag A. Kessler, Freising, Germany.

Kessler, H. G. & Fink, A. (1992) Physico-chemical effects of pasteurisation on cream properties. *Pasteurisation of Cream*, Document No. 271, pp. 11–17, International Dairy Federation, Brussels.

Klapwijk, P. M. (1992) Hygienic production of low-fat spreads and their application of HACCP during their development. *Food Control*, **1**(4), 183–189.

Koka, R. & Weimer, B. C. (2001) Influence of growth conditions on heat-stable phospholipase activity in *Pseudomonas*. *Journal of Dairy Research*, **68**, 109–116.

Kon, H. & Saito, Z. (1997) Factors causing temperature activation of lipolysis in cow's milk. *Milchwissenchaft*, **52**, 435–440.

Kornacki, J. & Flowers, R. (1998) Microbiology of butter and related products. *Applied Dairy Microbiology*, (eds. E. H. Marth & J. L. Steele), pp. 109–130, Marcel Dekker, New York.

Kosinski, E. (1996) Raw material quality. *UHT Cream*, Bulletin 315, pp. 12–16, International Dairy Federation, Brussels.

Law, B. A. (1984) *Advances in the Microbiology and Biochemistry of Cheese and Fermented Milk*, (eds. F. L. Davies & B. A. Law), pp. 187–253, Elsevier Applied Science, London.

Lewis, M. J. (1989) UHT cream processing. *Cream Processing Manual*, (ed. J. Rothwell), pp. 1–4, The Society of Dairy Technology, Huntingdon.

Linklater, P. M. (1968) Fundamentals of cheese manufacture. *Symposium: The Technology of Some Soft, Semi-Hard and Grating Cheese Varieties*, pp. 9–17, The Australian Dairy Produce Board, Melbourne.

Lucey, J. A. (2002) Acid and acid/heat coagulated cheese. *Encyclopedia of Dairy Sciences*, (eds. H. Roginski, J. W. Fuquay & P. F. Fox), Vol. 1, pp. 350–363, Academic Press, London.

Lucey, J. (2003) A closer look at cream cheese. *Dairy Pipeline*, **15**(1), 1–8.

Lucey, J. A., Tamehana, M., Singh, H. & Munro, P. A. (2001) Effect of heat treatment on the physical properties of milk gels made with both, rennet and acid. *International Dairy Journal*, **11**, 559–565.

Lyck, S., Nilsson, L.-E. & Tamime, A. Y. (2006) Miscellaneous fermented milk products. *Probiotic Dairy Products*, (ed. A. Y. Tamime), pp. 217–236, Blackwell Publishing, Oxford.

Manners, J., Tomlinson, N. & Jones, M. (1987) *Butter and Related Products*, Victorian College of Agriculture and Horticulture,, Werribee, Australia.

Marshall, V. M. E. & Law, B. A. (1984) The physiology and growth of dairy lactic-acid bacteria. *Advances in the Microbiology and Biochemistry of Cheese and Fermented Milk*, (eds. F. L. Davies & B. A. Law), pp. 67–98, Elsevier Applied Science, London.

McDowell, A. K. R. (1969) Storage of chilled cream in relation to butter quality. *Journal of Dairy Research*, **36**, 225–232.

Mortensen, G., Bertelsen, G., Mortensen, B. & Stapelfeldt, H. (2004) Light-induced changes in packaged cheeses – a review. *International Dairy Journal*, **14**, 85–102.

Muir, D. D. & Banks, J. M. (2003) Factors affecting the shelf life of milk and milk products. *Dairy Processing, Improving Quality*, (ed. G. Smit), pp. 185–207, Woodhead Publishing, New York.

Muir, D. D. & Kjaerbye, H. (1996) Quality aspects of UHT cream. *UHT Cream*, Document No. 315, pp. 25–34, International Dairy Federation, Brussels.

Munro, G. L., Grieve, P. A. & Kitchen, B. J. (1984) Effects of mastitis on milk yield, milk composition, processing properties and yield and quality of milk products. *Australian Journal of Dairy Technology*, **39**, 7–16.

Murphy, M. F. (1990) Microbiology of butter. *Dairy Microbiology*, Vol. 2: *The Microbiology of Milk Products*, (ed. R. K. Robinson), 2nd edn., pp. 109–130, Elsevier Science, London.

Neaves, P. & Williams, P. (1999) Microbiological surveillance and control in cheese manufacture. *Technology of Cheese-making*, (ed. B. A. Law), pp. 251–280, Academic Press, Sheffield.

Nelson, J. A. & Trout, G. M. (1965) *Judging Dairy Products*, 4th edn., Owen Publishing Co., Milwaukee.

Nichols, B. W. (1989) Processed oil and fat products. *Food Chemistry*, **33**, 27–31.

O'Donnell, E. (1989) Storage, handling and distribution. *Cream Processing Manual*, (ed. J. Rothwell), pp. 42–51, The Society of Dairy Technology, Huntingdon.

Ogden, L. V. (1993) Sensory evaluation of dairy products. *Dairy Science and Technology Handbook*, Vol. 1, *Principles and Properties*, (ed. Y. H. Hui), pp. 157–276, VCH Publishers, New York.

Pettersen, M. K., Eie, T. & Nilsson, A. (2005) Oxidative stability of cream cheese stored in thermoformed trays as affected by packaging material, drawing depth and light. *International Dairy Journal*, **15**, 355–362.

Powell, I. B., Broome, M. C. & Limsowtin, G. K. Y (2002) Starter cultures: general aspects. *Encyclopedia of Dairy Sciences*, (eds. H. Roginski, J. W. Fuquay & P. F. Fox), pp. 261–268, Academic Press, London.

Rajinder Nath, K. (1993) Cheese. *Dairy Science and Technology Handbook*, Vol. 2: *Product Manufacturing*, (ed. Y. H. Hui), pp. 161–256, VCH Publishers, New York.

Rinken, T. & Riik, H. (2006) Determination of antibiotic residues and their interaction in milk with lactate biosensor. *Journal of Biochemical and Biophysical Methods*, **66**, 13–21.

Robertson, G. L. (2005) *Food Packaging: Principles and Practice*, 2nd edn., CRC Press (member of Taylor & Francis Group), Boca Raton, Florida.

Robinson, R. K. & Wilbey, R. A. (1998) *Cheesemaking Practice – R. Scott*, 3rd edn., Aspen Publishers, Gaithersburg, Maryland.

Rønn, B. B., Hyldig, G., Wienberg, L., Qvist, K. B. & Laustsen, A. (1998) Predicting sensory properties from rheological measurements of low-fat spreads. *Food Quality and Preference*, **9**(4), 187–196.

Ronsivalli, L. J. & Vieira, E. R. (1992) *Elementary Food Science*, 3rd edn., Van Nostrand Reinhold, New York.

Rosenthal, I. (1991) *Milk and Dairy Products: Properties and Processing*. VCH Publishers, New York.

Rothwell, J. (1989a) Introduction. *Cream Processing Manual*, (ed. J. Rothwell), pp. 1–4, The Society of Dairy Technology, Huntingdon, England.

Rothwell, J. (1989b) Pasteurisation and homogenisation. *Cream Processing Manual* (ed. J. Rothwell), pp. 26–33, The Society of Dairy Technology, Huntingdon, England.

Rothwell, J., Jackson, A. C. & Faulks, B. (1989a) Trouble shooting. *Cream Processing Manual*, (ed. J. Rothwell), pp. 120–124, The Society of Dairy Technology, Huntingdon, England.

Rothwell, J., Jackson, A. C. & Faulks, B. (1989b) Modification and control of cream viscosity. *Cream Processing Manual*, (ed. J. Rothwell), pp. 83–87, The Society of Dairy Technology, Huntingdon, England.

Rousseau, D. & Marangoni, A. G. (1999) The effects of interesterification on physical and sensory attributes of butterfat and butterfat – canola oil spreads. *Food Research International*, **31**(5), 381–388.

Sanchez, C., Beauregard, J. L., Chassagne, M. H., Bimbenet, J. J. & Hardy, J. (1994a) Rheological and textural behaviour of double cream cheese. Part I: effect of curd homogenization. *Journal of Food Engineering*, **23**, 579–594.

Sanchez, C., Beauregard, J. L., Chassagne, M. H., Bimbenet, J. J. & Hardy, J. (1996) Effects of processing on rheology and structure of double cream cheese. *Food Research International*, **28**(6), 547–552.

Sanchez, C., Beauregard, J. L., Chassagne, M. H., Duquenoy, A. & Hardy, J. (1994b) Rheological and textural behaviour of double cream cheese. Part II: effect of curd cooling rate. *Journal of Food Engineering*, **23**, 595–608.

Smith, G. (1989) Sterilised cream. *Cream Processing Manual*, (ed. J. Rothwell), pp. 66–73, The Society of Dairy Technology, England.

Tamime, A. Y. (1990) Microbiology of starter cultures. *Dairy Microbiology*, Vol. 2, *The Microbiology of Milk Products*, (ed. R. K. Robinson), 2nd edn., pp. 131–202, Elsevier Science, London.

Tamime, A. Y. (2002a) Microbiology of starter cultures. *Dairy Microbiology Handbook*, (ed. R. K. Robinson), 3rd edn., pp. 261–366, John Wiley & Sons Inc., New York.

Tamime, A. Y. (ed.) (2006) *Fermented Milks*, Blackwell Publishing, Oxford.

Tamime, A. Y. & Marshall, V. M. E. (1997) Microbiology and technology of fermented milks. *Microbiology and Biochemistry of Cheese and Fermented Milk*, (ed. B. A. Law), 2nd edn., pp. 57–152, Blackie Academic & Professional, London.

Tossavainen, O., Pyykkönen, P., Vastamäki, P. & Huotari, H. (1996) Effect of milk protein products on the stability of model low-fat spread. *International Dairy Journal*, **6**, 171–184.

Tuorila, H., Matuszewska, I., Helleman, U. & Lampi, A. (1989) Sensory and chemical characterisation of fats used as spreads on bread. *Food Quality and Preference*, **1**(415), 157–162.

Vanderghem, C., Danthine, S., Blecker, C. & Deroanne, C. (2007) Effect of proteose-peptone addition on some physio-chemical characteristics of recombined dairy creams. *International Dairy Journal*, **17**, 889–895.

Van Slyke, L. L. (1949) *Cheese: A Treatise on the Manufacture of American Cheddar Cheese and some other Varieties*, Orange Judd Publishing Co., New York.

Varnam, A. H. & Sutherland, J. P. (2001) *Milk and Milk Products; Technology, Chemistry and Microbiology*. Aspen Publishers, Gaithersburg, Maryland.

Walstra, P., Wouters, J. T. M. & Geurts, T. J. (2006) *Dairy Science and Technology*, 2nd edn., CRC Press (member of Taylor & Francis Group), Boca Raton, Florida.

de Wit, R. & Nieuwenhuijse, H. (2008) Kinetic modelling of the formation of sulphur-containing flavour components during heat-treatment of milk. *International Dairy Journal*, **18**, 539–547.

Index